T0210661

# Communications
# in Computer and Information Science     601

*Commenced Publication in 2007*
Founding and Former Series Editors:
Alfredo Cuzzocrea, Dominik Ślęzak, and Xiaokang Yang

## Editorial Board

More information about this series at http://www.springer.com/series/7899

Vladimir Vishnevsky · Dmitry Kozyrev (Eds.)

# Distributed Computer and Communication Networks

18th International Conference, DCCN 2015
Moscow, Russia, October 19–22, 2015
Revised Selected Papers

 Springer

*Editors*

Vladimir Vishnevsky
V.A. Trapeznikov Institute of Control
  Sciences
Russian Academy of Sciences
Moscow
Russia

Dmitry Kozyrev
V.A. Trapeznikov Institute of Control
  Sciences
Russian Academy of Sciences
Moscow
Russia

ISSN 1865-0929          ISSN 1865-0937  (electronic)
Communications in Computer and Information Science
ISBN 978-3-319-30842-5      ISBN 978-3-319-30843-2  (eBook)
DOI 10.1007/978-3-319-30843-2

Library of Congress Control Number: 2016932356

Printed on acid-free paper

This Springer imprint is published by Springer Nature
The registered company is Springer International Publishing AG Switzerland

# Preface

This volume contains a collection of revised selected full-text papers presented at the 18th International Conference on Distributed Computer and Communication Networks (DCCN-2015), held in Moscow, Russia, October 19–22, 2015.

The conference is a continuation of traditional international conferences of the DCCN series, which took place in Bulgaria (Sofia, 1995, 2005, 2006, 2008, 2009, 2014), Israel (Tel Aviv, 1996, 1997, 1999, 2001), and Russia (Moscow, 1998, 2000, 2003, 2007, 2010, 2011, 2013) in the last 18 years. The main idea of the conference is to provide a platform and forum for researchers and developers from academia and industry from various countries working in the area of theory and applications of distributed computer and communication networks, to exchange their expertise, and to discuss the perspectives of development and collaboration in this area. The content of this volume is related to the following subjects:

1. Computer and communication networks architecture optimization
2. Control in computer and communication networks
3. Performance and QoS evaluation in wireless networks
4. Modeling and simulation of network protocols
5. Queueing and reliability theory
6. Wireless IEEE 802.11, IEEE 802.15, IEEE 802.16, and UMTS (LTE) networks
7. FRID technology and its application in intellectual transportation networks
8. Protocols design (MAC, Routing) for centimeter and millimeter wave mesh networks
9. Internet and Web applications and services
10. Application integration in distributed information systems
11. Big data in communication networks

The DCCN 2015 conference received 126 submissions from 164 authors in 16 different countries. From these, 94 submissions were accepted and presented at the conference, 38 of which were recommended by session chairs and selected by the Program Committee for the proceedings, yielding an overall acceptance rate of 40 %.

All the papers selected for the proceedings are given in the form presented by the authors. These papers are of interest to everyone working in the field of computer and communication networks.

We thank all the authors for their interest in DCCN, the members of the Program Committee for their contributions, and the reviewers for their peer-reviewing efforts.

February 2016                                                                                 Vladimir Vishnevsky

# Organization

DCCN 2015 was organized by the Russian Academy of Sciences (RAS), V.A. Trapeznikov Institute of Control Sciences of RAS, the Research and Development Company Information and Networking Technologies, and the Institute of Information and Communication Technologies (Bulgarian Academy of Sciences).

## Executive Committee

### General Co-chairs

| | |
|---|---|
| S.N. Vasilyev | ICS RAS, Russia |
| V.M. Vishnevsky | ICS RAS, Russia |

### Steering Committee

| | |
|---|---|
| V.V. Rykov | Gubkin Russian State University of Oil and Gas, Russia |
| D.A. Aminev | ICS RAS, Russia |
| T. Atanasova | IICT-BAS, Bulgaria |
| S.N. Kupriyahina | ICS RAS, Russia |

### Publicity Chair

| | |
|---|---|
| D.V. Kozyrev | ICS RAS, Russia |

## International Program Committee

| | |
|---|---|
| A.M. Andronov | Riga Technical University, Latvia |
| L.I. Abrosimov | Moscow Power Engineering Institute, Russia |
| Mo Adda | University of Portsmouth, UK |
| T.I. Aliev | ITMO University, Russia |
| G.P. Basharin | Peoples' Friendship University of Russia, Russia |
| A.S. Bugaev | Moscow Institute of Physics and Technology, Russia |
| E. Gelenbe | University of Paris, France |
| A. Gelman | IEEE Communications Society, USA |
| A.N. Dudin | Belarusian State University, Belarus |
| V.V. Devyatkov | Bauman University, Russia |
| D.-J. Deng | National Changhua University of Education, Taiwan |
| A.S. Ermakov | Kazakh National Technical University after K.I. Satpayev, Kazakhstan |
| V.S. Zhdanov | Higher School of Economics, Russia |
| Ju.P. Zaychenko | Kyiv Polytechnic Institute, Ukraine |
| V.A. Ivnitsky | Moscow State University of Railway Engineering, Russia |

| G. Kotsis | Johannes Kepler University Linz, Austria |
| E.A. Kucheryavy | Tampere University of Technology, Finland |
| L. Lakatosh | Budapest University, Hungary |
| E. Levner | Holon Institute of Technology, Israel |
| A.N. Latkov | Riga Technical University, Latvia |
| P.P. Malcev | Institute of Microwave Semiconduct Electronics of the RAS, Russia |
| S.D. Margenov | Institute of Information and Communication Technologies at the Bulgarian Academy of Sciences (IICT-BAS), Bulgaria |
| E.V. Morozov | Institute of Applied Mathematical Research of the Karelian Research Centre RAS, Russia |
| G.K. Mishkoy | Academy of Sciences of Moldova, Moldavia |
| A.A. Nazarov | Tomsk State University, Russia |
| A.N. Nazarov | OJSC Intellect Telecom, Russia |
| S.A. Nikitov | Kotel'nikov Institute of Radio-engineering and Electronics of the RAS, Russia |
| D.A. Novikov | V.A. Trapeznikov Institute of Control Sciences of the RAS, Russia |
| M. Pagano | Pisa University, Italy |
| S.L. Portnoy | Moscow State Institute of Radio-engineering, Electronics and Automation, Russia |
| E.Ja. Rubinovich | V.A. Trapeznikov Institute of Control Sciences of the RAS, Russia |
| V.V. Rykov | Gubkin Russian State University of Oil and Gas, Russia |
| K.E. Samuylov | Peoples' Friendship University of Russia, Russia |
| V.S. Sgurev | Institute of Information Technologies, Bulgarian Academy of Sciences (IIT-BAS), Bulgaria |
| P. Stanchev | Kettering University, USA |
| S.N. Stepanov | OJSC "Intellect Telecom", Russia |
| H. Tijms | Vrije University of Amsterdam, The Netherlands |
| T. Tozer | York University, UK |
| M.A. Fedotkin | Lobachevsky National Research University of Nizhni Novgorod, Russia |
| T. Czachorski | Institute of Informatics of Polish Academy of Sciences, Poland |
| V.A. Vasenin | Moscow State University, Russia |

## Support and Sponsors

Russian Academy of Science
IEEE Communication Society, Moscow Department
Research and Development Company Information and Networking Technologies, Moscow, Russia
Russian Scientific and Technical Journal *Electronika: Science, Technology, Business*

# Contents

# Autonomous Cars. History. State of Art. Research Problems

Peter Stanchev[1,2](✉) and John Geske[2]

[1] Institute of Mathematics and Informatics,
Bulgarian Academy of Sciences, Sofia, Bulgaria
[2] Kettering University, Flint, USA
{pstanche,geske}@kettering.edu

**Abstract.** Vehicle-to-infrastructure communications is the wireless exchange of critical safety and operational data between vehicles and highway infrastructure. The following research questions have to be answered by the V2I Communications: What safety applications are effective and have validated benefits?; What minimum infrastructure is needed for maximum benefit?; Can signal phase and timing, and mapping information be transmitted over a car network via a universal architecture?; What degree of market penetration is required for effectiveness?; Are there unique applications for specialty vehicles (transit bus, commercial vehicles, light rail, etc.)?

In the paper we present the history of the autonomous cars and explore different issues relevant to V2V, analyzing the research conducted so far, the technological solutions available for addressing the safety problems. The used communication technology is highlighted. Research problems and corresponding approaches are shown.

**Keywords:** Autonomous cars · Vehicle-to-vehicle communications · Connected vehicle · Safety applications · Vehicle awareness device · Communication security · V2V technology · V2I technology

## 1 Autonomous Cars. Definitions

Autonomous cars, or cars that run without human control, have been developed over the past several decades, starting in 1977. Currently, we have autonomous cars still in experimental and development stage that have driven autonomously thousands of miles. According to World Health Organization, each year, approximately 1.2 million lives are lost due to traffic accidents worldwide, and around 50 million people suffer car accidents. Self-driving cars promise to reduce this number of deaths and injuries. Some autonomy systems are already being used in the cars such as: Cruise Control, Anti-Lock Brakes. Some systems are just starting to be used: Stability and Traction Control, Pre-Accident Systems, Traffic Jam Assist, Self-Parking Systems. In 2015, car manufacturers will introduce new feature — wireless broadcast of vehicle operational data. This will

© Springer International Publishing Switzerland 2016
V. Vishnevsky and D. Kozyrev (Eds.): DCCN 2015, CCIS 601, pp. 1–10, 2016.
DOI: 10.1007/978-3-319-30843-2_1

be used both for vehicle-to-vehicle communications (V2V) as well as for vehicle-to-infrastructure communications (V2I). The Society of Automotive Engineers (SAE) introduces J2735 standard, supports interoperability of vehicular applications by use of standardized messages. Autonomous cars use equipment such as: Radar sensors, Cameras, Image-processing software, GPS Units, Accelerometer, Ultrasound Sensor, Wheel Sensor, Laser range Finder. The car will become a "networked computer on wheels". GM has been testing the technology with two different platforms, one being a mobile transponder about the size of a GPS unit while the other is a smartphone application that is tied to the vehicle's display unit. Both platforms use Dedicated Short-Range Communication (DSRC) to transfer data between devices and have a communication range of about one-quarter of a mile.

An autonomous car, also known as a driverless car, self-driving car and robotic car, is an automated or autonomous vehicle capable of fulfilling the main transportation capabilities of a traditional car. As an autonomous vehicle, it is capable of sensing its environment and navigating without human input. Advanced control systems interpret sensory information to identify appropriate navigation paths, as well as obstacles and relevant signage. By fully autonomous cars, we mean that the vehicle is able to completely manage itself from point A to point B, without any human intervention whatsoever. Autonomous cars need to do basically two things to find their way and drive: (a) the complete map of its surrounding area, and (b) its relative position. An autonomous car uses equipment such as: radar sensors, cameras, image-processing software, GPS units, accelerometer, ultrasound sensor, wheel sensor, laser range finder.

The main benefits of autonomous cars are: reduced accidents, traffic reduction, higher safe speeds, reduce traffic police, car insurance premiums will decrease, time saving, reduced number of cars needed, improved transportation of goods, impacts on economy. In 2050, at a time when the manual driving will for so long be history and our manual driving of cars will seem like horse carriages today.

In the United States, the National Highway Traffic Safety Administration (NHTSA) has proposed a formal classification system:

- Level 0: The driver completely controls the vehicle at all times.
- Level 1: Individual vehicle controls are automated, such as electronic stability control or automatic braking.
- Level 2: At least two controls can be automated in unison, such as adaptive cruise control in combination with lane keeping.
- Level 3: The driver can fully cede control of all safety-critical functions in certain conditions. The car senses when conditions require the driver to retake control and provides a "sufficiently comfortable transition time" for the driver to do so.
- Level 4: The vehicle performs all safety-critical functions for the entire trip, with the driver not expected to control the vehicle at any time. As this vehicle would control all functions from start to stop, including all parking functions, it could include unoccupied cars.

## 2   Autonomous Cars. History

Some steps in developing autonomous cars are [10]:

Around **1478** Leonardo Da Vinci sketch a pre-programmed clockwork cart. This cart would have been powered by large coiled clockwork springs, propelling it over 130 feet. The clever control mechanism could have taken the vehicle through a predetermined course.

In **1964–71** the Stanford Artificial Intelligence Laboratory cart was build. It was based on techniques for navigating through an unfamiliar environment with artificial intelligence and machine vision. The cart wandered into a nearby road and survived unscathed.

In **1977** Tsukuba Mechanical Engineering Lab, Japan developed computerized driverless car. It achieved speeds of up to 20 miles per hour, by tracking white street markers with machine vision.

In **1980s** Ernst Dickmanns and his group at University Bundeswehr Munich build robot cars, using saccadic vision, probabilistic approaches such as Karman filters, and parallel computers.

In **1986** teams from the Robotics Institute at the School of Computer Science, Carnegie Mellon University developed Navlab 1, one of the first autonomous vehicle.

**1996** ARGO team from Universities of Parma and Pavia, Italy drove their Lancia Thema test bed car 1200 miles around Italy. 94 % of the time was in autonomous mode.

In **2005**: DARPA's American "grand challenge" begins. The top car, with a max speed of 40 km per hour, to complete the 211 Km desert course was the Volkswagon of Stanford, which finished in 6 hours and 54 min.

Systems that are already in use are leading to autonomous cars are: Cruise Control, Anti-Lock Brakes. The systems that are just starting to be used now or in the near future are:

- Stability and Traction Control: These are the systems that use different sensors in order to determine the car parameters.
- Pre-Accident Systems: These are the systems that sense an imminent crash and prepare the car just before it.
- Traffic Jam Assist: which relieves drivers from the tiring work of stop and go traffic.
- Improved Cruise Control: In addition to the regular cruise control, using radar sensor placed in front of the car, the system can sense the car in front and will adjust the speed accordingly.
- Self-Parking Systems: is the self-parking ability.

Over the 6 years since Google started the project for autonomous cars, the been involved in 11 minor accidents (light damage, no injuries) during those 1.7 million miles of autonomous and manual driving with our safety drivers behind the wheel, and not once was the self-driving car the cause of the accident.

The history of V2V communications research began under the Vehicle Infrastructure Integration Initiative in 2003 [5]. It continue the Automated

Highway System (AHS) research of the 1990s. The mandate was issue by the Intermodal Surface Transportation Efficiency Act. The goal was to have a fully automated roadway or test track in operation by 1997. In 1997 twenty AHS-equipped vehicles demonstrating hands and feet off driving on I-15 in San Diego, California. U.S. Department of Transportation introduced the Intelligent Vehicle Initiative in 1997. It was authorized in the 1998 as Transportation Equity Act for 21st Century. In November 2003 it was announced allocating 75 MHz of spectrum at 5.9 GHz for research purposes to improve transportation mobility. The basic concept of operations was that V2V and vehicle-to- infrastructure (V2I) communication could support safety and mobility applications.

## 3    Autonomous Cars. Communication Systems

Today, most the Google car and the new Mercedes do not rely on V2V but only or mostly on their sensors. By the time a common V2V protocol is established, most driverless cars will be able to drive completely autonomously. V2V communication systems, are designed to prevent crashes in a number of scenarios such as:

- **Intersection assist**. When you approach an intersection, it alerts you if another vehicle is traveling at such a speed on a cross street that it could run a red light or stop sign and hit your car in the side.
- **Left-turn assist**. When in an intersection, it alerts you if there's not enough time to make a left-hand turn because of oncoming vehicles.
- **Do-not-pass warning**. When driving on a two-lane road, the system warns you when a vehicle coming in the opposite direction makes it unsafe to pass a slower-moving vehicle.
- **Advance warning of a vehicle braking ahead**. The system emits an alert when a vehicle that's two or more cars ahead in the same lane — and possibly out of sight — hits the brakes unexpectedly.
- **Forward-collision warning**. A warning will sound if the system detects that you're traveling at a speed that could cause you to hit a slower-moving vehicle in the rear.
- **Blind-spot/lane-change warning**. When traveling on a multi-lane road, this illuminates a warning light when a car is positioned in your blind spot.

Nearly every automaker — including Audi, Volkswagen, BMW, Ford, General Motors, Honda and Toyota — is developing some form of V2V technology. German automakers have launched a pilot program that combines V2V with V2I technology, allowing cars to communicate with each other and with traffic lights. GM is studying the possibility V2V systems also could recognize pedestrians by picking up their cellphone's wireless signal and alerting drivers to an impending collision.

**Vehicular communication systems** are a type of network in which vehicles and roadside units are the communicating nodes. There are several standards. The most popular are: **IEEE 1609**. It is a family of standards which deals with issues such as management and security of the network:

- 1609.1 -Resource Manager: This standard provides a resource manager for WAVE, allowing communication between remote applications and vehicles.
- 1609.2 -Security Services for Applications and Management Messages
- 1609.3 -Networking Services: This standard addresses network layer issues in WAVE.
- 1609.4 -Multi-channel Operation: This standard deals with communications through multiple channels.
- SAE J2735 SE which contains the Concept of Operations and the Software Requirements Specification for the J2735 SE standard, which covers the information exchange between applications in conjunction with wireless communications related to the next generation integrated transportation system, specifically, the interface to connected vehicles an infrastructure.

V2V communications includes a wireless network where cars send messages to each others on frequencies 5.9 GHz. The range is up to 300 m or 1000 feet or about 10 s at highway speeds. V2V communications include:

- Vehicle speed
- Vehicle position and heading (direction of travel)
- On or off the throttle (accelerating, driving, slowing)
- Brakes on, anti-lock braking
- Lane changes
- Stability control, traction control engaged
- Windshield wipers on, defroster on, headlamps on in daytime (raining, snowing)
- Brakes on, anti-lock braking
- Gear position (a car in reverse might be backing out of a parking stall)

In V2V network, every car, smart traffic signal could send, capture and retransmit signals. Five to ten hops on the network would gather traffic conditions a mile ahead. V2V warnings might come as different signals or cars brake and steer around hazards. Eight automakers: GM, Ford, Toyota, Hyundai/Kia, Honda, Volkswagen/Audi, Mercedes-Benz and Nissan/Infiniti, and 2500 vehicles been taking part in a University of Michigan V2V project along 73 lane-miles of roadway in Ann Arbor. GM suggests V2V will be effective when a quarter of cars on the road are equipped.

Some projects and research in the area are listed below.

AutoNet2030 (co-operative Systems in Support of Networked Automated Driving by 2030) is a European project connecting two domains of intensive research: cooperative systems for Intelligent Transportation Systems and Automated Driving. The research issues re as follows: how can all these vehicles with different capabilities most efficiently cooperate to increase safety and fluidity of the traffic system? What kind of information should be exchanged? Which organization (e.g. centralized or distributed) is the best? [2, 4]

[7] proposed model which can describe the network-wide spatiotemporal propagation of information while factoring the constraints arising from traffic flow dynamics and V2V communications and with the integrated multi-layer

framework. The proposed model can describe the interdependencies among information flow, traffic flow and V2V communication events by simultaneously tracking the dynamics of information flow and traffic flow. The captures information flow propagation wave that can characterize how the density, speed, and locations of the vehicles lead to the dynamics of information flow. [3] addresses the design, sensing, decision making, and acting infrastructure and several experimental tests that have been carried out to evaluate both platforms and proposed algorithms. The communication and security aspects are also investigated.

In [1] multimedia-based ad-hoc networking (VANET) is tested to fulfill demands in a vehicular environment, and the need to evaluate the current standards. In [6] set of algorithms that determine the crossing order are fed with information about surrounding vehicles: actual and further GPS position, speed and an identification number. In [8] a novel distributed intrusion detection system designed for a vehicular ad hoc network by combining static and dynamic detection agents that can be mounted on central vehicles, and a control center where the alarms about possible attacks on the system are communicated. The proposed DIDS can be used in both urban and highway environments for real time anomaly detection with good accuracy and response time. In [9] a vision-based multi-object tracking system for checking the plausibility of V2V communication is presented. The system is addressing the challenge of fusing relative sensor observations as provided by a Mobil Eye vision-system with time-delayed absolute GNSS-based measurements from Cooperative Awareness Messages (CAMs) as provided by V2V.

[11] evaluate the effects of buildings on the vehicle-to-vehicle performance at urban intersections based on a profound simulation campaign. Due to the two dimensional nature of intersection topologies, we investigate the performance of V2V communication by analyzing packet delivery ratios and packet drop rates with respect to sender and receiver's position under varying node density and intersection layout. [12] provide an overview of ITS activities and status worldwide. It is divided into 4 main parts: standards/specifications applicable for ITS, spectrum allocation and channel plans, 802.11p PHY details, and test and measurement solutions to aid in design and verification of ITS devices and systems.

The purpose of [5] is to assess the readiness for application of V2V communications, a system designed to transmit basic safety information between vehicles to facilitate warnings to drivers concerning impending crashes. The United States Department of Transportation and NHTSA have been conducting research on this technology for more than a decade. A suggested V2V Security System Design for Deployment and Operations diagram in the report is giving in Fig. 1. The schema presents a full deployment model. This diagram shows also in dotted lines the initial deployment model where there is no Intermediate CA (certificate authority) and the Root CA talks to the MA (misbehavior authority), PCA (pseudonym certificate authority), and ECA (enrollment certificate authority). SCMS stands for Security Credentials Management System and CRL stands for certificate revocation list.

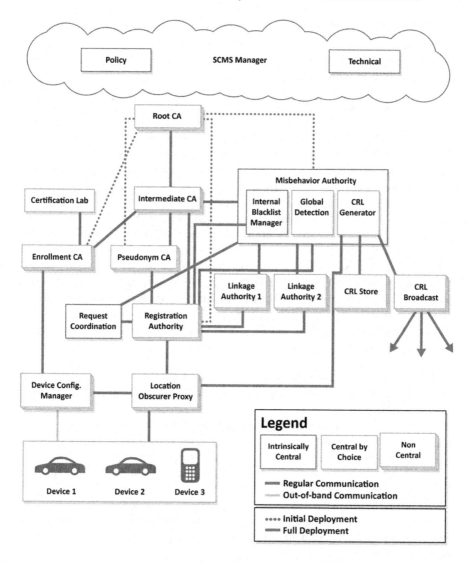

**Fig. 1.** V2V security system design for deployment and operations

Some issues for establishing V2V communications using [5] are:

- Additional protocols that enabled interoperability between devices have to be developed that would support a larger, widespread technology roll-out.
- More research would help the agency develop objective performance tests that would ensure consistent operation that is helpful to drivers.
- Based on preliminary information, NHTSA (National Highway Traffic Safety Administration) currently estimates that the V2V equipment would cost

approximately $341 to 350$ per vehicle in 2020. The communications costs will range from $3 to 13$ per vehicle.

- Based on preliminary estimates, the estimated total annual costs range from $0.3 to 2.1$ billion in 2020.
- In terms of safety impacts, an annual basis potentially prevent 25,000 to 592,000 crashes, save 49 to 1,083 lives, avoid 11,000 to 270,000 MAIS 1–5 injuries, and reduce 31,000 to 728,000 property-damage-only crashes by the time V2V technology had spread through the entire fleet.
- Given that Wi-Fi use is growing exponentially, opening the 5.8–5.9 GHz part of the spectrum could result in many more devices transmitting and receiving information on the same or similar frequencies, which could potentially interfere with V2V communications.
- V2V device certification issues: This means that auto manufacturers (and V2V device manufacturers), attempting to comply with a potential V2V mandate, could have a significant testing obligation to guarantee interoperability among their own devices and devices produced by other manufacturers.
- V2V and safety application standards need to be objective and practicable, meaning that technical uncertainties are limited, that tests are repeatable, and so forth.
- In order to function safely, a V2V system needs security and communications infrastructure to enable and ensure the trustworthiness of communication between vehicles.
- Auto manufacturers repeatedly have expressed to the agency their concern that V2V technologies will increase their liability as compared with other safety technologies.
- Privacy: The system will not collect financial information, personal communications, or other information linked to individuals. The system will enroll V2V enabled vehicles automatically, without collecting any information that identifies specific vehicles or owners. The system will not provide a pipe into the vehicle for extracting data.
- One potential issue with consumer acceptance is maintenance. If the security system is designed to require consumers to take action to obtain new security certificates – depending on the mechanism needed to obtain the certificates – consumers may find the required action too onerous.

Further research to move toward deployment has been identified and will be conducted to address the following:

- The impact of spectrum sharing with U-NII devices;
- Development of performance requirements for DSRC devices;
- Development of performance requirements for safety applications;
- The potential establishment of device certification and compliance procedures;
- The ability to mitigate V2V communication congestion:
- Incorporation of GPS positioning advancements to improve V2V relative positioning;
- Remedies to address false positive warnings from V2V safety applications;

- Driver-vehicle interface performance to enhance crash avoidance warning effectiveness;
- An appraisal of consumer acceptance of the technology;
- Evaluation of V2V system privacy risks.
- An assessment of the security system to ensure a trusted and a safe V2V system.

## 4 Conclusions

Further tests and developments have to be conducted for the V2V applications and combined with the sensor-based systems. The current test procedures should be modified to reflect a greater range of speeds and a greater variety of road geometry configurations. Research in V2V safety has to identify the performance of V2V safety applications. Many questions have to be answered.

## References

1. Akbar, M., Khan, M., Khaliq, K., Qayyum, A., Yousaf, M.: Evaluation of IEEE 802.11n for multimedia application in VANET, science direct. Procedia Comput. Sci. **32**, 953–958 (2014)
2. EU 7th Framework Programme, Project number 610542, Specifications for the enhancement to existing LDM and cooperative communication protocol standards. http://www.autonet2030.eu/wp-content/uploads/2015/02/D3.2-Specifications-cooperative-communication-protocol-standards-draft-for-approval.pdf
3. Fernandes, L., Souza, J., Pessin, G., Shinzato, P., Sales, D., Mendes, C., Prado, M., Klaser, R., Magalhaes, A., Hata, A., Pigatto, D., Branco, K., Grassi Jr., V., Osorio, F., Wolf, D.: CaRINA intelligent robotic car: architectural design and applications. J. Syst. Archit. **60**(4), 372–392 (2013)
4. Fortelle, A., Qian, X., Diemer, S., Grégoire, J., Moutarde, F., Bonnabel, S., Marjovi, A., Martinoli, A., Llatser, I., Festag, A., Katrin, S.: Network of Automated Vehicles: The Autonet2030 Vision, ITS World Congress, Detroit (2014)
5. Harding, J., Powell, G., R., Yoon, R., Fikentscher, J., Doyle, C., Sade, D., Lukuc, M., Simons, J., Wang, J.: Vehicle-to-vehicle communications: Readiness of V2V technology for application. (Report No. DOT HS 812 014). National Highway Traffic Safety Administration, Washington, DC, August 2014
6. Hernandez-Jayo, U., De-la-Iglesia, I., Lacoume, I., Mammu, A.: I-CROSS: intersection crossing warning application based on V2V communications. In: IEEE Vehicular Networking Conference, pp. 206–209 (2013)
7. Kim, Y., Peeta, S., He, X.: An analytical model to characterize the spatiotemporal propagation of information under vehicle-to-vehicle communications,. In: IEEE 17th International Conference on Intelligent Transportation Systems (ITSC), 8–11 October 2014, Qingdao, pp. 1142–1147 (2014)
8. Maglaras, A.: A novel distributed intrusion detection system for vehicular Ad Hoc networks. (IJACSA) Int. J. Adv. Comput. Sci. Appl. **6**(4), 101–106 (2015)
9. Obst, M., Hobert, L., Reisdorf, P.: Multi-sensor data fusion for checking plausibility of V2V communications by vision-based multiple-object tracking. In: IEEE Vehicular Networking Conference, pp. 143–150 (2014)

10. RIDE, The autonomous vehicle network of the future. http://mikechiafulio.com/RIDE/index.htm
11. Tchouankem, H., Zinchenko, T., Schumacher, H.: Impact of buildings on vehicle-to-vehicle communication at urban intersections. In: 12th Annual IEE Consumer Communications and Networking Conference (CCNC), pp. 198–204 (2015)
12. Ward L., Simon M.: Intelligent Transportation Systems Using IEEE 802.11p. http://www.rohde-schwarz-usa.com/rs/rohdeschwarz/images/1MA152_ITS_using_802_11p.pdf

# On Hyper-local Web Pages

D. Namiot[1](✉) and M. Sneps-Sneppe[2]

[1] Lomonosov Moscow State University, Moscow, Russia
dnamiot@gmail.com
[2] Ventspils University College, Ventspils, Latvia
manfreds.sneps@gmail.com

**Abstract.** In this paper, we discuss one approach for development and deployment of web sites (web pages) devoted to the description of objects (events) with a precisely delineated geographic scope. This article describes the usage of context-aware programming models for web development. In our paper, we propose mechanisms to create mobile web applications which content links to some predefined geographic area. The accuracy of such a binding allows us to distinguish individual areas within the same indoor space. Target areas for such development are applications for Smart Cities and retail.

**Keywords:** Browsers · Computer networks · Context awareness · HTML5 · Indoor communication

## 1 Introduction

Our paper deals with mobile web presentations of location-based services. How can we present some local (attached to a certain geographical location) information to mobile users? We are talking about programming (creating) mobile web sites, which content pages correspond to the current location of the mobile user. The traditional scheme is very straightforward. We have to determine the user's location and then create a dynamic web page, the issuance of which is clearly defined by specific geographical coordinates. For example, geo-location is a part of HTML5 standard [1].

As soon as web application obtained (as per user permission, of course) geo-coordinates, it can build a dynamic web page, which content depends on the current location (content is associated with obtained location). Technically, we can render our dynamic page on the client side (right in the browser), when application requests data from server via some asynchronous calls (AJAX) [2], or right on our server (in some CGI-script). In both cases obtained location info is used as a parameter either to AJAX script or to CGI script. For some of the applications (classes of applications), we may use several location-related datasets (e.g., so-called geo-fence [3]), but the common principles are similar. It is so called Location Based Services (LBS) [4].

© Springer International Publishing Switzerland 2016
V. Vishnevsky and D. Kozyrev (Eds.): DCCN 2015, CCIS 601, pp. 11–18, 2016.
DOI: 10.1007/978-3-319-30843-2_2

There are different methods for obtaining location information for mobile users [5]. Not all of them use GPS (GLONAS) positioning actually. Alternative approaches use Wi-Fi, Cell ID, collaborative location, etc. [6]. The above-mentioned geo-location in HTML5 has been a wrapper (interface) for location service. For the most of LBS, their top-level architecture is standard. LBS use obtained location info as a key for any database (data store) with location-dependent data. Location info is actually no more than a key for linking physical space (location) and virtual (e.g., coupon for the store). Only a small number of services actually use the coordinates. The typical example is indoor location based services. The paradigm *Location first* requires a digital map for an indoor space. This map should be created prior to the deployment, and it should be supported in an actual state during service's life time. On the other hand, there is a direction, called context-aware computing. In context-aware computing (ubiquitous computing) services can use other information (not related to geographic coordinates) as the "characteristics" of a user's location. Simplistically, the context is any additional information on the geographical location [7,8]. In this case, additional information (context), with the presence of certain metrics can serve as a unique (up to a certain approximation, of course) feature of a user's location. Or, in other words, we can substitute geo-location with context identification. Why might it be necessary? The typical example is indoor LBS [9]. Traditional geo-positioning can be difficult and positioning accuracy may be insufficient to distinguish the position of the mobile subscriber within the same premises. And yet, it is the distinction between positions within the same space (buildings) may be important for all kinds of services (for example, the buyer is located on the first or second floor of the hall).

Actually, it is a starting point for new approaches in LBS architecture, when the stage with obtaining (detecting) location info could be completely eliminated. Indeed, if location info is no more than a key for some database, then why do not replace geo-keys (e.g., latitude and longitude) with context-related IDs? It is sufficient to identify context and use this identification to search data.

The rest of the paper is organized as follows. In Sect. 2, we describe context identification. In Sect. 3, we describe how this identification could be used in web programming. In Sect. 4, we discuss the generic approaches for incorporating sensing information into web pages.

## 2   Network Proximity

One of the widely used methods for the identification of context is the use of wireless network interfaces of mobile devices (Wi-Fi, Bluetooth). The reasons for this are straightforward. On the first hand, these interfaces are supported in all modern smart-phones. Secondly, for obvious reasons, monitoring of network interfaces is directly supported and executed by the mobile operating systems. Therefore, a survey of network interfaces on the application level can be simplified and not cause additional power consumption, as compared with, for example, a specially organized monitoring for the accelerometer. Information received through the network interface is used to estimate the proximity of the mobile user to the elements of the network infrastructure (network proximity [10]). Note, that other mobile devices can act as these elements too (e.g., Wi-Fi access point, opened right onto mobile phone [11]). The classical form for collecting data about Wi-Fi devices are so-called Wi-Fi fingerprints sets [12]. Wi-Fi fingerprints are digital objects that describe availability (visibility) for network nodes. Their primary usage is navigation related tasks. The alternative approach lets users directly associate some data chunks with existing (or artificially created) network nodes. In other words, it is a set of user generated links between network nodes and some content that could be used by those in proximity to networks nodes. This approach is presented in SpotEx project and associated tools [13,14]. SpotEx lets users create a set of rules (logical productions) for linking network elements and available content. A special mobile application (context-aware browser) is based on the external set of rules (productions, if-then operators). The conditional part of the each rule includes predicates with the following objects:

– identity for Wi-Fi network (name, MAC-address)
– RSSI (signal strength),
– time of the day (optionally),

In other words, it is a set of operators like this:

*IF AccessPointIsVisible ('Cafe') THEN { show content for Cafe }*

Block { *show content for Cafe* } is some data (information) snippet presented in the rule. Each snippet has got a title (text) and some HTML content (it could be simply a link to any external site for example). Snippets could present coupons/discount info for malls, news data for campuses, etc. The context-aware browser (mobile application) maps current network environment against existing database, detects relevant rules (fires them) and builds a dynamic web page. This web page is presented to a mobile user in proximity. In fact, even the name of the application (context-aware browser) suggests the movement of this functionality in a mobile browser. This would eliminate the separate rule base as well as the special (separate) application. In fact, the standard mobile browser should play a role of this application. Rules for the content (data snippets) must be specified directly on the mobile web pages. And data snippets itself are HTML code chunks anyway. As applied implementations, we can mention, for example, Internet of Things applications [15,16]. The usage is very transparent. Data snippets (data, presented to mobile users) depends on visibility for some Wi-Fi access points. It lets us specify the positions for mobile users inside of some building (campus, etc.) Mobile users will see different information for different positions. And this approach does not use geo-coordinates at all. The next interesting direction is EU project FI-WARE [17]. Integration with the FI-CONTENT platform is one of the nearest goals.

## 3   Information Services

Technically, for the reuse of information about network proximity, we can talk about the two approaches. At the first hand, the implementation of a mobile browser can follow the same ideology that supports geo-coding in HTML5 [18]. A function from browser's interface

navigator.geolocation.getCurrentPosition()

accepts as a parameter some user-defined callback (another function). The callback should be called as soon as geo-location is completed. Obtained data should be passed as parameters. Note, that the whole process is asynchronous. By the analogue with the above-mentioned model, a mobile browser can add a new interface function. E.g., *getNetworks()* this function will accept a user-defined callback for accumulating network information (current fingerprint). A good candidate for data model is JSON. The browser will pass fingerprint as a JSON array to a user-defined callback. Each element from this array describes one network and contains the following information:

- SSID - name for access point
- MAC - MAC-address
- RSSI - signal strength

Note, that scanning networks is an asynchronous process in mobile OS. So, callback pattern is a good fit for this. Firefox OS is closest in ideology to this approach [19]. Also, Firefox OS offers Bluetooth API [20]. It has got the similar ideology, but there is no general unifier (e.g., even fields for objects are different). It should be possible, of course, to create some unified wrapper (shell), which will give a general list of networks. The biggest problem (we are not mentioning here the own prevalence and popularity for Firefox OS) is the status for both APIs. Wi-Fi API has just been scheduled yet. At the same time, the Bluetooth API exists, but it is declared preferred (privileged). Privileged APIs can be used by the operating system only. So, it could not be used in applications. The reason for this solution is security. API combines both network scanning and network connection (data exchange). It is the wrong design by our opinion. APIs functionality should be separated. The above-mentioned SpotEx approach is not about the connectivity. Mobile OS should use two separate APIs: one for scanning (networks poll) and one for connecting. Polling for networks does not require data exchange. So, scanning API is safe, and it should not be privileged. It is simple - we should have *WiFiManager* interface (as is, and it could be privileged), and *WiFiScan* with only one function *getNetworks()*:

```
<script>
function callback_function(json_data ) { ... }
WiFiScan.getNetworks(callback_function);
</script>
```

The callback function can loop over an array of existing networks IDs and show (hide) HTML div blocks with data related (associated) to the existing (visible) networks. Actually, it is a fundamental question. Traditionally, wireless networks on mobile phones are used as networks. But they are sensors too. The fact that some network node is reachable (visible) is a separate issue. And it could be used in mobile applications even without the ability to connect to that node. It is the main idea behind SpotEx, and it is the feature (option) we suggest to embed into mobile browsers. How can we present our rules for network proximity? As per our suggestion, each data snipped should be presented as a separate div block in HTML code. E.g., the above-mentioned example looks so:

```
<div id="Cafe_rule">
show content for Cafe
</div>
```

We can use CSS styles to hide/show this block. And this CSS visibility attribute depends on the visibility of Wi-Fi (Bluetooth) nodes. Of course, CSS visibility could be changes in JavaScript. So, our rules could be implemented in JavaScript code. We can directly present the predicates in our code, or describe their parts in CSS too. HTML5 custom attributes are good candidates for new attributes [21]. It means also, that adding some set of rules to existing web page looks like as adding (including) some JavaScript code (JavaScript file).

In general, this approach could change the paradigm of designing mobile web sites. It eliminates the demand to make separate versions for local sites or events. It is enough to have one common site with local offers (events, etc.) placed in hidden blocks. Blocks will be visible to mobile users in a proximity of some network nodes. Local blocks visibility depends on the network nodes visibility and so, it depends on the current location of mobile users. E.g., for the above-mentioned example, mobile users opened Cafe site being physically present in the proximity of Cafe, will see different (additional) data compared with any regular mobile visitor.

Of course, single data source (just one web site) support simplifies (makes it cheaper) the maintenance during life time. Web Intents [22] present the next interesting model for this approach. The Web Intents formation is a client framework (everything is executed in the browser) for the monitoring (polling) and building services interaction within the application. Interactions include data exchange and transfer of control. Web Intents form the core architecture of Android OS [23], but their future status is still unknown after some initial experiments from Google. We should note in this context a similar (by its concept) initiative from Mozilla Labs - Web Activities [24]. But the further status of this initiative is also unclear.

The next possible toolbox is seriously underrated in our opinion. It is a local web server. The first implementation, as far as we know, refers to the Nokia [25]. In our opinion, this is one of the most promising areas for communicating with phone sensors. The next possible idea resembles in some ways the old projects with WAP (Wireless Access Protocol). In this case, a mobile device used some intermediate server (WAP Gateway) for access to internet resources. This intermediate server should be able to collect sensing information (including network sensors). Internet service will get sensing info from our proxy.

## 4   A Generic Approach for Web Sensing

In this part, we would like to discuss the more generic approach (approaches) for embedding sensing information into web pages. As a workaround and prototype for this development, we can present a custom *WebView* for Android. On Android platform is possible to access from JavaScript to Java code for a web page, loaded into *WebView* control. Java code will provide a list of nearby network nodes (calculate the network fingerprint). The key moment here is the need for an asynchronous call from JavaScript, because scanning for wireless networks in Java is the asynchronous process. Let us describe this approach a bit more detailed. On Android side we activate JavaScript interface:

```
public void onCreate(Bundle savedInstanceState) {
super.onCreate(savedInstanceState);
WebView webView = new WebView(this);
setContentView(webView);
WebSettings settings = webView.getSettings();
settings.setJavaScriptEnabled(true);
webView.addJavascriptInterface(new MyJavascriptInterface(), "Network"); }
```

Now we can describe our Java code for getting network fingerprint. As a parameter, we will pass a name for callback function in JavaScript.

```
@JavascriptInterface
public void getNetworks(final String callbackFunction) { }
```

We skip the code for network scanning and demonstrate the final part only. As soon as a fingerprint in obtained, we can present it as JSON array and invoke our callback:

```
webView.loadUrl("javascript:" + callbackFunction + "(' " + data + " ')");
```

And on our web page, we can describe our callback function and call Java code:

```
function f_callback(json) { }
Network.getNetworks("f_callback");
```

This approach lets us proceed network proximity right in JavaScript (in other words, right on the web page). Actually, by the similar manner we can work with other sensors too. It is so-called Data Program Interface [26]. We would like to see something similar as a standard feature in the upcoming versions of Android.

## 5   Conclusion

The paper discusses the use of information about the network environment to create dynamic web pages. We propose several approaches to the implementation of a mobile browser that can handle data on a network (network proximity) to provide users with information tied to the current context. Also, we considered possible implementation details. The basic idea is to separate the functional for scanning network information and real data exchange.

# References

1. Holdener, A.T.: HTML5 Geolocation. O'Reilly Media Inc., Sebastopol (2011)
2. Namiot, D., Sneps-Sneppe, M.: Where are they now – safe location sharing. In: Andreev, S., Balandin, S., Koucheryavy, Y. (eds.) NEW2AN/ruSMART 2012. LNCS, vol. 7469, pp. 63–74. Springer, Heidelberg (2012)
3. Namiot, D., Sneps-Sneppe, M.: Geofence and network proximity. In: Balandin, S., Andreev, S., Koucheryavy, Y. (eds.) NEW2AN/ruSMART 2013. LNCS, vol. 8121, pp. 117–127. Springer, Heidelberg (2013)
4. Prasad, M.: Location Based Services, GIS Development, pp. 3–35 (2002)
5. Tabbane, S.: Location management methods for third generation mobile systems. IEEE Commun. Mag. **35**(8), 72–78 (1997). 10.1109/35.606034
6. Namiot, D.: Context-aware browsing - a practical approach. In: 6th International Conference on Next Generation Mobile Applications, Services and Technologies (NGMAST), pp. 18–23 (2012). http://dx.doi.org/10.1109/NGMAST.2012.13
7. Schilit, G., Theimer, B.: Disseminating active map information to mobile hosts. IEEE Netw. **8**(5), 22–32 (1994). 10.1109/65.313011
8. Namiot, D., Sneps-Sneppe, M.: Context-aware data discovery. In: 2012 16th International Conference on Intelligence in Next Generation Networks (ICIN), pp. 134–141 (2012). http://dx.doi.org/10.1109/ICIN.2012.6376016
9. Kolodziej, K., Danado, J.: In-building positioning: modeling location for indoor world. In : Proceedings of the 15th International Workshop on Database and Expert Systems Applications, pp. 830–834. IEEE (2004). http://dx.doi.org/10.1109/DEXA.2004.1333579
10. Sharma, P., Xu, Z., Banerjee, S., Lee, S.: Estimating network proximity and latency. ACM SIGCOMM Comput. Commun. Rev. **36**(3), 39–50 (2006). 10.1145/1140086.1140092
11. Namiot, D.: Network proximity on practice: context-aware applications and Wi-Fi proximity. Int. J. Open Inf. Technol. **1**(3), 1–4 (2013)
12. Cheng, Y.C., Chawathe, Y., LaMarca, A., Krumm, J.: Accuracy characterization for metropolitan-scale Wi-Fi localization. In: Proceedings of the 3rd International Conference on Mobile Systems, Applications, and Services, pp. 233–245. ACM (2005). http://dx.doi.org/10.1145/1067170.1067195
13. Namiot, D., Schneps-Schneppe, M.: About Location-aware Mobile Messages: Expert System Based on WiFi Spots. In: 5th International Conference on Next Generation Mobile Applications, Services and Technologies (NGMAST), pp. 48–53. IEEE (2011). http://dx.doi.org/10.1109/NGMAST.2011.19
14. Namiot, D., Sneps-Sneppe, M.: Wi-Fi proximity as a service. In: 1st International Conference on Smart Systems, Devices and Technologies, SMART, pp. 62–68 (2012)
15. Schneps-Schneppe, M., Namiot, D.: Open API for M2M applications: what is next?. In: 8th Advanced International Conference on Telecommunications, AICT, pp. 18–23 (2012)
16. Namiot, D., Schneps-Schneppe, M.: Smart cities software from the developer's point of view. In: 6th International Conference on Applied Information and Communication Technologies (AICT), LUA Jelgava, Latvia, pp. 230–237 (2013)
17. Castrucci, M., Cecchi, M., Priscoli, F.D., Fogliati, L., Garino, P., Suraci, V.: Key concepts for the future internet architecture. In: Future Network & Mobile Summit (FutureNetw), pp. 1–10. IEEE (2011). http://dx.doi.org/10.1145/1067170.1067195

18. Aghaee, S., Cesare, P.: Mashup development with HTML5. In: Proceedings of the 3rd and 4th International Workshop on Web APIs and Services Mashups, pp. 10. ACM (2010). http://dx.doi.org/10.1145/1944999.1945009

19. Amatya, S., Kurti, A.: Cross-platform mobile development: challenges and opportunities. In: Trajkovik, V., Anastas, M. (eds.) ICT Innovations 2013. Advances in Intelligent Systems and Computing, vol. 231, pp. 219–229. Springer, Heidelberg (2014)

20. Paul, A., Steglich, S.: Virtualizing devices. In: Crespi, N., Magedanz, T., Bertin, Emmanuel (eds.) Evolution of Telecommunication Services. LNCS, vol. 7768, pp. 182–202. Springer, Heidelberg (2013)

21. Cazenave, F., Quint, V., Roisin, C.: Timesheets. js: when SMIL meets HTML5 and CSS3. In: Proceedings of the 11th ACM Symposium on Document Engineering, pp. 43–52. ACM (2011). http://dx.doi.org/10.1145/2034691.2034700

22. Zheng, C., Shen, W., Ghenniwa, H.H.: An intents-based approach for service discovery and integration. In: IEEE 17th International Conference on Computer Supported Cooperative Work in Design (CSCWD), pp. 207–212. IEEE (2013). http://dx.doi.org/10.1109/CSCWD.2013.6580964

23. Chin, E., Felt, A.P., Greenwood, K., Wagner, D.: Analyzing inter-application communication in android. In: Proceedings of the 9th International Conference on Mobile Systems, Applications, and Services, pp. 239–252. ACM (2011). http://dx.doi.org/10.1145/1999995.2000018

24. Firtman, M.: Programming the Mobile Web, pp. 619–624. O'Reilly Media Inc, Sebastopol (2013)

25. Oliveira, L., Ribeiro, A.N., Campos, J.C.: The mobile context framework: providing context to mobile applications. In: Streitz, N., Stephanidis, C. (eds.) DAPI 2013. LNCS, vol. 8028, pp. 144–153. Springer, Heidelberg (2013)

26. Namiot, D., Sneps-Sneppe, M.: On software standards for smart cities: API or DPI. In: Proceedings of the 2014 ITU Kaleidoscope Academic Conference: Living in a Converged World-Impossible Without Standards?, pp. 169–174. IEEE (2014)

# Analysis of Two-Server Queueing Model with Phase-Type Service Time Distribution and Common Phases of Service

Chesoong Kim[1]([✉]), Alexander Dudin[2], Sergey Dudin[2], and Olga Dudina[2]

[1] Sangji University, Wonju, Kangwon 220-702, Korea
dowoo@sangji.ac.kr
[2] Belarusian State University, 4, Nezavisimosti Ave., 220030 Minsk, Belarus
dudin@bsu.by, dudin85@mail.ru, dudina_olga@email.com

**Abstract.** We consider two server queueing system with an infinite buffer. Customers arrive to the system according to the Markovian Arrival Process. Service time of a customer has a phase-type distribution. The servers use the same equipment (phases of $PH$) for customers processing. So, if service of a customer transits to the phase, at which another server is currently providing the service, the service of the customer is suspended until the phase will become available. Behavior of the system is described by the multi-dimensional Markov chain. The generator of this Markov chain is derived. Expressions for computation of the main performance measures are derived.

**Keywords:** Markovian arrival process · Phase-type service time distribution · Interacting service processes

## 1 Introduction

The simplest queueing models suggest that customer's service time is exponentially distributed. This allows to avoid the necessity of taking into account the elapsed service time when the Markovian process describing behavior of the queue should be constructed. It is quite obvious that in many real world applications of queueing theory assumption about the exponential distribution does not hold true and more general distributions of service time have to be considered. If service time has an arbitrary distribution, it is mandatory to take into account the elapsed or residual service times, e.g., by introducing supplementary variables or considering the embedded Markov chains. This may lead to huge analytical difficulties in analysis of the Markovian process describing dynamics of the queue, especially when multi-server queue is under study. To avoid such difficulties, so called phase type ($PH$) distribution, as natural extension of previously well-known Erlangian and hyper-exponential distribution, was offered, see, e.g., [1]. Good property of this distribution is its generality. Generally speaking, any distribution can be approximated, in sense of a weak convergence, by the

V. Vishnevsky and D. Kozyrev (Eds.): DCCN 2015, CCIS 601, pp. 19–29, 2016.
DOI: 10.1007/978-3-319-30843-2_3

$PH$ type distribution, see, e.g., [3]. Random time having the $PH$ type distribution can be interpreted as the sequence of phases duration of each of which has exponential distribution. The total number of existing phases of service is finite. However, implementation of the phases during the service of a customer may be repeated random number of times. So, in concrete realization of the random time having the $PH$ type distribution the number of phases in a sequence is random.

Formal definition of $PH$ type distribution is given in the next section. For purposes of this paper, we slightly rephrase this definition as follows. There is a virtual network with nodes (phases), say, $\{1, \ldots, M\}$. Random time having the $PH$ type distribution with an irreducible representation $(\boldsymbol{\beta}, S)$ is the time during which some virtual customer stays in this network conditional on the fact that this virtual customer starts its staying in the network from the visit to the state $m$ of the network with probability $\boldsymbol{\beta}_m$, $m = \overline{1, M}$, then it makes transitions inside of the set $\{1, \ldots, M\}$ with intensities given by the entries of the matrix $S$ or leaves the network from any state $m$ with intensity, which is the $m$th component of column vector $\mathbf{S}_0 = -S\mathbf{e}$, where $\mathbf{e}$ denotes unit column vector. In brief, as it was already noted above, random time having the $PH$ type distribution consists of the random number of *virtual* phases, duration of which is exponentially distributed. This allows to replace keeping track of the **continuous** elapsed or residual service time by the keeping track of the **discrete** current phase of the service what greatly simplifies the analysis.

So, phases of service in definition of $PH$ type distribution may be just the virtual entities. However, in many real world situations, random time having $PH$ type distribution may represent the sequence of *real* phases. E.g., processing time of the query in data base consists of implementation of a sequence of input/output operations alternating with the use of CPU. Processing of a car in service station consists of a sequence of technological operations. Security control in airport includes screening of a luggage and passengers with possible additional personal individual passenger inspection. If the service is provided by the single server, no collisions occur. But, if there are several servers operating in parallel, collisions may occur because the servers use some common resources, e.g. memory and CPU, tables and indices of data base, different equipment of a car service station, screening devices and cabins for the personal individual inspection, etc.

Traditionally in analysis of multi-server queues with $PH$ type service time distribution, it is assumed that service processes in the servers are independent. To the best of our knowledge, the systems with interference of phases of service in different servers are not considered in literature. In this paper, we start research of such systems from a relatively simple model where there are only two servers, state spaces of the virtual networks, in terms of which $PH$ type service time distribution is interpreted, coincide and if some phase of the service by a server is required while this phase is busy by another server this phase of service is postponed until it will be released by another server. In some sense, in this model we somehow unite the problems considered in two different popular branches of operations research, queueing theory and scheduling theory, which consider the similar objects (service systems) but with emphasis to different aspects of the problem. Scheduling theory addresses the aspect how the required phases of service should be ordered for different jobs to provide the shortest common processing time and completely ignores the stochastic aspects (arrival of jobs at random moments and random duration of implementation of a service at each phase). Queueing theory accounts these stochastic aspects, but is not focused on proper ordering the phases of the service.

The mathematical model is described in detail in Sect. 2. Behavior of the system is described by the multidimensional continuous time Markov chain in Sect. 3. The infinitesimal generator of this Markov chain as the block structured matrix is presented here. This Markov chain belongs to the class of continuous-time asymptotically quasi-Toeplitz Markov chains. Using this fact, it is shown that, due to impatience of customers, this Markov chain is ergodic for any set of the system parameters. The problem of computation of the stationary distribution of this Markov chain is touched. In Sect. 4, formulas for computation of the main performance measures of the system are presented. Section 5 concludes the paper.

## 2    Mathematical Model

Queueing system with two servers and a buffer of infinite capacity is considered. The structure of the system under study is presented in Fig. 1.

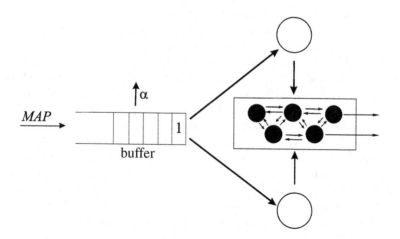

**Fig. 1.** Queueing system under study

Customers arrive at the system according to the Markovian arrival process ($MAP$). Arrivals in the $MAP$ are directed by an irreducible continuous time Markov chain $\nu_t$, $t \geq 0$, with the finite state space $\{0, 1, ..., W\}$. The sojourn time of the Markov chain $\nu_t$, $t \geq 0$, in the state $\nu$ has an exponential distribution with the parameter $\lambda_\nu$, $\nu = \overline{0, W}$. Here, notation such as $\nu = \overline{0, W}$ means that $\nu$ assumes values from the set $\{0, 1, ..., W\}$. After this sojourn time expires, with probability $p_k(\nu, \nu')$ the process $\nu_t$ transits to the state $\nu'$ and $k$ customers, $k = 0, 1$, arrive at the system. The intensities of transitions from one state to another, that are accompanied by the arrival of $k$ customers, are combined to the square matrices $D_k$, $k = 0, 1$, of size $W + 1$. The matrix generating function of these matrices is $D(z) = D_0 + D_1 z$, $|z| \leq 1$. The matrix $D(1)$ is an infinitesimal generator of the process $\nu_t$, $t \geq 0$. The stationary distribution vector $\boldsymbol{\theta}$ of this process satisfies the system of equations $\boldsymbol{\theta} D(1) = \mathbf{0}$, $\boldsymbol{\theta} \mathbf{e} = 1$. Here and throughout this paper, $\mathbf{0}$ is a zero row vector. In case if the dimension of a vector is not clear from the context, it is indicated as a lower index.

The average intensity $\lambda$ (fundamental rate) of the $MAP$ is defined by $\lambda = \boldsymbol{\theta} D_1 \mathbf{e}$.

The $MAP$ arrival process was introduced as a versatile Markovian point process ($VMPP$) by M.F. Neuts in the 70th. The original development of the $VMPP$ contained extensive notations; however these notations were greatly simplified in [4] and ever since this process bears the name Markovian arrival process. The class of $MAP$s includes many input flows considered previously, such as stationary Poisson ($M$), Erlangian ($E_k$), hyper-Markovian ($HM$), phase-type ($PH$), Markov modulated Poisson process ($MMPP$). Generally speaking, the $MAP$ is correlated, so it is ideal to model correlated and or bursty traffic in the modern telecommunication networks, see, e.g., [5–7].

The service time of a customer for the server has a $PH$ (phase-type) distribution with an irreducible representation $(\boldsymbol{\beta}, S)$. This service time traditionally is interpreted as the time until the underlying Markov process $m_t, t \geq 0$, with a finite state space $\{1, \ldots, M, M + 1\}$ reaches the single absorbing state $M + 1$ condition on the fact that the initial state of this process is selected among the states $\{1, \ldots, M\}$ according to the probabilistic row vector $\boldsymbol{\beta} = (\beta_1, \ldots, \beta_M)$. The transition rates of the process $m_t$ within the set $\{1, \ldots, M\}$ are defined by the sub-generator $S$, and the transition rates into the absorbing state (what leads to service completion) are given by the entries of the column vector $\mathbf{S}_0 = -S\mathbf{e}$. The mean service time is calculated as $b_1 = \boldsymbol{\beta}(-S)^{-1}\mathbf{e}$.

But here we assume that the service processes of customers by two servers are not independent. The state space of the underlying Markov processes for both the servers is the same. If there is no collision, these processes make transitions according to definition given above, independently of each other. However if the service of a customer by the $l$-th server, $l = 1, 2$ is at some phase, say, $m$ while the underlying service process of other customer by $l$'th, $l' = 1, 2$, $l' \neq l$, server should transit to this phase $m$, the $l'$-th server is stopped and the service is blocked until the $l$-th server finishes the $m$th phase of the service. The servers of the system are assumed to be identical and are enumerated in arbitrary order. However, if two servers need the same phase of the service, they are assumed being enumerated in order of occupation of the phase. Number 1 is appointed to the server that currently occupies the phase under the conflict. Number 2 is appointed to the server that currently waits for releasing this phase.

The customers from the buffer are impatient, i.e., the customer leaves the buffer and the system after an exponentially distributed waiting time described by the parameter $\alpha$, $0 < \alpha < \infty$.

## 3   Process of System States

It is easy to see that the dynamics of the system under study is described by the following regular irreducible multi-dimensional continuous-time Markov chain

$$\xi_t = \{i_t, r_t, \nu_t, n_t, m_t\}, \, t \geq 0,$$

where, during the epoch $t$, $t \geq 0$,

- $i_t$ is the number of customers in the system, $i_t \geq 0$;
- $r_t$ is an indicator that indicates whether some server is blocked or not: $r_t = 0$ corresponds to the case when a server isn't blocked and $r_t = 1$ otherwise;
- $\nu_t$ is the state of the underlying process of the $MAP$, $\nu_t = \overline{0, W}$;
- $n_t$ is the state of $PH$ service process at the first server, $n_t = \overline{1, M}$.
- $m_t$ is the state of $PH$ service process at the second server, $m_t = \overline{1, M}$, $m_t \neq n_t$.

The Markov chain $\xi_t$, $t \geq 0$, has the following state space:

$$\left( \{0, 0, \nu\} \right) \bigcup \left( \{1, 0, \nu, n\}, \, n = \overline{1, M} \right) \bigcup$$

$$\left( \{i, 0, \nu, n, m\}, \, i \geq 2, \, n = \overline{1, M}, \, m = \overline{1, M}, \, m \neq n \right) \bigcup$$

$$\left( \{i, 1, \nu, n\}, \, i \geq 2, \, n = \overline{1, M} \right), \, \nu = \overline{0, W}.$$

For further use throughout this paper, we introduce the following notation:

- $I$ is the identity matrix, and $O$ is a zero matrix of appropriate dimension;
- $\otimes$ and $\oplus$ indicate the symbols of Kronecker product and sum of matrices (see [8]), respectively;
- $\overline{W} = W + 1$;
- $I_{l_1, l_2}$, $l_1, l_2 = \overline{1, M}$, $l_1 \neq l_2$, is the matrix of size $(M - 1) \times (M - 1)$ with all zero entries except the entries $(I_{l_1, l_2})_{k,k}$, $k = \overline{0, M - 2}$, $k \neq l_2 - 2$, in the case $(l_1 < l_2)$ and $(I_{l_1, l_2})_{k,k}$, $k = \overline{0, M - 2}$, $k \neq l_2 - 1$, in the case $(l_1 > l_2)$ which are equal to 1;
- $\tilde{S}_l$, $l = \overline{1, M}$, is the square matrix of size $M - 1$ that is obtained from matrix $S$ by removing the $l - 1$-th column and the $l$-th row;

- $\mathbf{e}_{l_1,l_2}$, $l_1, l_2 = \overline{1,M}$, $l_1 \neq l_2$, is the column vector of size $(M-1)$ with all zero entries except the entry $(\mathbf{e}_{l_1,l_2})_{l_2-1}$ in the case $(l_1 > l_2)$ and $(\mathbf{e}_{l_1,l_2})_{l_2-2}$, in the case $(l_1 < l_2)$ which are equal to 1;
- $\mathbf{c}_{l_1,l_2}$, $l_1, l_2 = \overline{1,M}$, $l_1 \neq l_2$, is the row vector of size $(M-1)$ with all zero entries except the entry $(\mathbf{c}_{l_1,l_2})_{l_1-2}$ in the case $(l_1 > l_2)$ and $(\mathbf{c}_{l_1,l_2})_{l_1-1}$, in the case $(l_1 < l_2)$ which are equal to 1;
- $\boldsymbol{\beta}_l$, $l = \overline{1,M}$, - the row vector that obtained from the vector $\boldsymbol{\beta}$ by deleting $l-1$-th component;
- $\mathbf{a}_l$, $l = \overline{1,M}$, is the column vector of size $M-1$ that is obtained from the $l-1$-th column of the matrix $S$ by removing the $l-1$-th entry;
- $I_l^+$, $l = \overline{1,M}$, is the matrix of size $(M-1) \times M$ which obtained from the identity matrix of size $M-1$ by adding the zero column in position $l-1$;
- $\mathbf{S}_0^l$, $l = \overline{1,M}$ is a column vector of size $M-1$ which is obtained from the vector $\mathbf{S}_0$ by removing the $l-1$-th component.
- $\tilde{\mathbf{a}}_l$, $l = \overline{1,M}$, is a row vector of size $M$ with all zero components except the component $(\tilde{\mathbf{a}}_l)_{l-1}$ which is equal to 1;
- $B_l$, $l = \overline{1,M}$, is the matrix of size $(M-1) \times M(M-1)$ which obtained from the matrix $\mathrm{diag}\{\boldsymbol{\beta}_1, \ldots, \boldsymbol{\beta}_M\}$ by deleting the $l-1$-th row;
- $C_l$, $l = \overline{1,M}$, is the matrix of size $(M-1) \times M$ which obtained from the matrix $\mathrm{diag}\{\boldsymbol{\beta}_1, \ldots, \boldsymbol{\beta}_M\}$ by deleting the $l-1$-th row;

Let us enumerate the states of the Markov chain $\xi_t$, $t \geq 0$, in the direct lexicographic order of the components $r$, $k$, $\nu$, $\zeta$, $\eta$ and refer to the set of the states of the chain having values $(i,r)$ of the first two components of the Markov chain as a macro-state $(i,r)$.

Let $Q$ be the generator of the Markov chain $\xi_t$, $t \geq 0$, consisting of the blocks $Q_{i,j}$, which, in turn, consist of the matrices $(Q_{i,j})_{r,r'}$ of the transition rates of this chain from the macro-state $(i,r)$ to the macro-state $(j,r')$, $r, r' = 0, 1$. The diagonal entries of the matrices $Q_{i,i}$ are negative, and the modulus of the diagonal entry of the blocks $(Q_{i,i})_{r,r}$ defines the total intensity of leaving the corresponding state of the Markov chain $\xi_t$, $t \geq 0$.

Analysing all transitions of the Markov chain $\xi_t, t \geq 0$, during an interval of an infinitesimal length and rewriting the intensities of these transitions in the block matrix form we obtain the following result.

**Theorem 1.** *The infinitesimal generator $Q = (Q_{i,j})_{i,j \geq 0}$ of the Markov chain $\xi_t$, $t \geq 0$, has a block-tridiagonal structure:*

$$Q = \begin{pmatrix} Q_{0,0} & Q_{0,1} & O & O & \cdots \\ Q_{1,0} & Q_{1,1} & Q_{1,2} & O & \cdots \\ O & Q_{2,1} & Q_{2,2} & Q^+ & \cdots \\ \vdots & \vdots & \ddots & \ddots & \ddots \end{pmatrix}.$$

*The non-zero blocks $Q_{i,j}$, $i, j \geq 0$, have the following form:*

$$Q_{0,0} = D_0,$$

$$Q_{1,1} = D_0 \oplus S,$$

$$Q_{i,i} = \begin{pmatrix} Q_{i,i}^{(0,0)} & Q_{i,i}^{(0,1)} \\ Q_{i,i}^{(1,0)} & Q_{i,i}^{(1,1)} \end{pmatrix}, \, i > 1,$$

$$Q_{i,i}^{(0,0)} = D_0 \otimes I_{M(M-1)} + I_{\bar{W}} \otimes (\mathcal{S} + \mathrm{diag}\{\tilde{S}_1, \ldots, \tilde{S}_M\}) - (i-2)\alpha I_{\bar{W}M(M-1)}, \, i > 1,$$

$$\mathcal{S} = \begin{pmatrix} S_{1,1}I_{M-1} & S_{1,2}I_{1,2} & \cdots & S_{1,M}I_{1,M} \\ S_{2,1}I_{2,1} & S_{2,2}I_{M-1} & \cdots & S_{2,M}I_{2,M} \\ \vdots & \vdots & \cdots & \vdots \\ S_{M,1}I_{M,1} & S_{M,2}I_{M,2} & \cdots & S_{M,M}I_{M-1} \end{pmatrix},$$

$$Q_{i,i}^{(0,1)} = I_{\bar{W}} \otimes \left( \begin{pmatrix} \mathbf{0}^T & S_{1,2}\mathbf{e}_{1,2} & \cdots & S_{1,M}\mathbf{e}_{1,M} \\ S_{2,1}\mathbf{e}_{2,1} & \mathbf{0}^T & \cdots & S_{2,M}\mathbf{e}_{2,M} \\ \vdots & \vdots & \cdots & \vdots \\ S_{M,1}\mathbf{e}_{M,1} & S_{M,2}\mathbf{e}_{M,2} & \cdots & \mathbf{0}^T \end{pmatrix} + \mathrm{diag}\{\mathbf{a}_1, \ldots, \mathbf{a}_M\} \right), \, i > 1,$$

$$Q_{i,i}^{(1,0)} = I_{\bar{W}} \otimes \begin{pmatrix} \mathbf{0} & S_{1,2}\mathbf{c}_{1,2} & \cdots & S_{1,M}\mathbf{c}_{1,M} \\ S_{2,1}\mathbf{c}_{1,M} & \mathbf{0} & \cdots & S_{2,M}\mathbf{c}_{1,M} \\ \vdots & \vdots & \cdots & \vdots \\ S_{M,1}\mathbf{c}_{1,M} & S_{M,2}\mathbf{c}_{1,M} & \cdots & \mathbf{0} \end{pmatrix},$$

$$Q_{i,i}^{(1,1)} = D_0 \oplus \mathrm{diag}\{S_{1,1}, \ldots, S_{M,M}\} - (i-2)\alpha I_{\bar{W}M}, \, i > 1,$$

$$Q_{0,1} = D_1 \otimes \beta,$$

$$Q_{1,2} = \begin{pmatrix} Q_{1,2}^{(0,0)} & Q_{1,2}^{(0,1)} \end{pmatrix},$$

$$Q_{1,2}^{(0,0)} = D_1 \otimes \mathrm{diag}\{\beta_1, \ldots, \beta_M\},$$

$$Q_{1,2}^{(0,1)} = D_1 \otimes \mathrm{diag}\{\beta_1, \ldots, \beta_M\},$$

$$Q^+ = \begin{pmatrix} D_1 \otimes I_{M(M-1)} & O \\ O & D_1 \otimes I_M \end{pmatrix},$$

$$Q_{1,0} = I_{\bar{W}} \otimes S_0, \, Q_{2,1} = \begin{pmatrix} Q_{2,1}^{(0,0)} \\ Q_{2,1}^{(1,0)} \end{pmatrix}, \, i > 1,$$

$$Q_{2,1}^{(0,0)} = I_{\bar{W}} \otimes \left( \begin{pmatrix} \mathbf{S}_{01}I_1^+ \\ \vdots \\ (\mathbf{S}_0)_M I_M^+ \end{pmatrix} + \mathrm{diag}\{(\mathbf{S}_0)^l, \, l = \overline{1,M}\} \right),$$

$$Q_{2,1}^{(1,0)} = I_{\bar{W}} \otimes \begin{pmatrix} (\mathbf{S}_0)_1 \tilde{\mathbf{a}}_1 \\ \vdots \\ (\mathbf{S}_0)_M \tilde{\mathbf{a}}_M \end{pmatrix},$$

$$Q_{i,i-1} = \begin{pmatrix} Q_{i,i-1}^{(0,0)} & Q_{i,i-1}^{(0,1)} \\ Q_{i,i-1}^{(1,0)} & Q_{i,i-1}^{(1,1)} \end{pmatrix}, \, i > 2,$$

$$Q_{i,i-1}^{(0,0)} = I_{\bar{W}} \otimes \left( \begin{pmatrix} \mathbf{S_{01}}B_1 \\ \vdots \\ (\mathbf{S_0})_M B_M \end{pmatrix} + \mathrm{diag}\{(\mathbf{S_0})^l \boldsymbol{\beta}_l, \, l = \overline{1,M}\} \right) + (i-2)\alpha I_{\bar{W}M(M-1)},$$

$$Q_{i,i-1}^{(0,1)} = I_{\bar{W}} \otimes \left( \begin{pmatrix} (\mathbf{S_0})_1 C_1 \\ \vdots \\ (\mathbf{S_0})_M C_M \end{pmatrix} + \mathrm{diag}\{\mathbf{S_0}^l \boldsymbol{\beta}_l, \, l = \overline{1,M}\} \right),$$

$$Q_{i,i-1}^{(1,0)} = I_{\bar{W}} \otimes \mathrm{diag}\{(\mathbf{S_0})_l \boldsymbol{\beta}_l, \, l = \overline{1,M}\},$$

$$Q_{i,i-1}^{(1,1)} = I_{\bar{W}} \otimes \mathrm{diag}\{(\mathbf{S_0})_l \boldsymbol{\beta}_l, \, l = \overline{1,M}\} + (i-2)\alpha I_{\bar{W}M}.$$

**Corollary 1.** The Markov chain $\xi_t$, $t \geq 0$, belongs to the class of continuous-time asymptotically quasi-Toeplitz Markov chains ($AQTMC$), for definition and relevant information see paper [9].

Proof. It can be verified that the limits $Y_0$, $Y_1$ and $Y_2$

$$Y_0 = \lim_{i \to \infty} R_i^{-1} Q_{i,i-1}, \; Y_1 = \lim_{i \to \infty} R_i^{-1} Q_{i,i} + I, \; Y_2 = \lim_{i \to \infty} R_i^{-1} Q_{i,i+1}$$

exist and the matrix $Y_0 + Y_1 + Y_2$ is stochastic where the matrix $R_i$ is a diagonal matrix with the diagonal entries which are defined as the moduli of the corresponding diagonal entries of the matrix $Q_{i,i}$, $i \geq 0$. According to the definition given in [9] this means that the Markov chain $\xi_t$, $t \geq 0$, belongs to the class of $AQTMC$.

Let us analyze the properties of this Markov chain. This analysis includes derivation of conditions which should be imposed on the system parameters to guarantee existence of the stationary distribution of the states of the chain (ergodicity condition) and a procedure for computation of the stationary probabilities of the states.

As follows from [9], a sufficient condition for the existence of a stationary distribution of $AQTMC$ $\xi_t$, $t \geq 0$, is expressed in terms of the matrices $Y_0$, $Y_1$ and $Y_2$ defined above. This sufficient condition of ergodicity of Markov chain $\xi_t$, $t \geq 0$, is fulfillment of inequality

$$\mathbf{y}Y_0\mathbf{e} > \mathbf{y}Y_2\mathbf{e} \tag{1}$$

where the vector $\mathbf{y}$ is the unique solution to the system

$$\mathbf{y}(Y_0 + Y_1 + Y_2) = \mathbf{y}, \; \mathbf{y}\mathbf{e} = 1.$$

It is easy to verify that for the considered Markov chain the matrices $Y_0$, $Y_1$ and $Y_2$ have the following form:

$$Y_0 = I, \; Y_1 = O, \; Y_2 = O.$$

It is easy to see that here the ergodicity condition (1) is transformed to inequality $1 > 0$ which is true for all possible values of the system parameters.

Thus, the following limits (stationary probabilities) exist for any set of the system parameters:

$$\pi(i, r, \nu, n, m) = \lim_{t \to \infty} P\{i_t = i,\ r_t = r,\ \nu_t = \nu,\ n_t = n, m_t = m\},$$

$$i \geq 0,\ r = \overline{0, 1},\ \nu = \overline{0, W},\ n = \overline{1, M},\ m = \overline{1, M}.$$

Then let us form the row vectors $\boldsymbol{\pi}_i$ of the stationary probabilities as follows:

$$\boldsymbol{\pi}_0 = (\pi(0,0,0), \pi(0,0,1), \ldots, \pi(0,0,W)),$$

$$\boldsymbol{\pi}_1 = (\boldsymbol{\pi}(1,0,0), \boldsymbol{\pi}(1,0,1), \ldots, \boldsymbol{\pi}(1,0,W)),$$

where

$$\boldsymbol{\pi}(1,0,\nu) = (\pi(1,0,\nu,1), \pi(1,0,\nu,2), \ldots, \pi(1,0,\nu,M)),\ \nu = \overline{0, W}.$$

$$\boldsymbol{\pi}_i = (\boldsymbol{\pi}(i,0), \boldsymbol{\pi}(i,1)),\ i \geq 2,$$

where

$$\boldsymbol{\pi}(i,r) = (\boldsymbol{\pi}(i,r,0), \boldsymbol{\pi}(i,r,1), \ldots, \boldsymbol{\pi}(i,r,W)),\ r = 0,1,\ \ i \geq 2,$$

$$\boldsymbol{\pi}(i,0,\nu) = (\boldsymbol{\pi}(i,0,\nu,1), \boldsymbol{\pi}(i,0,\nu,2), \ldots, \boldsymbol{\pi}(i,0,\nu,M)),\ \nu = \overline{0, W},$$

$$\boldsymbol{\pi}(i,0,\nu,n) = (\pi(i,0,\nu,n,1), \pi(i,0,\nu,n,2), \ldots, \pi(i,0,\nu,n,M)),\ n = \overline{1, M},$$

$$\boldsymbol{\pi}(i,1,\nu) = (\pi(i,1,\nu,1), \pi(i,1,\nu,2), \ldots, \pi(i,1,\nu,M)),\ \nu = \overline{0, W}.$$

It is well known that the probability vectors $\boldsymbol{\pi}_i$, $i \geq 0$, satisfy the following system of linear algebraic equations:

$$(\boldsymbol{\pi}_0, \boldsymbol{\pi}_1, \ldots)Q = \mathbf{0},\quad (\boldsymbol{\pi}_0, \boldsymbol{\pi}_1, \ldots)\mathbf{e} = 1. \tag{2}$$

To solve system (2), we use the numerically stable algorithm for computation of the probability vectors $\boldsymbol{\pi}_i$, $i \geq 0$, developed in [10] which effectively uses information about the asymptotic behavior of the Markov chain $\xi_t$, $t \geq 0$, and the sparse structure of the generator $Q$.

## 4   Performance Measures

The average number $N_{buffer}$ of customers in the buffer is computed by

$$N_{buffer} = \sum_{i=3}^{\infty} (i-2)\boldsymbol{\pi}_i \mathbf{e}.$$

The average number $N_{busy}$ of busy servers at an arbitrary moment is computed by

$$N_{busy} = \sum_{i=1}^{\infty} \min\{i, 2\}\boldsymbol{\pi}_i \mathbf{e}.$$

The probability $P_{blocked}$ that a server is blocked at an arbitrary moment is computed by

$$P_{blocked} = \sum_{i=2}^{\infty} \pi(i,1)\mathbf{e}.$$

The probability $P_{loss}$ that an arbitrary customer will be lost (due to impatience) is computed by

$$P_{loss} = \frac{\alpha N_{buffer}}{\lambda}.$$

The average intensity $\lambda_{out}$ of flow of customers who receive service is computed by

$$\lambda_{out} = \lambda(1 - P_{loss}).$$

## 5   Conclusion

Two server queueing model with an infinite buffer and Markovian arrival process is analysed. Service times by two servers have phase type distribution with coinciding state spaces of underlying Markov chains. The phases of service times at two servers are implemented independently if the underlying Markov chains of services currently have different states (phases). If the required phases of service coincide, service by one of the servers is postponed until the phase will be released by the competitive server. Generator of multi-dimensional Markov chain describing behavior of the system is written down. Formulas for computation of the key performance measure of the system in terms of stationary probabilities of the Markov chain are presented. It is planned to extend the results to the cases when the number of servers is more than two, when the state spaces of underlying processes of service coincide partially, when the vectors and subgenerators defining irreducible representations of service times may we selected in such a way as to minimize possibility of conflicts, when the number of active servers can be dynamically changed, etc.

**Acknowledgements.** This research was supported by Basic Science Research Program through the National Research Foundation of Korea (NRF) funded by the Ministry of Education (Grant No.2014R1A1A4A01007517).

## References

1. Neuts, M.F.: Matrix-geometric Solutions in Stochastic Models. The Johns Hopkins University Press, Baltimore (1981)
2. Bocharov, P.P., D'Apice, C., Pechinkin, A.V., Salerno, S.: Queueing Theory. VSP, Utrecht-Boston (2004)
3. Asmussen, S.: Applied Probability and Queues. Springer, New York (2003)
4. Lucantoni, D.: New results on the single server queue with a batch Markovian arrival process. Commun. Stat. Stoch. Models **7**, 1–46 (1991)

5. Heyman, D.P., Lucantoni, D.: Modelling multiple IP traffic streams with rate limits. IEEE/ACM Trans. Netw. **11**, 948–958 (2003)
6. Klemm, A., Lindermann, C., Lohmann, M.: Modelling IP traffic using the batch Markovian arrival process. Perform. Eval. **54**, 149–173 (2003)
7. Chakravarthy, S.R.: The batch Markovian arrival process: a review and future work. In: Krishnamoorthy, A., Raju, N., Ramaswami, V. (eds.) Advances in Probability Theory and Stochastic Processes, pp. 21–29. Notable Publications Inc., New Jersey (2001)
8. Graham, A.: Kronecker Products and Matrix Calculus with Applications. Ellis Horwood, Cichester (1981)
9. Klimenok, V.I., Dudin, A.N.: Multi-dimensional asymptotically quasi-Toeplitz Markov chains and their application in queueing theory. Queueing Syst. **54**, 245–259 (2006)
10. Dudina, O., Kim, C., Dudin, S.: Retrial queueing system with Markovian arrival flow and phase-type service time distribution. Comput. Ind. Eng. **66**, 360–373 (2013)

# Optimization of Topological Structure of Broadband Wireless Networks Along the Long Traffic Routes

V.M. Vishnevsky[✉], Andrey Larionov, and R.V. Smolnikov

V. A. Trapeznikov Institute of Control Sciences of Russian Academy of Sciences,
Moscow, Russia
vishn@inbox.ru, Larioandr@gmail.com, rodion.smolnikov@gmail.com

**Abstract.** The paper deals with a relevant problem of optimal placement of base stations in broadband wireless networks along the long highways. This problem is formulated in terms of integer linear programming. The exact solution of the problem as well as the high-speed heuristic algorithm are presented.

**Keywords:** Topology structure synthesis · Broadband wireless networks · Linear and nonlinear programming · Heuristic algorithm · Network protocols

## 1 Introduction

Broadband wireless networks and communication channels have become recently one of the main directions of development of the telecommunications industry [1,2]. The development of networks of this class for creation of a modern infrastructure for multimedia data transmission along the long transport routes is one of the major problems of implementation of new transport routes and pipelines and exploiting the existing ones.

Creation of such communication infrastructure allows to: provide the operating control over the technical parameters of a route by the means of high-speed data transfer from sensors and data units to the Control Center; provide the security control over the route sections and strategically important objects using data from the video surveillance systems; provide the voice communication (IP-telephony) and transmission of multimedia information between the stationary and mobile objects on long highways as well as communication with the Control Center, etc.

Due to the high practical demands for high-performance communication networks along the long transport routes based on IEEE 802.11-2012 [3] hardware and software, in recent years a considerable number of studies on this class of

This research was financially supported by the Ministry of Education and Science of the Russian Federation in the framework of the applied research project №14.613.21.0020 of 22.10.2014 (RFMEFI61314X0020).

V. Vishnevsky and D. Kozyrev (Eds.): DCCN 2015, CCIS 601, pp. 30–39, 2016.
DOI: 10.1007/978-3-319-30843-2_4

networks have appeared [4–11]. Particularly, in [4] the problem of wireless network base stations deployment, maximizing the network coverage is investigated under constraints on the total network cost. The initial data for solution of the problem are the potential locations for deployment of the base stations, and the beforehand-collected statistics of traffic from fixed and mobile users. The close problem of the deployment of base stations, maximizing the coverage, is formulated in [5]. For the analytical description of the problem, it is modeled as a Maximum Coverage with Time Threshold Problem (MCTTP), and a genetic algorithm is used to solve it. Strategies for roadside units (RSUs) placement based on the road traffic characteristics, aiming at improving connectivity in vehicular ad hoc networks, are presented in [6]. To divide the coverage area of each RSU, the authors propose an Expansion and Coloration Algorithm (ECA). The average connectivity model for all vehicles in the network is established based on the results obtained from ECA. The RSUs placement problem is formulated as a combinatorial optimization problem, of which the objective is to maximize the average connectivity probability by searching for an optimal position combination of the given RSUs. Taking part of an actual urban road network as an example, the RSUs placement problem is calculated and the optimal placement scheme is evaluated. Simulation results show that the optimal placement scheme obtained from the proposed strategy leads to the best connectivity compared to uniform placement and hot-spot placement.

The problem of roadside units placement in IEEE 802.11p/WAVE (wireless access in vehicular environment) networks is studied in [7]. An analytical model is presented that allows to analyze the delay of data transmission in communication networks along highways. The cases of bound and unbound base stations are investigated. It is shown that only those deployment strategies are efficient in which roadside units are connected to each other within the line-of-sight range. Paper [8] designs a connectivity analysis scheme for the roadside-to- vehicle telematics network based on the real movement history of vehicle objects collected from taxi telematics system currently in operation, aiming at providing a useful guideline and information to build a telematics network. The implemented analyzer can locate the current and previous positions of all vehicles, decide whether it can be connected to an RSU, and calculate the duration of disconnected state, taking into account the transmission range, the number of RSUs, and RSU deployment. The RSU placement scheme can improve the network coverage exploiting the analysis result.

The results of simulation and measurement of the performance of a roadside unit placement scheme for the vehicular telematics network on the road network of Jeju city (South Korea) are presented in [9]. The calculated optimal topology of the backbone wireless network provides improvement of connectivity and reduction of the disconnection interval for the given number of roadside units, the transmission range, and the overlap ratio. A research problem of finding the optimal locations to place dissemination points (i.e. roadside infrastructure nodes for information dissemination) is considered in [10]. In this paper a novel approach is proposed for dissemination points placement in grid road networks without knowing trajectories. Based on the analysis of path number between two intersections, a probabilistic model is proposed to get the trajectories estimation of vehicles. The problem of the roadside unit placement in vehicular networks is studied in [11], where the authors focus on the highway-like scenario in which there may be multiple lanes with exits or intersections along the road. In the proposed model, each vehicle can access RSUs in two ways: (1) direct delivery, which occurs when the vehicle is in the transmission range of the RSUs, and (2) multi-hop relaying, which takes place when the vehicle is out of RSU transmission range. Both access patterns in this placement strategy are worked out and this placement problem is formulated via an integer linear programming model such that the aggregate throughput in the network can be maximized. The impact of wireless interference, vehicle population distribution, and vehicle speeds are also taken into account in the formulation. The performance of the proposed placement strategy is evaluated via NS-2 simulations to generate vehicle mobility patterns.

In this article is considered RSU optimal placement problem (discrete Case) with number of practical limitations. This limitations was founded while development broadband wireless network for Ring Road of Kazan City (M7 "Volga").

## 2   Formulation of the Problem

The problem is to choose placements of Road Side Units (RSU), where every RSU has its own connection coverage radius, radio transmission range and cost. Besides this set of RSU must has total cost less than a budget and the arrangement must cover as big area as possible. Consider a one-dimensional road, which is a straight line $\alpha[A, B]$ of length $a$. Consider set of points $X \subset \alpha$, and these points are possible placements for considered RSU. Consider C - a budget to be spent for net creation. $T = \{t_1 \ldots t_n\}$ - set of station types, where every type is characterized by three parameters:

- $r$ - coverage radius. It is a half of length of such segment *coverage segment*, containing a RSU in the middle of one, as every client inside the segment can communicate to the RSU and every other one can't.
- $R$ - transmission range. It is a half of length of such segment *connection segment*, containing a RSU in the middle of one, as every another RSU inside the segment can establish a connection with the RSU and every other one can't.
- $c$ - station cost.

Let give some definitions. Let the section $\alpha$ has orientation and $x_A < x_B$. Designate r(t), R(t), c(t) as coverage radius, transmission range and station cost of station type t respectively.

**Definition 1.** Internal arrangement *or just arrangement of RSU is called ending set of pairs* $P = \{(x_1, t_1), (x_2, t_2) \ldots (x_N, t_N)\}$, $x_1 < x_2 < \cdots < x_N$, *where* $x_i \in X$ - *possible placement,* $t_i \in T$ - *station type.*

**Definition 2.** Correct arrangement *of RSU is called an arrangement* $P = \{(x_1, t_1), (x_2, t_2) \ldots (x_N, t_N)\}$, *which:*

$$\forall i = \overline{1, N-1} : x_{i+1} - x_i \leq \min\{R(t_i), R_{(t_{i+1})}\}, \tag{1}$$

*Ie every neighbour stations can establish a connection with each other.*

Designate $x(p)$ $t(p)$ as coordinate and station type respectively in pair p.

**Definition 3.** Extended arrangement *of RSU is called ending set of such pairs* $\hat{P} = \{(x_A, t_0), (x_1, t_1) \ldots (x_N, t_N)(x_B, t_0)\}$, *where* $\{(x_1, t_1), (x_2, t_2) \ldots (x_N, t_N)\}$ *is an internal arrangement,* $t_0$ *- is special station type, which:*

- $c(t_0) = 0$, *RSU of this type has zero cost*
- $r(t_0) = 0$, *RSU of this type doesn't serve clients*
- $R(t_0) = |\alpha|$, *Ie if RSU in* $x_A$ *and in* $x_B$ *of type* $t_0$ *are situated inside transmission ranges of RSU in* $x_1$ *and* $x_n$ *respectively than connections is established. (Ends of* $\alpha$ *achievable by internal arrangement)*

*Type* $t_0$ *- is an device for connection with gateways.*

Obviously, an extended arrangement is correct if and only if its internal arrangement is correct.

**Definition 4.** Coverage $\Delta(P)$ *is called set of points, which consists of every points of* $\alpha$ *being situated inside coverage radius at least one of placed RSU:*

$$\Delta(P) = \{x \in \alpha, \exists p \in P : |x(p) - x| \leq r(t(p))\}, \tag{2}$$

Obviously, $\Delta(P)$ is a set of segments.

**Definition 5.** *Let* $P$ *- an arrangement. Coverage area* $\mathfrak{M}(\Delta(P))$ *is called summary length of set of segments defined in.*

**Definition 6.** Coverage percentage *is called ratio:*

$$\delta(P) = \frac{\mathfrak{M}(\Delta(P))}{\mathfrak{M}(\alpha)}, \tag{3}$$

Considering all definitions above, the general formulation of the problem is:

**Problem 7.** *Find such extended correct arrangement* $\hat{P}_0$ *as:*

$$\mathfrak{M}(\Delta(\hat{P}_0)) = \max(\mathfrak{M}(\Delta(\hat{P}))), \tag{4}$$

*with limitations*

$$\begin{cases} \hat{P} - & an\ extended\ correct\ arrangement \\ \sum_{i=1}^{N} c(t(p_i)) \leq & C,\ \text{- total system cost is less than budget } C \end{cases}$$

Other words, telecommunications coverage area along segment of highway $\alpha$ is needed to be maximized considering limitations of total cost and connection availability as described in Definition 2. That problem is too difficult to solve, so *discrete case* will be considered as being of the most practical interest. The discrete case has additional limitations for set X:

Let $X = \{x_1 \ldots x_m\}$, $\forall i = \overline{0 \ldots m+1} : x_i \in \alpha$, where m - ending number and every $x_i \in \alpha$, so X is a discrete set of points.

The formulation with last limitation may be used for modeling of real problem, for example, when there are number of power transmission towers along highway.

## 3    Formulation Discrete Problem in Terms of Linear Programming

**Input Parameters:**

**X**=$\{x_0, x_1 \ldots x_m, x_{m+1}\}$, - set of possible placement points and $x_0 = A$, $x_{m+1} = B$.

**T**=$\{t_0 \ldots t_n\}$, - set of available station types and $t_0$ is the special type, described in Definition 3.

**C** - budget size for purchase RSUs.

**Boolean Variables are Introduced:**

$$\mathbf{e_i} = \begin{cases} 1, & \text{If in point } i \text{ is situated an RSU} \\ 0, & \text{If point } i \text{ is empty} \end{cases} , \forall i = \overline{0 \ldots m+1}, \tag{5}$$

$$\mathbf{v_{i,j}} = \begin{cases} 1, & \text{If in point } i \text{ is situated an RSU of type } j \\ 0, & \text{If point } i \text{ is empty} \\ & \text{or type installed RSU different.} \end{cases} , \forall i = \overline{0 \ldots m+1}, \tag{6}$$

$$\mathbf{u_{i,j}} = \begin{cases} 1, & \text{If stations are situated in points} \\ & i \text{ and } j \text{ respectively and} \\ & \text{establish a connection between each other.} \\ 0, & \text{in other case.} \\ & \text{In addtion, the station is considered not} \\ & \text{to connect itself. } u_{i,j} = 0 \text{ if } i = j. \end{cases} , \begin{cases} \forall i = \overline{0 \ldots m+1}, \\ \forall j = \overline{0 \ldots m+1}, \end{cases} \tag{7}$$

$$\mathbf{z_{i,j,k}} = \begin{cases} 1, & \text{If a station of type } j \text{ are situated in point } i \\ & \text{and establish a connection with} \\ & \text{station in point } k \\ 0, & \text{In other case} \end{cases} , \begin{cases} \forall i = \overline{0 \ldots m+1}, \\ \forall j = \overline{1 \ldots n}, \\ \forall k = \overline{0 \ldots m+1}, \end{cases} \tag{8}$$

**Real Variables are Introduced:**

$\rho_i^+$ – segment, *left* end of one is in point $i$

$\rho_i^-$ – segment, *left* end of one is in point $i$

This sets of segments are real coverage segment of an RSU:

$\rho_i^+$ and $\rho_i^-$ are covered only by RSU in point $i$.

if $e_i = 0$ (i.e. point $i$ is empty) than $\rho_i^{+-} = 0$.

$\theta_i$ – cost of station being situated in point. $i$ $e_i = 0 \Rightarrow \theta_i = 0$.

**The Objective Function:**

$$\Phi = \sum_{i=0}^{m+1} (\rho_i^+ + \rho_i^-), \tag{9}$$

Maximum of $\Phi$ is sought. It ensures the widest area of highway coverage.

**Limitations of Correctness of Arrangement:**

**Condition 1.** *There isn't more than one RSU in a point:*

$$\forall i = \overline{0 \dots m+1} : \sum_{j=1}^{n} v_{i,j} = y_i, \tag{10}$$

**Condition 2.** *Points A and B has RSUs of special type $t_0$*

$$v_{0,0} = 1,$$
$$v_{m+1,0} = 1, \tag{11}$$

This limitation has come from Definition 3.

**Condition 3.** *There aren't another points except A and B where is installed RSU of type $t_0$:*

$$\forall i = \overline{1 \dots m}, v_{i,0} = 0, \tag{12}$$

**Cost Limitations:**

**Condition 4.** *Cost of RSU in point i:*

$$\forall i = \overline{0 \dots m+1}, \sum_{j=1}^{n} c(t(j)) v_{i,j} = \tilde{c}_i, \tag{13}$$

**Condition 5.** *Total cost of the system is less than budget size:*

$$\sum_{i=0}^{m+1} \theta_i \leq C, \tag{14}$$

As it follows from $4$ the sum $\sum\limits_{i=0}^{m+1} \theta_i$ is exactly total cost of set all installed RSU.

**Limitations of Coverage:**

**Condition 6.** *Coverage segments $\rho_i^+ \; \rho_i^-$ must be less than coverage radius of RSU in point i:*

$$\forall i = \overline{0 \dots m+1}, \sum_{j=1}^{n} r(t(j))v_{i,j} \geq \rho_i^+,$$

$$\forall i = \overline{0 \dots m+1}, \sum_{j=1}^{n} r(t(j))v_{i,j} \geq \rho_i^-,$$

(15)

Ie real coverage segment is everywhere equal ar less than declared by RSU type.

**Condition 7.** *Every segments $\rho_i^+ \; \rho_j^-$ don't have more than one mutual points:*

$$\forall i = \overline{0 \dots m+1}, \forall j = \overline{0 \dots m+1}, x_i + \rho_i^+ \leq x_j - \rho_j^-, \tag{16}$$

In terms of coverage area it means that the segments not intersect each other.

**Limitations RSUs Links:**

**Condition 8.** *Every station except RSU in point B has only one RSU with bigger coordinate value which establish a connection to:*

$$\forall i = \overline{0 \dots m+1}, \sum_{j=i+1}^{k+1} u_{i,j} = y_i, \tag{17}$$

**Condition 9.** *Every station except RSU in point A has only one RSU with less coordinate value which establish a connection to:*

$$\forall i = \overline{0 \dots m+1}, \sum_{j=0}^{i-1} u_{i,j} = y_i, \tag{18}$$

**Condition 10.** *Every connection is reciprocal:*

$$\forall i = \overline{0 \dots m+1}, j = \overline{0 \dots m+1}, u_{i,j} = u_{j,i}, \tag{19}$$

That limitations means every RSU except of RSUs in points $A$ and $B$ has exactly two neighbours one - on the left and one on the right.

**Condition 11.** *Relation between variables $z_{i,j,k}$ and $u_{i,k}$, $v_{i,j}$:*

$$\forall i = \overline{0 \dots m+1}, \forall j = \overline{1 \dots n}, \forall k = \overline{0 \dots m+1}, z_{i,j,k} = u_{i,k}v_{i,j}, \tag{20}$$

This equation matches with $z_{i,j,k}$ description.

**Condition 12.** *RSU of type $j$ installed in point $i$ establish a connection with RSU in point $k$ only if the last is inside of transmission range of first:*

$$\forall i = \overline{0 \ldots m+1}, \forall j = \overline{1 \ldots n}, \forall k = \overline{0 \ldots m+1}, z_{i,j,k}(R(t_k) - |x_i - x_k|) \geq 0, \ (21)$$

Limitation 11 is not linear. But according to [12] it may be replaced by two linear ones:

**Lemma 1.** *Let $x_j$ and $z$ - boolean variables. Than equality:*

$$\begin{cases} \sum_{i=1}^{J} x_i - z \leq J - 1 \\ - \sum_{i=1}^{J} x_i + J \cdot z \leq 0 \end{cases} \Leftrightarrow \prod_{i=1}^{J} x_i = z, \qquad (22)$$

Using the lemma, *Condition* 11 becomes identically equal to:

**Condition 13.**

$$\begin{cases} u_{i,k} + s_{i,j} - z_{i,j,k} \leq 1, \\ 2z_{i,j,k} - u_{i,k} - s_{i,j} \leq 0, \end{cases} \begin{cases} \forall i = \overline{0 \ldots m+1}, \\ \forall j = \overline{1 \ldots n}, \\ \forall k = \overline{0 \ldots m+1}, \end{cases} \qquad (23)$$

Formulation of discrete problem of optimization of topological structure of broadband wireless networks along the long traffic routes in terms of linear programming:

**Problem 8.** *Find set $\rho_i^+(0), \rho_i^-(0), i = \overline{0 \ldots m+1}$ which:*

$$\Phi(\rho_i^+(0), \rho_i^-(0)) = \max(\Phi(\rho_i^+, \rho_i^-)), \qquad (24)$$

*with number of limitations 1-13 (except 12).*

# 4  Solution of Discrete Problem in Terms of Linear Programming

To solve the problem of Integer linear programming Problem 8 can be used a variety of methods including: Branch and bound method, implemented by, inter alia, in the form of the application package GLPK. For the exact solution of the problem of low dimension can be used restricted brute-force algorithm. The definition "the problem of Low dimension" refers to input parameters, in which the brute-force method works within a reasonable time. Branch and bound method and other popular methods for solving the mixed integer linear programming tasks are now implemented in such packages of applied programs as FortMP, Gurobi, Parma Polyhedra Library, GLPK, etc. This problem of Integer linear programming (discrete task) solved by using package GLPK (GNU Linear Programming Kit). GLPK prefers because it is a popular cross- platform system

**Table 1.** Calculation of Problem 8 by Branch and bound method

| | Number of RSU types | | |
|---|---|---|---|
| Number of possible placements | T = 4 | T = 5 | T = 6 |
| X = 18 | 4,5 | 9,2 | 16,9 |
| X = 28 | 86,4 | 614,2 | - |
| X = 54 | 241,5 | - | - |

for solution of linear programming problems using the full power of advanced algorithms. The GLPK has the ability to solve problems in many different methods, such as simplex method, branch and bound method and many others. In this case- the case of the mixed problem of linear programming (Mixed Integer Problem)-branch and bound method is used. For each ongoing experiment GLPK-project consists of two text files: file with a description of the model (common for all experiments) and model data file generated by Java application from user input. Both files are written in the programming language MathProg (GNU MathProg Language)-main language for modeling of linear programming solvers (for example, glpsol).

Results of the problem solution by Branch and bound method are presented in Table 1. Time characteristics expressed in seconds. Dashes means that too much time took calculation of problem with appropriate dimension (reasonable time is 10 min).

GLPK successfully copes with Problem 8 of dimensions $0 < |P| < 50$, $0 < |T| < 5$. It needs to note, that parameter $|T|$ has more influence to calculation time than parameter $|P|$. Heuristic algorithm based on gradient descent is developed for solving bigger dimension problems.

## 5   Conclusion

Practically important and theoretically interesting problem has been formulated in this article. General results are:

The problem of optimization of topological structure of broadband wireless networks along the long traffic routes has been formulated. Exact solution has been got using GLPK packages. Heuristic algorithm for high dimension cases has been researched. Calculation utilities have been developed.

## References

1. Vishnevsky, V.M., Portnoi, S.L., Shakhnovich, I.V.: WiMAX Encyclopaedia. Way to 4G, p. 470. Tekhnosfera, Moscow (2010)
2. Vishnevsky, V.M., Semenova, O.V.: Polling Systems: Theory and Applications for Broadband Wireless Networks, p. 317. Academic Publishing, London (2012)

3. IEEE Std. - IEEE Standard for Information technology. Telecommunications and information exchange between Local and metropolitan area networks. Specific requirements Part 11: Wireless LAN Medium Access Control (MAC) and Physical Layer (PHY) Specifications, 3 March 2012 (2012)

4. Brahim, M.B., Drira, W., Filali, F.: Roadside units placement within city-scaled area in vehicular ad-hoc networks. In: 3rd International Conference on Connected Vehicles and Expo (ICCVE 2014), Vienna, Austria, 3–7 November 2014 (2014)

5. Cavalcante, E.S., Aquino, A.L., Pappa, G.L., Loureiro, A.A.: Roadside unit deployment for information dissemination in a VANET: an evolutionary approach. In: 14th Annual Conference Companion on Genetic and Evolutionary Computation (GECCO 2012), NY, USA, pp. 27–34, May 2012

6. Liu, H., Ding, S., Yang, L., Yang, T.A.: Connectivity-based strategy for roadside units placement in vehicular ad hoc networks. Int. J. Hybrid Inf. Technol. **7**, 91–108 (2014)

7. Reis, A.B., Sargento, S., Neves, F., Tonguz, O.K.: Deploying roadside units in sparse vehicular networks: what really works and what does not. IEEE Trans. Veh. Technol. **63**, 2794–2806 (2014)

8. Lee, J.: Design of a network coverage analyzer for roadside-to-vehicle telematics networks. In: Ninth ACIS International Conference on Software Engineering, Artificial Intelligence, Networking, and Parallel/Distributed Computing, 6–8 August 2008 (2008)

9. Lee, J., Kim, C.M.: A roadside unit placement scheme for vehicular telematics networks. In: Kim, T., Adeli, H. (eds.) AST/UCMA/ISA/ACN 2010. LNCS, vol. 6059, pp. 196–202. Springer, Heidelberg (2010)

10. Xie, B., Xia, G., Chen, Y., Xu, M.: Roadside infrastructure placement for information dissemination in urban ITS based on a probabilistic model. In: Hsu, C.-H., Li, X., Shi, X., Zheng, R. (eds.) NPC 2013. LNCS, vol. 8147, pp. 322–331. Springer, Heidelberg (2013)

11. Wu, T.-J., Liao, W., Chang, C.-J.: A cost-effective strategy for road-side unit placement in vehicular networks. IEEE Trans. Commun. **60**, 2295–2303 (2012)

12. Vishnevsky, V.M., Lyakhov, A.I., Portnoy, S.L., Shakhnovich, I.V.: Broadband wireless networks of information transfer. Shirikopolosnye besprovodnye seti peredachi informacii, pp. 435–436. Technosfera, Moscowpp (2005)

# Control Strategies of Subscription Notification Delivery in Smart Spaces

Dmitry G. Korzun[1]([✉]), Michele Pagano[2], and Andrey Vdovenko[1]

[1] Petrozavodsk State University, Petrozavodsk, Russia
{dkorzun,vdovenko}@cs.karelia.ru
[2] University of Pisa, Pisa, Italy
m.pagano@iet.unipi.it

**Abstract.** The paper studies performance of the subscription operation in smart spaces for the case when the notification delivery to a client is subject to losses. We consider an active control of the check interval for the client to adapt to the observable loss rate. The adaptive strategy takes into account the number of lost notifications. The performance is analyzed and compared with other strategies under simplified assumptions about the loss distribution.

**Keywords:** Smart spaces · Subscription · Mobile clients · Control strategy · Performance · Simulation

## 1 Introduction

Information sharing in a smart space employs the subscription operation for detection of content changes and for subsequent delivery of notifications to clients [10,13]. Both change detection and notification delivery are subject to losses in networked environments [14], especially with the emergency of Internet of Things (IoT).

In existing solutions for smart spaces, the major role is played by information brokers [5,9]. In this paper, we study a solution closer to the Internet philosophy, in which the control of notification delivery is partially delegated to clients. The latter actively request the broker for new notifications, in addition to default passive waiting for incoming notifications from the broker. A key performance parameter is the check interval, whose length is adapted to the observable loss rate. This issue resembles the "classical" congestion control in Transmission Control Protocol (TCP) [1–3], by which the window size is reduced in case of losses and incremented otherwise.

Our study continues research [8,14]. We consider an active control of notification delivery for subscription operation for the case of notification losses. The problem was initially stated in [14]. In our solution, the client follows an adaptive strategy controlling the check interval based on the number of notifications lost in the latest window. This adaptive strategy was early experimented in [8]. In this extended paper we show that the strategy is a generalization of the

© Springer International Publishing Switzerland 2016
V. Vishnevsky and D. Kozyrev (Eds.): DCCN 2015, CCIS 601, pp. 40–51, 2016.
DOI: 10.1007/978-3-319-30843-2_5

TCP algorithm of additive–increase/multiplicative–decrease (AIMD). We perform simulation experiments to compare the performance with other strategies. Several different notification loss distributions and performance metrics are used to make this comparison more comprehensive.

The rest of the paper is organized as follows. Section 2 states the notification loss problem for subscription operation in smart spaces. Section 3 introduces strategies that a client can additionally use for controlling the notification delivery. Section 4 describes the results of our simulation experiments to evaluate and compare the performance of the considered control strategies. Finally, Sect. 5 concludes the paper.

## 2    Subscription Notification Delivery in Smart Spaces

The pub/sub model is widely used for organizing multi-agent interactions in distributed systems [4]. In this study we consider this model for the case of smart spaces [10, 13].

In general, a smart space forms a sparse-connected multi-agent system deployed in a networked computing environment, typically with access to the Internet [7]. Such an environment consists of various digital devices, including the growing family of IoT devices. Software agents run on the devices and interact over the shared information content. This type of interaction involves, in parallel and asynchronously, a lot of informational sources and destinations. Information sharing makes the interaction indirect, based on a semantic information broker (SIB) [9, 13]. The latter implements a shared information storage, serving requests from agents on read/write operations. In other words, SIB acts as an information hub maintaining knowledge of the whole environment and enabling the agents to construct digital services over this cooperative knowledge.

The subscription operation specifies a persistent query from an agent (a subscription client) to the SIB (a subscription server) for a particular part of the shared content [10]. Whenever the specified part is changed, the agent should receive the subscription notification. Changes are due to parallel activity of other agents, which act as publishers in this interaction (note that an agent may combine the roles of publisher and subscriber). SIB monitors subscriptions of all clients and maps all incoming content changes to the specified interests. Therefore, changes are controlled on the SIB side, and corresponding notifications are sent to the clients. SIB acts as a passive receiver, and we call such subscription notifications passive [14].

We employ Smart-M3 as a reference software platform for creating smart spaces [5, 9]. For each subscription, the SIB maintains a network connection (e.g., a TCP connection) established by the client's request [10, 11]. Knowing the set of all subscriptions, the SIB regularly checks that they are alive, removing the subscription if its network connection is lost.

Smart-M3 follows the best effort style in subscription notification delivery. A notification *should* be sent to a client if a related change in the content has happened. Some notifications can be unsent by SIB due to its overload or internal operability faults. SIB does not check delivery for already sent notifications, and a new notification can be sent although the underlying network connection is broken on the client side.

The above properties do not ensure the dependable notification delivery even if reliable network protocols are used, such as TCP. A possible solution is for a client to have an additional mechanism reducing the number of undelivered notifications. The obvious way is augmenting the passive notification delivery with an active control strategy that the client performs individually on its own.

Consider the following model to formalize the key properties of the subscription notification loss problem in smart spaces. Let $i = 1, 2, \ldots$ be the event-based time evolution on the client side, where $i$ is the index of notification events. An event $i$ is either a passive notification (i.e., received from SIB) or an explicit check of the notification delivery (made by the client within its active control).

Denote by $t_i$ and $k_i$ the time elapsed and the number of losses occurred between $i$ and $i + 1$, respectively. Assume that some initial value $t_0$ is always defined. The values for $k_i$ are non-negative integers. We consider the following distributions to model the notification losses.

1. Let the time elapsed between consecutive losses follow a uniform distribution $\mathcal{U}\{0, \xi t_0\}$. Hence, the average number of losses in any check interval is proportional to its length $t_i$
2. Let $k_i$ follow a Poisson process of parameter $\lambda t_i$. Hence, the number of losses during $t_i$ has the probability mass function

$$\mathbb{P}(k_i = k) = \frac{(\lambda t_i)^k}{k!} e^{-\lambda t_i} \qquad (1)$$

3. Let $k_i$ follow a two-state alternated Poisson process: the loss rate is $\lambda_1$ and $\lambda_2$ with probability $p_1$ and $p_2 = 1 - p_1$, respectively. The assumptions $0 < p_2 < p_1$ and $\lambda_1 \ll \lambda_2$ describe that the network typically operates with moderate losses ($\lambda_1$), while from time to time the network suffers from high losses ($\lambda_2$), e.g., due to burst overload.

## 3   Control Strategies

The notification loss problem is similar to the packet loss problem in TCP congestion control [1–3]. The additive–increase/multiplicative–decrease (AIMD) algorithm is a feedback control algorithm used in the TCP congestion avoidance mechanism. The AIMD algorithm calculates the size $\omega(\tau)$ of congestion window at time $\tau$, after the initial slow start phase.

Denote by $\tau_i$ the end of the $i^{\text{th}}$ TCP round, whose duration is then given by $t_i = \tau_i - \tau_{i-1}$ and by $\omega(\tau)$, where the latter is expressed in MSS units (maximum segment size). The following equation describes the process [2]:

$$\omega_{i+1} = \omega(\tau_i + 0) = \begin{cases} \omega(\tau_i) + 1 & \text{if no loss,} \\ \left\lfloor \dfrac{\omega(\tau_i)}{\kappa} \right\rfloor & \text{if losses are observed on } [\tau_{i-1}, \tau_i], \end{cases} \qquad (2)$$

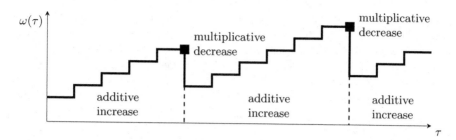

**Fig. 1.** Typical stepwise evolution of TCP congestion window.

The stepwise behavior of $w(\tau)$ is schematically depicted in Fig. 1.

The standard case is $\kappa = 2$, and the multiplicative reduction is applied as soon as a loss is detected (by triple duplicate acknowledgment). Although in TCP modeling the analysis is often carried out at round level, actual TCP implementations react at the arrival of each acknowledgement and not just at the end of round. Moreover, handling of multiple losses for congestion window is still an open issue and highly impacts on TCP performance, especially in wireless and satellite networks [3]. Consequently, the AIMD algorithm cannot be applied directly for the case of subscription notification losses.

For the case of subscription in smart spaces, let us introduce an analogue of the AIMD algorithm to perform active control of notification delivery. If the client has observed no losses during $t_{i-1}$ (i.e., $k_{i-1} = 0$) then this observation indicates that the system state is "good". To save resources the client can increase additively the check interval, i.e. $t_i = t_{i-1} + \delta$ for a fixed parameter $\delta > 0$. As in the AIMD algorithm, the increment is conservative, since a high increase of $t_i$ might determine a clear risk for suffering a burst of losses.

On the contrary, if the client has observed losses, i.e., $k_{i-1} > 0$, it must reduce $t_i$ to decrease the number of losses in near future. The reduction is multiplicative, since the client is interested in fast achieving $k_i = 0$. Moreover, the reaction should take into account the amount of losses and the previous value of the control interval. We apply the multiplicative average

$$t_i = \alpha t_{i-1} + (1 - \alpha)\frac{t_{i-1}}{k_{i-1} + 1},$$

where $0 \le \alpha < 1$ is a fixed parameter (cf. Fig. 1 for $\kappa = k_{i-1} + 1$ and $\alpha = 0$).

We yield the recurrent system that describes an *adaptive strategy* [14], by which the check interval $t_i$ is reduced (multiplicative decrease) in case of losses and incremented (additive increase) otherwise:

$$t_i = \begin{cases} t_{i-1} + \delta & \text{if } k_{i-1} = 0, \\ \dfrac{1 + \alpha k_{i-1}}{k_{i-1} + 1} t_{i-1} & \text{if } k_{i-1} > 0. \end{cases} \tag{3}$$

Note that (3) is valid only for active control of subscription notifications. When a passive notification $i$ is delivered, then the value of $t_i$ is not set by the client.

This adaptive strategy refines the AIMD algorithm of TCP, i.e., providing a TCP-like control of notification delivery. One can interpret $t_i$ in (3) as TCP congestion window. Fixed $\delta = 1$ means the additive increment by one full-sized segment. In the AIMD algorithm, $\alpha = 0$. TCP congestion implies $k_{i-1} = 1$, and $\kappa = k_{i-1} + 1 = 2$ halves the value of $t_i$, as it is described in (3).

In order to evaluate and compare the performance, we consider alternative control strategies and present some auxiliary analytical results. The simplest approach is the strategy of a *constant check interval*, i.e., $t_i = t_0$ for $i = 1, 2, \ldots$. The mean and variance of the random variable $K$, which describes the number of losses, depend only on the loss distributions introduced in Sect. 2. In the case of Poissonian losses, we have:

$$\mathbb{E}\left[K\right] = \text{Var}\left[K\right] = \lambda t_0. \tag{4}$$

Since in many networked systems, randomness can improve the performance, an interesting strategy is *random selection* of the check interval. Intuitively, let $t_i$ be chosen at random around $t_0$. In particular, we consider a continuous uniform distribution $\mathcal{U}\left\{t_0 - \Delta, t_0 + \Delta\right\}$ with $0 < \Delta < t_0$. In the case of Poissonian losses, applying the laws of total expectation and variance (e.g., see [15]), we have

$$\mathbb{E}\left[K\right] = \lambda t_0, \quad \text{Var}\left[K\right] = \mathbb{E}_T\left[\text{Var}\left[K|T\right]\right] + \text{Var}_T\left[\mathbb{E}\left[K|T\right]\right] = \lambda t_0 + \frac{1}{3}\lambda^2\Delta^2. \tag{5}$$

The mean value is the same as in (4), while the variance is increased due to the variability of the interval length. Roughly speaking, in this case the randomness does not change the average performance indexes, but increases the probability that higher number of notifications can be lost.

Finally, we consider a semi-adaptive approach, in which the check interval is halved in case of losses and set to the initial (reference) value $t_0$ otherwise:

$$t_i = \begin{cases} t_{i-1}/2 & \text{if } k_{i-1} > 0, \\ t_0 & \text{if } k_{i-1} = 0. \end{cases} \tag{6}$$

We shall refer this strategy as *multiplicative–decrease*. The evolution of the check interval can be described by a discrete-time Markov chain [6]. State $i$ corresponds to a check interval of length $t_i = 2^{-i}t_0$ and the only possible transitions from state $i$ are back to state 0 (no loss is experienced) with probability $p_i$ and to the following state $i + 1$ (the check interval is halved in case of losses) with probability $q_i = 1 - p_i$, as shown in Fig. 2.

Due to the particular structure of the Markov chain, the state probabilities can be easily written as a function of $\pi_0$ (local balance equations):

$$\pi_{n+1} = \pi_0 \prod_{i=0}^{n} q_i \tag{7}$$

with the additional normalization condition

$$\sum_{n=0}^{\infty} \pi_n = 1 \quad \Rightarrow \quad \pi_0\left[1 + \sum_{n=0}^{\infty} \prod_{i=0}^{n} q_i\right] = 1. \tag{8}$$

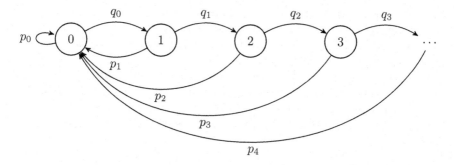

**Fig. 2.** Markov chain for the multiplicative–decrease strategy.

Note that the sum converges under reasonable assumptions over the loss processes. (It is enough to assume that the loss probability goes to zero as $n \to \infty$.) The state probabilities permit to find all the relevant statistics, such as the average check interval

$$\mathbb{E}[T] = \sum_{n=0}^{\infty} \pi_n t_n = t_0 \sum_{n=0}^{\infty} \pi_n 2^{-n}.$$

The numerical values of $p_n$ and $q_n = 1 - p_n$ depend on the loss distribution. In the case of Poissonian losses, we have

$$\begin{cases} p_0 = \mathbb{P}(K = 0 | T = t_0) = e^{-\lambda t_0}, \\ p_n = \mathbb{P}(K = 0 | T = t_n) = p_0^{2^{-n}}, \quad n \geq 1. \end{cases}$$

In the case of uniformly distributed losses, the transition probability $q_n = 1 - p_n$ is proportional to the length of the check interval $t_n = t_0 2^{-n}$, i.e., $q_n = q_0 2^{-n}$. Then (7) is rewritten as

$$\pi_n = \pi_0 q_0^n \left(\frac{1}{2}\right)^{n(n-1)/2}$$

Taking into account the specific values of $p_n$ and $q_n$ as well as the normalization condition (8), it is possible to calculate the state probabilities (at least numerically). In the special case in which the loss probabilities do not depend on the duration of the check interval (i.e., $q_n = q \,\forall n$), the close-form expression for the state probabilities is $\pi_n = (1 - q)q^n$ with the average check interval

$$\mathbb{E}[T] = \sum_{n=0}^{\infty} \pi_n t_n = t_0 \frac{1-q}{1-q/2} > t_0,$$

and the variance

$$\mathrm{Var}[K] = \mathbb{E}[T^2] - \mathbb{E}^2[T] = t_0^2 \frac{q(1-q)/4}{(1-q/4)(1-q/2)^2}.$$

The assumption that the number of losses does not depend on the length of the check interval might be unrealistic. Nevertheless, it provides a lower bound for the average (and an upper bound for the variance) w.r.t. the more realistic assumptions described in Sect. 2.

Moreover, a practical implementation does not use very short check intervals. This property corresponds to truncating the Markov chain at state $N$, remaining in that state in case of further losses (returning to state 0 with the probability $p_N$ as usual). Hence, (7) is still valid for all the states up to $N-1$, while

$$\pi_N = \pi_0 \frac{1}{1-q_N} \prod_{i=0}^{N-1} q_i,$$

and the normalization condition (8) involves just $N+1$ terms.

## 4  Simulation Experiments

In the experiments, we use the three notification loss distributions (see Sect. 2). The key simulation parameters are summarized in Table 1. In all four control strategies, we take $t_0 = 20$ s. According to the selected parameters of the loss distributions, it means that every 20 s one notification is lost on average. The simulation parameters of the four control strategies are presented in Table 2.

The average check interval is exactly $t_0 = 20$ for the strategies of constant check interval and random selection. The adaptive strategy starts from $t_0$ and then forgets about $t_0$, balancing in accordance with the observed number of losses. The multiplicative–decrease strategy attempts to be below $t_0$ when losses are observed, leading to shorter check interval on average.

Experimental behavior of the four control strategies for the given three losses distributions is shown in Figs. 3, 4, and 5. From the visually comparison, it is immediately clear that the adaptive strategy makes the check interval longer compared with the other strategies. From the point of view of the performance, the adaptive strategy saves the resources, reducing the number of network requests from the client to SIB.

Consider quantitative performance indexes. The basic metrics are the average number of losses $k_{avg}$ and the average length of the check interval $t_{avg}$:

$$k_{avg} = \frac{1}{n} \sum_{i=1}^{n} k_i, \quad t_{avg} = \frac{1}{n} \sum_{i=1}^{n} t_i.$$

**Table 1.** Simulation parameters of the notification loss distributions.

| Uniform losses | Poissonian losses | Two-state alternated Poissonian losses |
|---|---|---|
| $\xi = 0.1$ | $\lambda = 0.05$ | $p_1 = 0.8$ and $\lambda_1 = 0.0375$, $p_2 = 0.2$ and $\lambda_2 = 0.1$ |

**Table 2.** Simulation parameters of the control strategies for notification delivery.

| Adaptive | Constant check | Random selection | Multiplicative–decrease |
|---|---|---|---|
| $\alpha = 0.3,\ \delta = t_0 = 20$ | $t_i = t_0 = 20$ | $a = 10,\ b = 30$ | $t_0 = 20$ |

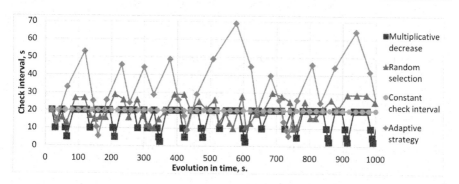

**Fig. 3.** Experimental evolution of the check interval: uniform losses.

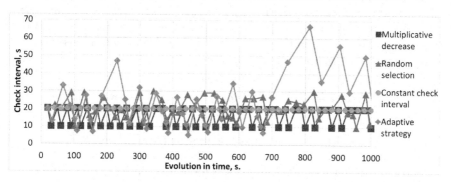

**Fig. 4.** Experimental evolution of the check interval: Poissonian losses.

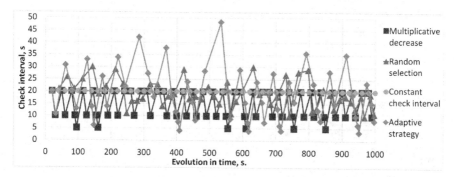

**Fig. 5.** Experimental evolution of the check interval: 2-state alternated Poissonian losses.

**Table 3.** Performance comparison based on the metrics $Q_{tot}$ and $Q_{avg}$.

| Loss distribution | Metric | Multiplicative–decrease | Random selection | Constant check interval | Adaptive strategy |
|---|---|---|---|---|---|
| Uniform losses | $k_{avg}/t_{avg}$ | 0.60/14.15 | 0.78/19.24 | 1.04/20.00 | 1.09/30.36 |
| | $Q_{tot}$ | 0.043 | 0.041 | 0.052 | 0.035 |
| | $Q_{avg}$ | 0.053 | 0.045 | 0.052 | 0.040 |
| Poissonian losses | $k_{avg}/t_{avg}$ | 0.76/15.60 | 1.10/20.38 | 1.13/20.00 | 1.13/25.18 |
| | $Q_{tot}$ | 0.048 | 0.054 | 0.057 | 0.044 |
| | $Q_{avg}$ | 0.038 | 0.057 | 0.057 | 0.046 |
| Two-state alternated Poissonian losses | $k_{avg}/t_{avg}$ | 1.01/14.25 | 1.66/18.81 | 1.54/20.00 | 1.33/20.81 |
| | $Q_{tot}$ | 0.071 | 0.088 | 0.077 | 0.063 |
| | $Q_{avg}$ | 0.060 | 0.087 | 0.077 | 0.067 |

Intuitively, the client is interested in reducing the losses (i.e., $k_{avg} \to \min$) for high values of the check interval (i.e., $t_{avg} \to \max$). Since these two requirements are in opposition, in analogy with the concept of power in computer networks defined as the ratio between the throughput and the delay [12], we use the ratio among the two, such as the following metrics:

$$Q_{tot} = \frac{k_{avg}}{t_{avg}} \to \min , \qquad Q_{avg} = \frac{1}{n}\sum_{i=1}^{n} \frac{k_i}{t_i} \to \min .$$

The performance comparison based on these metrics is shown in Table 3. The experiments indicate that the control by the multiplicative–decrease strategy and adaptation strategy achieves the lowest loss level. The efficiency of multiplicative decrease is due to frequent checks, which consumes resources of the client and the network. On the other hand, the adaptive strategy tries to increase the check interval length (on the risk of higher losses).

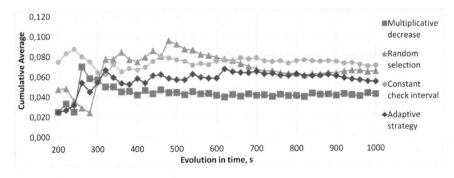

**Fig. 6.** Experiemental behavior of the cumulative average: Poissonian losses, $\beta = \gamma = 1$.

**Table 4.** Performance comparison based on the metric $C_{\mathrm{cum}}(n; \beta, \gamma)$.

| Loss distribution | Cost | | Strategy performance in terms of $C_{\mathrm{cum}}(n; \beta, \gamma)$ | | | |
| | $\beta$ | $\gamma$ | Multiplicative–decrease | Random selection | Constant check interval | Adaptive strategy |
|---|---|---|---|---|---|---|
| Uniform losses | 0.5 | 1.5 | 0.016 | 0.010 | 0.009 | 0.007 |
| | 0.75 | 1.25 | 0.028 | 0.021 | 0.022 | 0.016 |
| | 1 | 1 | 0.053 | 0.045 | 0.052 | 0.041 |
| | 1.25 | 0.75 | 0.102 | 0.098 | 0.126 | 0.109 |
| | 1.5 | 0.75 | 0.204 | 0.215 | 0.307 | 0.308 |
| Poissonian losses | 0.5 | 1.5 | 0.006 | 0.009 | 0.010 | 0.008 |
| | 0.75 | 1.25 | 0.015 | 0.023 | 0.024 | 0.0019 |
| | 1 | 1 | 0.038 | 0.054 | 0.057 | 0.046 |
| | 1.25 | 0.75 | 0.097 | 0.130 | 0.137 | 0.119 |
| | 1.5 | 0.75 | 0.251 | 0.329 | 0.340 | 0.325 |
| Two-state alternated Poissonian losses | 0.5 | 1.5 | 0.011 | 0.014 | 0.012 | 0.012 |
| | 0.75 | 1.25 | 0.025 | 0.034 | 0.030 | 0.028 |
| | 1 | 1 | 0.059 | 0.087 | 0.077 | 0.067 |
| | 1.25 | 0.75 | 0.146 | 0.224 | 0.201 | 0.170 |
| | 1.5 | 0.75 | 0.370 | 0.594 | 0.530 | 0.453 |

For further performance comparison, we consider the cumulative average

$$C_{\mathrm{cum}}(j; \beta, \gamma) = \frac{1}{j} \sum_{i=1}^{j} \frac{k_i^\beta}{t_i^\gamma}, \quad j = 1, 2, \ldots, n, \quad \beta + \gamma = 2, \quad \beta, \gamma > 0$$

Parameters $\beta$ and $\gamma$ allow defining the performance tradeoff between losses and check interval. Clearly, the case of $Q_{\mathrm{tot}}$ and $Q_{\mathrm{avg}}$ corresponds to $\beta = \gamma = 1$, i.e., losses and check interval have the same cost influence to the performance. In particular, $C_{\mathrm{cum}}(n; 1, 1) = Q_{\mathrm{avg}}$. Experimental behavior of $C_{\mathrm{cum}}(j; 1, 1)$ for the case of Poissonian losses is shown in Fig. 6.

If $\beta < 1 < \gamma$ then longer check intervals are preferable, i.e., many active checks degrade the performance more significantly than a high number of occurred losses. If $\beta > 1 > \gamma$ then less loss rate is preferable, i.e., many losses degrade the performance more significantly than a high rate of check requests. Our experiment results with varying $\beta$ and $\gamma$ are summarized in Table 4. When notification losses are considered less expensive then the multiplicative–decrease strategy becomes more efficient.

## 5    Conclusion

This paper has studied the problem of efficient notification delivery for the subscription operation in smart spaces. We considered different assumptions on the notification loss distribution in wireless networked environments. We introduced several control strategies, including a proposal for a TCP-like control for adaptation of the check interval to the observable loss rate. A client can implement such a strategy individually to increase the delivery efficiency. The presented strategies are provided with initial analytical estimates and simulation experiments for performance comparison.

**Acknowledgments.** This research is financially supported by the Ministry of Education and Science of the Russian Federation: project # 14.574.21.0060 (RFMEFI57414X0060) of Federal Target Program "Research and development on priority directions of scientific-technological complex of Russia for 2014–2020" and project # 1481 of the basic part of state research assignment for 2014–2016. The work of M. Pagano is partially supported by the PRA 2016 research project 5GIOTTO funded by the University of Pisa.

# References

1. Allman, M., Paxson, V., Blanton, E.: TCP Congestion Control. RFC 5681 (Draft Standard), September 2009. http://www.ietf.org/rfc/rfc5681.txt
2. Bogoiavlenskaia, O.: Discrete model of TCP congestion control algorithm with round dependent loss rate. In: Balandin, S., Andreev, S., Koucheryavy, Y. (eds.) NEW2AN/ruSMART 2015. LNCS, vol. 9247, pp. 190–197. Springer, Heidelberg (2015)
3. Callegari, C., Giordano, S., Pagano, M., Pepe, T.: A survey of congestion control mechanisms in linux TCP. In: Vishnevsky, V., Kozyrev, D., Larionov, A. (eds.) DCCN 2013. CCIS, vol. 279, pp. 28–42. Springer, Heidelberg (2014)
4. Eugster, P.T., Felber, P.A., Guerraoui, R., Kermarrec, A.M.: The many faces of publish/subscribe. ACM Comput. Surv. **35**, 114–131 (2003)
5. Honkola, J., Laine, H., Brown, R., Tyrkkö, O.: Smart-M3 information sharing platform. In: Proceedings of the IEEE Symposium on Computers and Communications (ISCC 2010), pp. 1041–1046. IEEE Computer Society, June 2010
6. Kleinrock, L.: Queueing Systems: Theory, vol. 1. Wiley Interscience, New York (1975)
7. Korzun, D.: Service formalism and architectural abstractions for smart space applications. In: Proceedings of the 10th Central & Eastern European Software Engineering Conference in Russia (CEE-SECR 2014). ACM, October 2014
8. Korzun, D., Pagano, M., Vdovenko, A.: A TCP-like control of notification delivery for subscription operation in smart spaces. In: Proceedings of the 18th International Conference on Distributed Computer and Communication Networks: Control, Computation, Communications (DCCN-2015), pp. 10–18. ICS RAS, October 2015
9. Korzun, D.G., Kashevnik, A.M., Balandin, S.I., Smirnov, A.V.: The Smart-M3 platform: experience of smart space application development for internet of things. In: Balandin, S., Andreev, S., Koucheryavy, Y. (eds.) NEW2AN/ruSMART 2015. LNCS, vol. 9247, pp. 56–67. Springer, Heidelberg (2015)
10. Lomov, A.A., Korzun, D.G.: Subscription operation in Smart-M3. In: Balandin, S., Ovchinnikov, A. (eds.) Proceedings of the 10th Conference of Open Innovations Association FRUCT and 2nd Finnish-Russian Mobile Linux Summit, pp. 83–94. SUAI, November 2011

11. Morandi, F., Roffia, L., D'Elia, A., Vergari, F., Cinotti, T.S.: RedSib: a Smart-M3 semantic information broker implementation. In: Balandin, S., Ovchinnikov, A. (eds.) Proceedings of the 12th Conference of Open Innovations Association FRUCT and Seminar on e-Tourism, pp. 86–98. SUAI, November 2012

12. Peterson, L.L., Davie, B.S.: Computer Networks: A Systems Approach. The Morgan Kaufmann Series in Networking, 5th edn. Morgan Kaufmann Publishers Inc., San Francisco (2012)

13. Smirnov, A., Kashevnik, A., Shilov, N., Oliver, I., Balandin, S., Boldyrev, S.: Anonymous agent coordination in smart spaces: state-of-the-art. In: Balandin, S., Moltchanov, D., Koucheryavy, Y. (eds.) ruSMART 2009. LNCS, vol. 5764, pp. 42–51. Springer, Heidelberg (2009)

14. Vdovenko, A.S., Korzun, D.G.: Active control by a mobile client of subscription notifications in smart space. In: Proceedings of the 16th Conference Open Innovations Framework Program FRUCT, pp. 123–128. IEEE, ITMO University, October 2014

15. Weiss, N., Holmes, P., Hardy, M.: A Course in Probability. Pearson Addison Wesley, New York (2005)

# Modeling and Performance Analysis of Interconnected Servers in a Cloud Computing System with Dynamic Load Balancing

Udo R. Krieger[✉]

Otto-Friedrich-Universität, Fakultät WIAI,
An der Weberei 5, 96047 Bamberg, Germany
udo.krieger@ieee.org

**Abstract.** We consider the efficiency of dynamic resource pooling and allocation in a cloud computing system offering infrastructure-as-a-service (IaaS). We assume that the demand for service computing by virtual machines (VMs) follows a Poisson load pattern and that the response times of the provided computing services can be classified into several service categories that are governed by exponential service time patterns. A hierarchical, dynamic, class-dependent balancing policy based on a least-loading scheme is applied to provide a uniform utilization among the servers. It is derived from cascaded mutual overflow routing using information on the utilization of VM clusters of similar type on adjacent servers within this resource pool. Regarding the allocation of virtual machines of these different service types to the user demand by a pool of physical servers, we derive a Markovian loss model with adaptive routing induced by cascaded mutual overflow as effective, state-dependent load balancing policy. We determine its basic performance characteristics applying a Markovian fixed-point model. Based on the latter we gain insight on the power of the proposed dynamic load balancing policy among service classes.

**Keywords:** Cloud computing · IaaS · Performance analysis · Randomized load balancing · Mutual overflow system

## 1 Introduction

In recent years modern Web services have been provided by cloud computing systems that are hosted by powerful infrastructures in modern data centers like Microsoft Windows Azure or Amazon Web Services. The latter environment constitutes an example of an infrastructure-as-a-service (IaaS) system where users can deploy their virtualized service computing systems on the physical resources of the infrastructure provider. Pooling the virtualized resources offers the chance to follow efficiently the dynamically changing demand curves of the Web services advertised by a service provider, and to satisfy the scalability, elasticity, and resilience requirements of service-oriented computing.

© Springer International Publishing Switzerland 2016
V. Vishnevsky and D. Kozyrev (Eds.): DCCN 2015, CCIS 601, pp. 52–60, 2016.
DOI: 10.1007/978-3-319-30843-2_6

Effective load balancing and resource allocation schemes are an important ingredient of IaaS systems. Recently, new randomized resource assignment policies based on sampling the utilization of physical servers such as the power-d scheme have been studied (cf. [2–4]). Following Mukhopadhyay et al. [3], we can argue that the resource allocation of virtual machines (VMs) on an interconnected cluster of physical servers may be considered as a randomized routing among loss systems hosting the VMs as service units. Then a utilization-oriented resource allocation policy that first senses the status of the physical servers, allocates VMs and tries to optimize the load assignment subject to loss minimization and uniformization constraints can be translated into a state-dependent routing policy of VM requests to the coupled loss systems (cf. [2,3,5]).

Related research [3] has revealed that randomized power-d load balancing schemes for interconnected loss systems are very powerful mechanisms, but their technical realization requires some overhead. Therefore, we propose a much simpler load balancing mechanism that is derived from classical mutual overflow routing (MOR) with state-dependent load splitting (cf. [1]). Its superior performance has already been revealed in the context of circuit-switched networking and guided a worldwide deployment in the switching equipment of two big European manufacturers.

We suppose that a cluster of interconnected physical machines is given which hosts groups of VMs as virtualized computing resources. The latter are divided into different service classes. VM groups of the same class on two neighboring servers are coupled by a MOR scheme. In this way we construct a hierarchical binary load balancing tree among groups of VMs of the same class. Considering a binary tree component, the load is balanced is such a way that a VM request is first offered to the least loaded component with the option to overflow to the other one in case of a fully loaded structure. In this way an effective adaptive local load balancing scheme can be extended to a tree structure.

In this paper we first analyze the derived performance model of a basic component of this IaaS system and describe a fixed-point model of the underlying coupled loss systems with dynamic load balancing in Sect. 2. Then we investigate its basic performance metrics in Sect. 3. Finally, some general conclusions are drawn.

## 2  Performance Modeling of Dynamic Load Balancing Among Virtualized Server Units in Cloud Computing

We consider a dynamic resource allocation of virtual machines hosted on the interconnected physical servers that is governed by a dynamic load balancing scheme. The latter uses both information about the utilization of the $C$ virtual machines of $M$ different service types and the number of different machines on the interconnected cluster of the $N = 2^k, k \in \mathbb{N}$, physical servers. Motivated by the approach of Mazumdar et al. [3] and references therein, we assume that the server $i \in \mathbb{S} = \{1, \dots, N\}$ accomodates $C_{ij}$ virtual machines as service units of $M$ different service types $j \in \mathbb{T} = \{1, \dots, M\}$. Then we suppose that the resulting number $C_{\cdot j} = \sum_{i \in \mathbb{S}} C_{ij}$ of virtualized servers of a certain type $j \in \mathbb{T}$ can be arranged such that $C_{\cdot 1} \leq C_{\cdot 2} \leq \dots \leq C_{\cdot M}$ holds.

## 2.1  Dynamic Load Balancing by Mutual Overflow Routing

We propose to apply a dynamic load balancing scheme among two adjacent virtualized clusters of the same type $j \in \mathbb{T}$ on pairs $(i, i+1), i = 2l - 1 \in \{1, \ldots, N-1\}, l = 1, \ldots N/2$ of adjacent servers that is derived from a mutual overflow scheme (see Fig. 1, cf. [1]). Then the scheme can be applied in a hierarchical way to the compound of two clustered servers $(i, i+1)$ and $(i+2, i+3)$ as single load balancing block of the mutual overflow scheme and so on. In this way a hierarchical, binary load balancing tree is formed among the service units of each service type.

The offered traffic to service category $j \in \mathbb{T}$ is resulting from a splitting of the overall load to all interconnected servers with rate $\lambda_S$ by a splitting ratio $p_j^{(T)}$. Each server $i \in \mathbb{S}$ gets a conditional splitting ratio $p_i^{(S|T=j)}$ of the overall traffic of type $j$. We assume that the offered load is determined by a Poisson stream, hence, all traffic of service demand for virtual machines of a certain class $j \in \mathbb{T}$ at different servers $i \in \mathbb{S}$ is governed by Poisson processes with rates $\lambda_{ij} = \lambda_S \cdot p_j^{(T)} \cdot p_i^{(S|T=j)}$.

## 2.2  Performance Model of a Two-Server System

Let us now consider two adjacent servers $(i, i+1), i \in \{1, \ldots, N-1\}$ within the server farm as building block of the service infrastructure with hierarchical, dynamic load balancing among a certain service class. For simplicity we assume $i = 1$. We call the latter service computing system 1 and 2, respectively (see Fig. 1). Then we look at a fixed service category $j \in \mathbb{T}$, e.g. $j = 1$, and apply the mutual overflow scheme as basic load balancer among the $N_1 = C_{1j}$ and $N_2 = C_{2j}$ virtual machines on system 1 and 2, respectively, that are serving the incoming service requests of type $j$. In this case we interpret the system of parallel virtual machines on each server as fully available trunk groups 1 and 2 with Poisson arrival streams and rates $\lambda_1, \lambda_2$ as offered traffic 1 and 2, and exponential service times. Without loss of generality, we assume a common service time with rate $\mu = 1$. If all virtual machines in both service groups are busy, an arriving VM service request is lost in this combined server system. In the binary tree structure this portion of the traffic will overflow to the neighboring block of the server cascade. Thus, a coupled system of two isolated Erlang loss models can be used to describe the basic performance behavior of the coupled virtualized server cluster of fixed type $j$ (cf. [1]).

We propose to use both the information on the relative server capacities $N_1, N_2$ and the VM utilizations $\rho_1, \rho_2$ on the different servers 1 and 2 to allocate the incoming requests to the least loaded server. This state-dependent dynamic routing policy within the mutual overflow system between system 1 and 2 is modelled by an adaptive routing with a random splitting of the offered Poisson traffic of class $j$ with common rate:

$$\lambda = \lambda_{ij} + \lambda_{i+1j} = \lambda_S \cdot p_j^{(T)} \cdot (p_i^{(S|T=j)} + p_{i+1}^{(S|T=j)}) \quad i = 1, j \in \mathbb{T}.$$

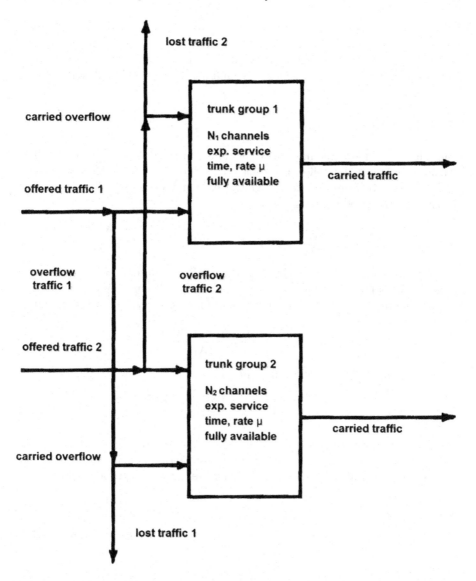

**Fig. 1.** Principle of mutual overflow routing between two coupled loss systems of virtualized machines that are interpreted as trunk group 1 and 2 (cf. [1]).

Due to the dynamic MOR-type load balancing subject to the server capacities and utilization states, we assume that this offered traffic with rate $\lambda$ is randomly split up into two portions for system 1 and 2 with rates $\lambda_1$ and $\lambda_2$. Then the described policy can be modelled by the following splitting probabilities

$$p = \frac{\frac{N_1}{N_1+N_2}(1 - \rho_1(N_1))}{1 - \rho(N_1 + N_2)}, \quad 1 - p = \frac{\frac{N_2}{N_1+N_2}(1 - \rho_2(N_2))}{1 - \rho(N_1 + N_2)} \tag{1}$$

of the Poisson stream with rate $\lambda$ offered to the consecutive virtualized server groups on system 1 and 2. Here $\rho_1(N_1) = E(X_1)/N_1$, $\rho_2(N_2) = E(X_2)/N_2$ denote the utilization of a single virtual machine on server 1 and 2, respectively, where the state variable $X_k$ records the number of active virtual machines on server $k \in \{1, 2\}$.

$$\rho(N_1 + N_2) = E(X_1 + X_2)/(N_1 + N_2) = \frac{N_1}{N_1 + N_2} \cdot \rho_1(N_1) + \frac{N_2}{N_1 + N_2} \cdot \rho_2(N_2)$$

is the utilization of a single machine of the same type on two adjacent servers that are coupled by mutual overflow load balancing. Here we assume that the systems 1 and 2 will not be fully utilized. The latter case may be simply incorporated by adopting the modified splitting probability

$$p = \frac{\frac{N_1}{N_1 + N_2}(1 - \rho_1(N_1) + \varepsilon)}{1 - \rho(N_1 + N_2) + \varepsilon}, \quad 1 - p = \frac{\frac{N_2}{N_1 + N_2}(1 - \rho_2(N_2) + \varepsilon)}{1 - \rho(N_1 + N_2) + \varepsilon}$$

for sufficiently small $\varepsilon > 0$. It yields a static splitting according to the relative number of VMs for fully utilized service blocks in the coupled VM clusters.

Then the resulting fresh Poisson stream of type $j$ to system 1 has the rate

$$\lambda_1 = \lambda \cdot p = \lambda \cdot \frac{\frac{N_1}{N_1 + N_2}(1 - \rho_1(N_1))}{1 - \rho(N_1 + N_2)} \tag{2}$$

and that one offered to system 2 has the rate

$$\lambda_2 = \lambda \cdot (1 - p) = \lambda \cdot \frac{\frac{N_2}{N_1 + N_2}(1 - \rho_2(N_2))}{1 - \rho(N_1 + N_2)} \tag{3}$$

The flow analysis of the interacting server systems coupled by the adaptive routing which is induced by the mutual overflow load balancing now yields the following offered traffic rates $L_1$ and $L_2$ on system 1 and 2, respectively:

$$L_1 = L_1(p) = \lambda_1 + \lambda_2 \cdot B_2 = \lambda \cdot p + \lambda \cdot (1 - p) \cdot B_2 = \lambda[1 - (1 - p) \cdot (1 - B_2)] \tag{4}$$
$$L_2 = L_2(p) = \lambda_2 + \lambda_1 \cdot B_1 = \lambda \cdot (1 - p) + \lambda \cdot p \cdot B_1 = \lambda[1 - p \cdot (1 - B_1)] \tag{5}$$

Here the terms $B_1$ and $B_2$ denote the arrival-stationary blocking probabilities of system 1 and 2, respectively. If we make the simplifying assumption that the Markov-modulated overflow traffic is again approximated by Poisson streams, the latter coincide with the time-stationary blocking probabilities due to the PASTA property.

## 2.3   Fixed-Point Approximation of the Blocking Probabilities

The time-stationary blocking probabilities $B_1, B_2$ are determined by Erlang's formula $B = E(N, A)$ for a pure loss system with offered load $A$ and $N$ service units if we suppose that the overflow streams are Poisson flows. Then they are given by the following quantities

$$B_1 = f_1(N_1, L_1, B_2, p) = L_1^{N_1}/[N_1!(\sum_{i=0}^{N_1} L_1^i/i!)] = E(N_1, \lambda \cdot [1 - (1-p) \cdot (1 - B_2)])$$

$$B_2 = f_2(N_2, L_2, B_1, p) = L_2^{N_2}/[N_2!(\sum_{j=0}^{N_2} L_2^j/j!)] = E(N_2, \lambda \cdot [1 - p \cdot (1 - B_1)])$$

if we assume without loss of generality that the service rates of all classes are uniformly given by $\mu = 1$. In [1] it has been revealed that this approximation of the Markov-modulated overflow streams by Poisson flows with identical overflow rates yields a very accurate approximation of the blocking behavior.

For any fixed splitting probability $p \in (0, 1)$ this overall Erlang loss model yields a system of fixed-point equations $F = (f_1 \circ f_2, f_2 \circ f_1) : I \to I$ on the compact unit square $I = [0, 1]^2$ to determine the blocking probabilities $(B_1(p), B_2(p)) \in [0, 1]^2$ by

$$B_1 = B_1(p) = E\left(N_1, \lambda \cdot [1 - (1-p) \cdot (1 - E(N_2, \lambda \cdot [1 - p \cdot (1 - B_1)]))]\right) \quad (6)$$

$$B_2 = B_2(p) = E\left(N_2, \lambda \cdot [1 - p \cdot (1 - E(N_1, \lambda \cdot [1 - (1-p) \cdot (1 - B_2)]))]\right). \quad (7)$$

The existence of a fixed point $B^*(p) = (B_1^*(p), B_2^*(p)) \in [0, 1]^2$ is guaranteed by Brower's fixed-point theorem. In [1] it was shown that for fixed $p \in (0, 1)$ the independent offered Poisson streams with the positive rates $\lambda_1, \lambda_2$ in (2), (3) determine even a unique fixed point $B^*(p)$ due to the monotonicity of Erlang's loss formula. Then they can be computed by a simple power iteration, e.g. $B_1^{(n)} = [f_1 \circ f_2]^n(B_1^{(0)})$, $B_1^{(0)} = E(N_1, L_1(p))$. Both blocking terms $B_1, B_2$ arising as fixed point $B^*(p) = (B_1, B_2)$ in (6), (7) are coupled by the common splitting probability $p = g(B_1, B_2)$ in (1) which depends in a nonlinear manner on the individual server utilizations

$$\rho_1(N_1) = g_1(N_1, L_1, B_1) = E(X_1)/N_1 = \frac{1}{N_1} \cdot L_1 \cdot (1 - B_1)$$

$$\rho_2(N_2) = g_2(N_2, L_2, B_2) = E(X_2)/N_2 = \frac{1}{N_2} \cdot L_2 \cdot (1 - B_2)$$

and, hence, blocking probabilities in both loss systems 1 and 2, and on the server utilization

$$\rho(N_1 + N_2) = \frac{E(X_1 + X_2)}{(N_1 + N_2)} = \frac{N_1}{N_1 + N_2} \cdot \frac{L_1}{N_1} \cdot (1 - B_1) + \frac{N_2}{N_1 + N_2} \cdot \frac{L_2}{N_2} \cdot (1 - B_2)$$

in the overall system. Hence, the splitting probability $p \in (0, 1)$ in (1) is determined by the resulting fixed-point equation:

$$p = h(B_1, B_2, p) = \frac{\frac{N_1}{N_1 + N_2}(1 - \frac{L_1}{N_1}(1 - B_1))}{1 - [\frac{N_1}{N_1 + N_2}\frac{L_1}{N_1}(1 - B_1) + \frac{N_2}{N_1 + N_2}\frac{L_2}{N_2}(1 - B_2)]}$$

$$= \frac{N_1 - L_1(p)(1 - B_1(p))}{N_1 - L_1(p)(1 - B_1(p)) + N_2 - L_2(p)(1 - B_2(p))} \quad (8)$$

$$= \frac{N_1 - \lambda \cdot [1 - (1-p) \cdot (1 - B_2(p))](1 - B_1(p))}{(N_1 - \lambda \cdot [1 - (1-p) \cdot (1 - B_2(p))](1 - B_1(p))}{+ N_2 - \lambda \cdot [1 - p \cdot (1 - B_1(p))](1 - B_2(p)))} \quad (9)$$

The corresponding fixed-point model (4), (5), (6), (7) and (9) of the combined splitting-blocking model $X = (B_1(p), B_2(p), p) \in [0,1]^3$ is simple, but analytically complex due to the ratio structure of the term $p$. It can be solved by a power iteration $p^{(n)} = h(B_1, B_2, p^{(n-1)})$, $n \in \mathbb{N}$, $p^{(0)} = 0.5$ whose local convergence to a fixed point $X^* = (B_1^*(p^*), B_2^*(p^*), p^*)$ is guaranteed by Brower's fixed-point theorem. Starting with the outcome of our previous analysis [1], we can reveal the dependence on the splitting parameter $p^*$ by a

**Steady-State Representation Result.** *We consider the basic IaaS component of two servers that host $N_1$ and $N_2$ virtual machines (VMs) and serves Poisson arrival streams with offered loads $\lambda_1 = \lambda p^*$ and $\lambda_2 = \lambda(1 - p^*)$, respectively. They are coupled by mutual overflow routing combined with a least-load balancing scheme with splitting probability $p^*$ in (1). The steady-state distribution $\Pi = (\pi_{i,j}), \pi_{i,j} = \lim_{t\to\infty} \mathbb{P}(X(t) = (i,j))$ of the resulting ergodic loss model $X = (X_1, X_2) \in [0, N_1] \times [0, N_2] \subset \mathbb{N}^2$ that describes the number of active VMs in server 1 and 2 is determined by a perturbed product-form solution*

$$\pi_{i,j} = \sum_{k=0}^{N_1} c_k(p^*) g_i(\rho_k(p^*), \lambda_1, p^*) \cdot g_j(-\rho_k(p^*), \lambda_2, p^*)$$

$$+ \sum_{l=0}^{N_2} d_l(p^*) g_i(\gamma_l(p^*), \lambda_1, p^*) \cdot g_j(-\gamma_l(p^*), \lambda_2, p^*) \qquad (10)$$

*for $0 \leq i \leq N_1, 0 \leq i \leq N_2$. The parameters $\rho_k(p^*) \in (-\infty, -1), 1 \leq k \leq N_1$, are the distinct zeros of the Brockmeyer polynomial $g_{N_1}(1 + \rho(p^*), \lambda_1, p^*)$ and $\gamma_l(p^*) \in (1, \infty), 1 \leq l \leq N_2$, the distinct zeros of the Brockmeyer polynomial $g_{N_2}(1 - \gamma(p^*), \lambda_2, p^*)$ and $\rho_0(p^*) = \gamma_0(p^*) = 0$. The $N_1 + N_2$ coefficients $0 \leq k \leq N_1, 0 \leq l \leq N_2$ are the unique solution of a linear system determined by these Brockmeyer polynomials. All terms dependent in unique manner on the existing fixed-point $p^* \in [0, 1]$ of the non-linear system (9). They uniquely determine the IaaS blocking probability:*

$$\pi_{N_1, N_2} = E(N_1, \lambda_1) \cdot E(N_2, \lambda_2) + \sum_{k=1}^{N_1} c_k(p^*) g_i(\rho_k(p^*), \lambda_1, p^*) \cdot g_j(-\rho_k(p^*), \lambda_2, p^*)$$

$$+ \sum_{l=1}^{N_2} d_l(p^*) g_i(\gamma_l(p^*), \lambda_1, p^*) \cdot g_j(-\gamma_l(p^*), \lambda_2, p^*) \qquad (11)$$

Combining all these coupled loss models arising from the hierarchical binary load balancing tree, we are able to determine by these analytic means the vector of the server-dependent blocking probabilities $B_{ij}$ of class $j$ on server $i$ and the overall blocking probabilities $B_{i,*}, B_{*,j}$ of all servers and classes as fundamental performance metrics of the sketched interconnected service computing system.

## 3   Performance Study

We have investigated the loss and utilization performance of a basic building block of the proposed IaaS computing model. It consists of two loss systems

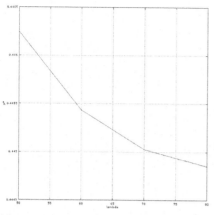

(a) Dependence relation of the splitting probability $p^* \in (0,1)$ on the overall rate $\lambda$ of the arriving Poisson traffic.

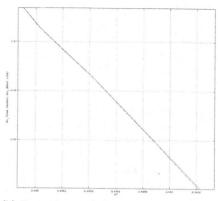

(b) Dependence of the utilizations $\rho_1, \rho_2$ of system 1 and 2 on the fixed-point splitting probability $p^* \in (0,1)$.

**Fig. 2.** Dependence relations in two interconnected loss systems with $C_1 = 20$ and $C_2 = 25$ VMs and dynamic MOR-based load balancing.

with $C_1 = 20$ and $C_2 = 25$ virtual machines of the same class coupled by mutual overflow routing with a least-load balancing policy in (1). In Fig. 2(a) we depict the dependence relation of the adaptive splitting probability $p^* \in (0,1)$ on the overall rate $\lambda$ of the arriving Poisson traffic for a heavily loaded system. In Fig. 2(b) we illustrate the dependence of the utilization $\rho_1$ of system 1 (blue, dashed line) and the corresponding utilization $\rho_2$ of system 2 (black, solid line) on the fixed-point of the splitting probability $p^* \in (0,1)$.

## 4   Conclusions

We have investigated the basic component of a new IaaS computing model with dynamic load balancing that is derived from mutual overflow routing (MOR) among two physical servers with virtualized computing resources and a state-dependent splitting of the offered traffic based on a least-load policy. We have first derived a fixed-point model for the MOR scheme on an underlying binary tree of Erlang loss systems that can reflect the state-dependent load balancing policy. Then we have developed an analysis method to compute the loss performance of the service system. The outcome has been demonstrated by a case study of a single class IaaS block illustrating the dependence relation of the traffic splitting and the utilization vector of the system.

# References

1. Krieger, U.R.: Analysis of a loss system with mutual overflow. In: Proceedings of ITC-Seminar, Peking, September 1988
2. Maguluri, S.T., Srikant, R., Ying, L.: Stochastic models of load balancing and scheduling in cloud computing clusters. In: Proceedings IEEE INFOCOM (2012)
3. Mukhopadhyay, A., Mazumdar, R., et al.: The power of randomized routing in heterogeneous loss systems. In: Proceedings of ITC, Ghent (2015)
4. Mitzenmacher, M.: The power of two choices in randomized load balancing. IEEE Trans. Parallel Distrib. Syst. **12**, 1094–1104 (2001)
5. Turner, S.R.E.: The effect of increasing routing choice on resource pooling. Probab. Eng. Inf. Sci. **12**, 109–124 (1998)

# Formalizing Set of Multiservice Models for Analyzing Pre-emption Mechanisms in Wireless 3GPP Networks

Konstantin Samouylov, Irina Gudkova$^{(\boxtimes)}$, and Ekaterina Markova

Peoples' Friendship University of Russia, Moscow, Russia
{ksam,igudkova,emarkova}@sci.pfu.edu.ru
http://aipt.sci.pfu.edu.ru/en

**Abstract.** Users of wireless 3GPP LTE networks are provided with a wide range of multimedia services with varying QoS requirements; due to this fact a problem of an effective network resources' distribution arises and, consequently, a task of the optimal RAC schemes development. According to the international standards, two types of services are defined within LTE networks – GBR services and non-GBR services. The GBR services generate streaming traffic and non-GBR services – elastic traffic the bit rate of which can dynamically change depending on the cell load. Also, the service priorities differ and are organized with the help of different mechanisms, e.g. service interruption mechanism and mechanism of bit rate degradation. The paper proposes a formal unique description of RAC schemes that is used to develop an example set of models realizing three possible pre-emption based scenarios in multiservice wireless networks.

**Keywords:** Wireless network · Radio admission control · RAC · Guaranteed bit rate · GBR · Non-GBR · Service interruption · Bit rate degradation

## 1 Introduction

Mobile 4G (4th generation) networks based on LTE (Long Term Evolution) technology [1] provide a wide range of multimedia services with varying quality of service (QoS) requirements. This is one of the most important trends in the modern telecommunication systems and networks modernization. A new classification of multiservice traffic identification has been proposed in LTE networks for reflecting the varying demands of users. According to the 3GPP (3rd Generation Partnership Project) consortium recommendations (TS 36.300, TS 23.401, TS 23.203), fifteen types of services are singled out in LTE. They are characterized by different service priorities and network resource requirements: guaranteed bit rate (GBR) services and non-GBR services.

The reported study was funded by RFBR according to the research project No. 16-37-00421 mol_a.

The GBR services are real time services, e.g. voice telephony, video telephony, or real time gaming, for which a minimum bit rate value is specified, i.e. guaranteed bit rate. Nevertheless, when free resources are available in the network, the instantaneous bit rate can exceed the minimum specified one, but can not exceed some threshold value known as the maximum bit rate (MBR).

The non-GBR services are those for which no minimum (guaranteed) bit rate value is specified, the instantaneous bit rate can vary depending on the cell load. It is determined with the help of the so-called aggregate maximum bit rate (AMBR), allowing to differentiate services by the service priority levels. So, the bit rate cannot exceed the maximum value specified for user equipments (UE-AMBR) or determined by network characteristics (access point name AMBR, APN-AMBR). Examples of such services include e-mail, web browsing, or interactive gaming.

Various radio resource management (RRM) mechanisms are used to guarantee QoS. One of these mechanisms is the radio admission control (RAC) [2,4–10] aimed at admitting or rejecting user requests for service taking into account the limited frequency bands and varying QoS requirements. For this purpose, the different priority levels (allocation and retention priority, ARP) are assigned to multimedia services and define the pre-emption RAC algorithms within the corresponding RAC schemes. According to 3GPP TS 23.203: "The range of the ARP priority level is 1 to 15 with 1 as the highest level of priority. The pre-emption capability information defines whether a service data flow can get resources that were already assigned to another service data flow with a lower priority level. The pre-emption vulnerability information defines whether a service data flow can lose the resources assigned to it in order to admit a service data flow with higher priority level. The pre-emption capability and the pre-emption vulnerability can be either set to 'yes' or 'no'." In accordance with this definition of ARP, it is evident that in case of the lack of resources already occupied by lower priority services a new arrived higher priority request could be accepted, at best, through service bit rate degradation (partial pre-emption) [4,6,7,9,10], or at worst, through service interruption (full pre-emption) [4,7,8]. Other mechanisms such as reservation [5,6], threshold [5], and probabilistic management [5,9] could also be applied to realize the priority-service discipline.

Since the number of services LTE users are interested in varies as well as the services themselves within this specified range varies, more than 200 RAC schemes could be set up considering all possible pairwise influences of users on each others due to various pre-emption algorithms. Taking into account this fact and the annual growth of mobile traffic, it seems that the increasing need for the development of a general model considering the overall possible priority mechanisms could be transformed into the need for a set of specific models reflecting the predefined required services and mechanisms. Nevertheless, a unique notation is needed to describe at least, firstly each service within a RAC scheme, and secondly, the pre-emption mechanism applied to each pair of services. This is exactly the main purpose of the paper, i.e. to propose a such notation (Sect. 2) and to illustrate its usage with an example set of specific models of RAC schemes (Sects. 3 and 4).

## 2 Formal Description of RAC Models

Thereby, each RAC scheme is characterized by the number $K \in \{1, \ldots, 15\}$ of services (Table 1), as well as by the pre-emption scenario that is realized with the use of bit rate degradation and service interruption mechanisms.

**Table 1.** Characteristics of LTE service types (TS 23.203)

| QCI | Resource type | Priority level | Examples of services |
|-----|---------------|----------------|----------------------|
| 1 | | 2 | Conversational voice |
| 2 | GBR | 4 | Conversational live streaming video |
| 3 | | 3 | Real time gaming |
| 4 | | 5 | Non-conversational buffered streaming video |
| 65 | | 0.7 | Mission critical user plane push to talk voice |
| 66 | | 2 | Non-mission-critical user plane push to talk voice |
| 5 | | 1 | IMS signalling |
| 7 | | 7 | Voice, live streaming video, interactive gaming |
| 6 | Non-GBR | 6 | Buffered streaming video, TCP-based applications | for multimedia priority services subscribers |
| 8 | | 8 | (e.g., www, e-mail, chat, | for premium subscribers |
| 9 | | 9 | FTP, P2P file sharing, progressive video, etc.) | for non-privileged subscribers |
| 69 | | 0.5 | Mission critical delay sensitive signalling |
| 70 | | 5.5 | Mission critical data |

Note that for GBR services increasing or decreasing bit rate from a GBR value to a MBR one does not affect the mean service time. At the same time, changing bit rate for non-GBR services results in changing the mean transfer time. So, two types of traffics can be singled out: streaming traffic [11, 12] that is characterized by fixed values of bit rate and time and elastic traffic [2] that is characterized by a fixed value of file size and a variable data transfer time [3]. Traffic can be transferred in two modes – unicast mode or multicast mode. For the last one two disciplines have been analyzed [4]: T1 – multicast session is closed when the first user who has opened the session leaves it, – and T2 – multicast session is closed when the last user leaves the session. T stands for the so-colled "transparent" service discipline in queuing theory.

Each of fifteen services could be described with the help of the following parameters:

$$S =< \text{priority level, resource type, data transfer mode} > .$$

Assume that any RAC scheme is defined by the following mechanisms: bit rate degradation (partial pre-emption) and service interruption (full pre-emption). If $k$ is the pre-emption capable service and $k'$ is the pre-emption vulnerable service, than the pre-emption algorithm could be defined by the following matrix: $\mathbf{P} = (\mathbf{p}(k, k'))_{k,k'=1,\ldots,K} = (d(k, k'), i(k, k'))_{k,k'=1,\ldots,K}$, where

$$d(k, k') = \begin{cases} 1, \text{ users of service } k' \text{ will be degraded,} \\ \quad \text{when a request for service } k \text{ arrives,} \\ 0, \text{ otherwise,} \end{cases}$$

$$i(k, k') = \begin{cases} 1, \text{ users of service } k' \text{ will be interrupted,} \\ \quad \text{when a request for service } k \text{ arrives,} \\ 0, \text{ otherwise.} \end{cases}$$

## 3   Example Set of Markov Models of RAC Schemes

Let us illustrate the proposed formal description by an example set of models namely three Markov models (Fig. 1) with different traffic types, namely: unicast streaming, multicast streaming, and elastic traffics, as well as some combinations of different pre-emption mechanisms – first of all, bit rate degradation and service interruption.

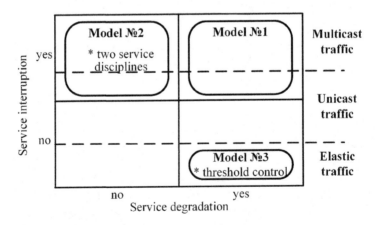

**Fig. 1.** Set of the RAC schemes.

Let us consider an LTE cell having the capacity of $C$ bps. The incoming requests for service generate unicast, multicast, and elastic traffics and arrive as

Poisson processes. Resource occupancy times by unicast and multicast traffics as well as elastic file sizes are exponentially distributed. Bit rates are $d = 1$ and $b$ bps for unicast and multicast traffics and $e$ bps corresponds to the MBR for elastic traffic. The system states of models will be described by vectors with the following components: $n$ – number of unicast users, $m$ – state of a multicast connection ($m = 1$, if at least one user is being served; $m = 0$, otherwise), $u$ – number of transferred elastic files.

## 3.1 Model of RAC Scheme with Multicast Traffic Degradation

Let us consider the first model (Fig. 2) [4]. Suppose that users are provided with two services that generate streaming traffic. The service that has a higher priority level ($S_1$) (e.g. video conference) generates multicast traffic, and the service that has a lower priority level ($S_2$) (e.g. video on demand) generates unicast traffic. The number of resources allocated for the establishment of a multicast connection is adaptively changed along a certain specified value set of $b_1 > ... > b_k > ... > b_K$ bps and is determined in accordance with the number of free resources – $\max \{b_k, \ k = 1, ..., K : \ b_k \leq C - n\}$.

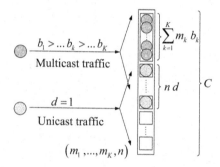

**Fig. 2.** Model of RAC scheme with multicast traffic degradation.

The RAC scheme is realized by two mechanisms of priority admission control – degradation of multicast bit rate and interruption of unicast users. So, the scheme could be defined in the following manner:

$$K = 2, \ S_1 = < 4, \text{GBR, multicast} >, S_2 = < 5, \text{GBR, unicast} >,$$

| P | 1 | 2 |
|---|---|---|
| 1 | $(0,0)$ | $(0,1)$ |
| 2 | $(1,0)$ | $(0,0)$ |

The admission control is realized in two stages. The first stage is the degradation of the number of resources occupied by multicast traffic down to the minimum value of $b_K$ bps. The second stage is the interruption of $b_K - (C - dn)$

unicast users in case of the lack of resources. The state of a multicast connection is described as $m_k \in \{0,1\}$, $k = 1, ..., K$: $m_k = 1$ if connection is established and occupies $b_k$ bps, $m_k = 0$ if connection does not occupy $b_k$ bps. The system state space of the corresponding Markov model has the following form

$$\mathcal{X}_1 = \{(\mathbf{m}, n): \quad \mathbf{m} = \mathbf{0},\ n = 0, \ldots, C,\quad \mathbf{m} = \mathbf{e}_1,\ n = 0, \ldots, C - b_1,$$
$$\mathbf{m} = \mathbf{e}_k,\ n = C - b_{k-1} + 1, \ldots, C - b_k,\ k = 2, \ldots, K\}.$$

## 3.2  Model of RAC Scheme with Unicast Traffic Interruption

Let us consider the second scheme of the set (Fig. 3) with two multicast traffic disciplines – T2 (e.g. real time gaming $\mathcal{S}_1$), and T1 (e.g. video conference $\mathcal{S}_2$), – and unicast traffic (e.g. video on demand $\mathcal{S}_3$).

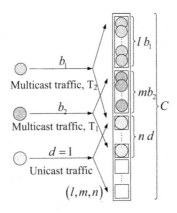

**Fig. 3.** Model of RAC scheme with unicast traffic interruption.

The RAC scheme realizes the mechanism of unicast traffic interruption in case of the lack of resources required for the establishment of a multicast T1 or T2 connection; the admission control is defined by the following services' parameters and matrix:

$$K = 3, \mathcal{S}_1 = < 3,\ \text{GBR, multicast} >,$$
$$\mathcal{S}_2 = < 4,\ \text{GBR, multicast} >, \mathcal{S}_3 = < 5,\ \text{GBR, unicast} >,$$

| **P** | 1 | 2 | 3 |
|---|---|---|---|
| 1 | (0,0) | (0,0) | (0,1) |
| 2 | (0,0) | (0,0) | (0,1) |
| 3 | (0,0) | (0,0) | (0,0) |

Let us denote as $l \in \{0,1\}$ the state of a multicast connection with discipline T2, as $b_1$ and $b_2$ bps the number of resources necessary for the establishment of multicast T1 and T2 connections, respectively. Then the state space has the following form

$$\mathcal{X}_2 = \{(l, m, n) \in \{0,1\} \times \{0,1\} \times \{0,1,...,C\} : b_1 m + b_2 l + dn \leq C\}.$$

## 3.3   Model of RAC Scheme with Elastic Traffic Degradation

Unlike the previous sections where we have proposed the models with streaming traffic, let us construct a scheme with two types of services that generate elastic traffic (Fig. 4), and with MBR thresholds $e_1 > e_2$ bps [10].

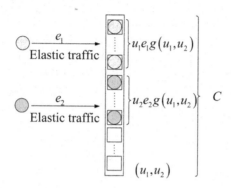

**Fig. 4.** Model of RAC scheme with elastic traffic degradation.

The number of resources occupied by files can dynamically vary from a maximum to a minimum value that is necessary to guarantee requirements for the mean transfer time and is realized through a threshold $U$ of the number of transferred files. In case of low cell load, the maximum number or resources is occupied for files transfer. If the load rises so that it is no longer possible to guarantee a MBR for each transferred file, the instantaneous bit rate is degraded proportionally to an individual MBR, in accordance with some coefficient of bit rate degradation. Because of this, the RAC scheme and corresponding state space of the Markov model have the following forms:

$$K = 2, \mathcal{S}_1 = < k,\ \text{non-GBR, elastic} >, \mathcal{S}_2 = < k,\ \text{non-GBR, elastic} >,$$

| **P** | 1 | 2 |
|---|---|---|
| 1 | (0,0) | (1,0) |
| 2 | (1,0) | (0,0) |

$$\mathcal{X}_3 = \{\mathbf{u} \geq 0 :\ u_1 + u_2 \leq U\}.$$

## 4   Numerical Analysis of Performance Measures

Let us consider a LTE network cell with the peak capacity equal to 100 Mbps. For our experiment, we have chosen the capacity of the cell with streaming or elastic traffic according to the data provided by the Cisco Systems company [13]. For example, in 2015, 58 % of the global mobile traffic is video traffic and 25 % is data traffic. It means that the capacity of the cell with streaming or elastic traffic is equal to 58 Mbps or to 25 Mbps respectively.

Based on Cisco Systems report, as well as on the requirements of the program Skype [14], which are recommended for providing video conference service to seven and more people, we can obtain initial data for numerical analysis (Table 2). In view of fact that the model No.1 and the model No.3, namely models with multicast and elastic traffic degradation, are the most general ones out of the models considered in Sect. 2, let us consider the numerical analysis for these models' performance measures – namely, mean bit rate of multicast streaming traffic and mean coefficient of the bit rate degradation for elastic traffic.

**Table 2.** Example of initial data for numerical analysis

| Parameter | Example of service | Traffic type | |
| --- | --- | --- | --- |
| | | Streaming | Elastic |
| $C$ | | 59 Mbps | 38 Mbps |
| $d$ | video on demand | 2 Mbps | – |
| $b$ | video conference | 8 Mbps, 6 Mbps, 4 Mbps | – |
| $e_1$ | service within 4G network | – | 6 Mbps |
| $e_2$ | service within 3G network | – | 3 Mbps |

According to the forecasts by Cisco Systems, mobile applications providing video services will generate most of global mobile data traffic by 2018, namely about 69 %. However, not only in the future, but also in the beginning of 2012, mobile video represents more than half of global mobile data traffic. In this regard, let us consider an example of a single cell supporting video conference and video on demand services. One of the main trends of 5G networks is the need to save resources to allocate for high quality video services. This is possible through the implementation of MBMS (Multimedia Broadcast and Multicast Service) subsystem.

To illustrate the behavior of the models' performance measure – mean bit rate $\bar{b}$ – let us summarize the initial data of the numerical example in the Table 3. Having such initial set {8 Mbps, 6 Mbps, 4 Mbps} of bit rates, we can distinguish seven variants of bit rate sets $(K; b_1, ..., b_K)$: (1; 4), (1; 6), (1; 8), (2; 6,4), (2; 8,4), (2; 8,6), (3; 8,6,4) (we omit units of measurement). The Fig. 5 shows also the combination of bit rates sets into three groups but already according to maximum bit rate $b_1$.

**Table 3.** Numerical data for model with multicast traffic degradation

| Traffic type | Service | Traffic rate | Initial set of requirements | Resource occupation |
|---|---|---|---|---|
| Unicast | video on demand | 90 % | 2 Mbps | 2 h |
| Multicast | video conference | 10 % | 8 Mbps, 6 Mbps, 4 Mbps | 1 h |

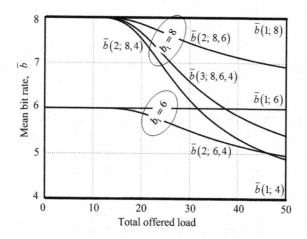

**Fig. 5.** Mean bit rates of video conference service.

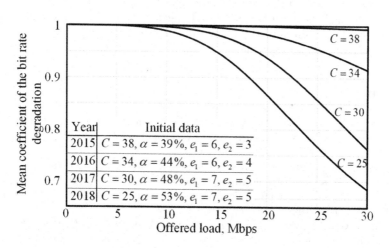

**Fig. 6.** Mean coefficient of the bit rate degradation for elastic traffic.

To analyze the RAC scheme for the model with elastic traffic, let us define maximum bit rate values $e_1$ and $e_2$ as the minimum rate values based on "group of devices" and "technology" criterion (e.g. 4G and 3G, respectively). Note that the ITU-T G.1010 contains information on the preferred (15 s) and acceptable (60 s) data transfer time values for a 10 MB file, which allows us to define a limit on the minimum elastic traffic bit rate and, consequently – on the maximum number of the transferred data blocks $U$. Suppose, in this case the ratio $\alpha$ is between the volume of traffic generated within a 4G network and the total volume of traffic generated within the 2G and 3G networks. Then, for the annual range of 2015-2018 [13], according to the Cisco Systems data, the values of $\alpha$ can be obtained. Using the obtained initial data, let us conduct the analysis of the mean coefficient of the bit rate degradation. The Fig. 6 shows that the growth of the total offered load results in decreasing the mean degradation factor.

## 5 Conclusion

The paper discusses the principle of RAC schemes construction within multi-service networks with the pre-emption based admission control. A set of three Markov models of RAC schemes with unicast streaming, multicast streaming, and elastic traffics was proposed. Different scenarios of priority admission control were analyzed, that are based, first and foremost, on the mechanism of bit rate degradation and service interruption of users with lower priority levels. In addition to the results described in the paper, a unique input data for numerical experiments is developed for further performance analysis.

Currently, the understanding and specifications of the fifth-generation networks are formed. In these specifications there are new types of services, generating traffic, which does not always fit into the existing classification of LTE network services, for example, machine-to-machine traffic. Thereby, it is assumed that in future releases of 3GPP specifications the number of services will be expanded again, and the formal description of RAC schemes suggested in the paper could be also applied.

## References

1. Stasiak, M., Glabowski, M., Wisniewski, A., Zwierzykowski, P.: Modelling and Dimensioning of Mobile Wireless Networks: From GSM to LTE. Willey, Chichester (2010)
2. Samouylov, K., Gudkova, I.: Recursive computation for a multi-rate model with elastic traffic and minimum rate guarantees. In: IEEE International Congress on Ultra Modern Telecommunications and Control Systems ICUMT-2010, pp. 1065–1072. IEEE Press, Moscow (2010)
3. Basharin, G., Gaidamaka, Y., Samouylov, K.: Mathematical theory of teletraffic and its application to the analysis of multiservice communication of next generation networks. Autom. Control Comput. Sci. **47**(2), 62–69 (2013)

4. Borodakiy, V., Gudkova, I., Samouylov, K., Markova, E.: Modelling and performance analysis of pre-emption based radio admission control scheme for video conferencing over LTE. In: 6th International Conference ITU Kaleidoscope: Living in a Converged World - Impossible Without Standards?. K-LCW-2014, pp. 53–59. ITU Press, St. Petersburg (2014)

5. Ghaderi, M., Boutaba, R.: Call admission control in mobile cellular networks: a comprehensive survey. Wirel. Commun. Mob. Comput. **6**(1), 69–93 (2006)

6. Gudkova, I.A., Samouylov, K.E.: Modelling a radio admission control scheme for video telephony service in wireless networks. In: Andreev, S., Balandin, S., Koucheryavy, Y. (eds.) NEW2AN/ruSMART 2012. LNCS, vol. 7469, pp. 208–215. Springer, Heidelberg (2012)

7. Khabazian, M., Kubbar, O., Hassanein, H.: A fairness-based pre-emption algorithm for LTE-advanced: In: 10th IEEE Global Telecommunications Conference GLOBECOM-2012, pp. 5320–5325. IEEE Press, Anaheim (2012)

8. Kwan, R., Arnott, R., Trivisonno, R., Kubota, M.: On pre-emption and congestion control for LTE systems. In: 72nd Vehicular Technology Conference VTC2010-Fall, pp. 6–8. IEEE Press, Ottawa (2010)

9. Liao, H., Wang, X., Chen, H.H.: Adaptive call admission control for multi-class services in wireless networks. In: IEEE International Conference on Communications ICC-2008, pp. 2840–2844. IEEE Press, Beijing (2008)

10. Shorgin, S.Y., Samouylov, K.E., Gudkova, I.A., Markova, E.V., Sopin, E.S.: Approximating performance measures of radio admission control model for non real-time services with maximum bit rates in LTE. In: AIP Conference Proceedings, vol. 1648, pp. 1–4. AIP Publishing, USA (2015). doi:10.1063/1.4912508

11. Gudkova, I., Plaksina, O.: Performance measures computation for a single link loss network with unicast and multicast traffics. In: Balandin, S., Dunaytsev, R., Koucheryavy, Y. (eds.) ruSMART 2010. LNCS, vol. 6294, pp. 256–265. Springer, Heidelberg (2010)

12. Iversen, V.: Teletraffic Engineering and Network Planning. Technical University of Denmark, Lyngby (2011)

13. Cisco Systems: Cisco visual networking index: Global Mobile Data Traffic Forecast Update, 20132019: usage: White paper. Cisco Systems, 40 (2015)

14. Skype 2015: How much bandwidth does Skype need?. https://support.skype.com/en/faq/FA1417/how-much-bandwidth-does-skype-need

# Methods of Performance Evaluation of Broadband Wireless Networks Along the Long Transport Routes

Vladimir Vishnevsky[1], Alexander Dudin[2], Dmitry Kozyrev[1,3(✉)], and Andrey Larionov[1]

[1] V.A. Trapeznikov Institute of Control Sciences
of Russian Academy of Sciences, Moscow, Russia
vishn@inbox.ru, kozyrevdv@gmail.com, larioandr@gmail.com
[2] Belarusian State University, Minsk, Belarus
dudin@bsu.by
[3] Peoples' Friendship University of Russia, Moscow, Russia

**Abstract.** This paper presents the description of the method of performance evaluation and assessment of main characteristics of broadband wireless networks with linear topology, which is based on a model of stochastic multiphase queueing systems with correlated MAP input flows and a cross-traffic. The results of analytical calculations for the networks of small dimension are given. A simulation model for assessment of performance characteristics of large-scale wireless networks with linear topology is developed.

**Keywords:** Wireless networks · Stochastic multi-stage systems · Markov arrival process · Performance evaluation · Analytical modeling and simulation

## 1 Introduction

One of the major problems in the development of new and operation of existing transport routes (railroads, highways, gas and oil pipes) is the creation of a modern wireless communication infrastructure based on the international IEEE 802.11–2012 standard [1]. This standard regulates the creation of high-speed communication channels and wireless networks that operate under the control of IEEE 802.11n and IEEE 802.11s protocols, on the basis of which one can effectively implement the wireless networks along the long transport routes. These networks can provide not only the backbone for high-speed transmission of multimedia information by deploying the base stations on high-rise buildings and towers along the transport routes, but also an operation communication between the fixed and mobile users (cars, trains, road signs, weight control points and transport security control points, traffic lights control points, etc.) as well [2,3].

This research was financially supported by the Ministry of Education and Science of the Russian Federation in the framework of the applied research project №14.613.21.0020 of 22.10.2014 (RFMEFI61314X0020).

V. Vishnevsky and D. Kozyrev (Eds.): DCCN 2015, CCIS 601, pp. 72–85, 2016.
DOI: 10.1007/978-3-319-30843-2_8

The deployment and development of wireless networks along extended routes requires solving a number of complex organizational and technical tasks under tight restrictions on the use of frequency, economic, and hardware resources. In this regard, it seems to be increasingly urgent to solve the problem of optimal allocation of base stations along the long transport routes, which is one of the most important problems in designing the broadband wireless networks of this class. Its solution is aimed both at the realization of high speed backbone network and at the maximum network coverage of the route to provide connectivity for mobile users, as well as to minimize interference and time delays when transmitting the multimedia data over the network. Numerous papers are devoted to the solution of this problem [4–8,10,12].

In current paper we present a new approach to performance evaluation of wireless networks with linear topology based on the model of stochastic multi-stage system with a Markov arrival process (MAP) and a cross-traffic. This paper is the development of studies initiated in [9,11], with regard to the development of effective computational algorithms and conduction of simulations that enable to evaluate the performance characteristics of large-scale networks.

## 2 Description of the Model of a Wireless Network with a Linear Topology

A broadband wireless network along the long transport routes is a set of base stations, connected with each other with high-speed wireless communication channels. An adequate mathematical model for such a network is a multi-stage queueing system with a MAP input flow, PH-distribution of service time at the stages of the system and a cross-traffic (Fig. 1).

The term "service" of each message refers to a number of technical processes in a real network, the duration of which is a random variable. Thus the messages are processed by one or several program components and the processing time depends on the current load of the central processing unit (CPU) and the memory of a base station, on the number of concurrently served messages, the number of CPU cores and other parameters. The messages then come to the output device and are transmitted through the network over one or more serial communication channels to the next base station. The transmission time via communication channels is also a random variable, since it is influenced by the background traffic, the implemented networking technologies, the setup of traffic profiling, and many other factors.

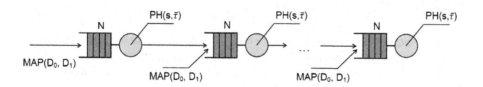

**Fig. 1.** Tandem queueing model

In general case, such processing time should be modeled by PH-distributions: each phase corresponds to a certain technical process (processing by software component, transmission through communication lines). PH-distribution is described by a continuous-time Markov chain with $M + 1$ states in which the zero state is "absorbing" - after arriving to this state a message is considered to be processed. Denote by $PH(\overline{\tau}, S)$ the phase-type distribution of time till absorption at the $0$ phase in a continuous-time Markov chain with a discrete state space $\{0, 1, \ldots, M\}$, steady-state distribution $(\tau_0, \overline{\tau})$, where $\tau_0 = 1 - \overline{\tau} \overline{1}_M$, and infinitesimal generator $T$:

$$T = \begin{pmatrix} 0 & 0 \\ -S\overline{1}_M & S \end{pmatrix}, \quad S \in \mathbb{R}^{M \times M}$$

$$\overline{1}_M = [1 \ \ldots \ 1]^T \in \mathbb{R}^{M \times 1}$$

Each base station of the wireless network transmits not only messages from the previous base station, but the messages from mobile users as well (the cross-traffic). We assume that the input flows of messages and the cross-traffic flow are MAP-flows, which allows to take into account the complex and correlated nature of data flows in wireless communication networks.

In the Markov chain, which controls the MAP, all transitions are divided into observable and hidden. When an observable transition occurs, the arrival process generates a package and, if the initial and the final states differ, then the change of states occurs. Under a hidden transition only the change of states is done. Hidden and observable transition rates are represented by the matrices $D_0$ and $D_1$ respectively:

$$D_0 = \begin{cases} \lambda_{ij}^{(0)}, & i \neq j \\ -\lambda_i, & i = j \end{cases}, \quad i, j = \overline{1, W}$$

$$D_1 = \left\{ \lambda_{ij}^{(1)} \right\}, \quad i, j = \overline{1, W}$$

$$\lambda_i = \lambda_{ii}^{(1)} + \sum_{\substack{j=1 \\ j \neq i}}^{W} \left( \lambda_{ij}^{(0)} + \lambda_{ij}^{(1)} \right)$$

where $\lambda_{ij}^{(0)}$ — hidden transition rates, $\lambda_{ij}^{(1)}$ — observable transition rates, and $\lambda_i$ — total rate of leaving a state or generation of a message without changing the state. The sum of matrices $D_0$ and $D_1$ is the infinitesimal generator of the Markov chain:

$$D = D_0 + D_1$$

An elementary check, considering the definition of $\lambda_i$, shows that the sum of all elements for each row of the generator $D$ is equal to zero.

Stationary probabilities $\overline{\theta} \in \mathbb{R}^W$ of the Markov process are calculated from the balance equations and the normalization condition:

$$\begin{cases} \overline{\theta}D & = \overline{\mathbf{0}}_W \\ \overline{\theta}\overline{\mathbf{1}}_W & = 1 \end{cases} \tag{1}$$

where $\overline{\mathbf{0}}_W = \|0\ 0\ \ldots 0\| \in \mathbb{R}^W$ — row vector of zeros and $\overline{\mathbf{1}}_W = \|1\ 1\ \ldots 1\|^T \in \mathbb{R}^W$ — column vector of ones. Using the obtained stationary probability distribution, we can calculate the average rate of messages, generated by the MAP-flow, as a mean value of a random variable equal to a cumulative rate of observable transitions from the specified state:

$$\lambda = \sum_{i=0}^{W} \left[ \theta_i \sum_{j=0}^{W} \lambda_{ij}^{(1)} \right] = \overline{\theta}D_1 \overline{\mathbf{1}}_W \tag{2}$$

Since the base stations of the wireless network have buffers of limited capacity, it is necessary to take into consideration in the mathematical model the restrictions on queue lengths at servers of each stage.

As the result, a multi-stage queuing system $MAP/PH/1/N \rightarrow \bullet/PH/1/N \rightarrow \cdots \rightarrow \bullet/PH/1/N$ is described, which adequately describes the functioning of a broadband wireless network with a linear topology.

## 3 Properties and Characteristics of a $MAP/PH/1/N$ System

Before proceeding to calculation of the stationary characteristics of a $MAP/PH/1/N$ system, we formulate the following two theorems:

**Theorem 1.** Let the input flow $X = MAP(D_0^{(X)}, D_1^{(X)})$, $D_0^{(X)}, D_1^{(X)} \in \mathbb{R}^{W \times W}$ arrive at the system $MAP/PH/1/N$, and let the service time has a phase-type distribution $Y = PH(S, \overline{\tau})$, $S \in \mathbb{R}^{M \times M}, \overline{\tau} \in \mathbb{R}^{1 \times M}$. Then the output flow of the system will be a MAP flow $Z = MAP(D_0^{(Z)}, D_1^{(Z)})$, which transition rate matrices $D_{0,1}^{(Z)} \in \mathbb{R}^{(WM(N+2)) \times (WM(N+2))}$ have forms:

$$D_0^{(Z)} = \begin{pmatrix} \tilde{D}_0 & B_1 & 0 & 0 & \ldots & 0 & 0 & 0 \\ 0 & R_0 & \tilde{D}_1 & 0 & \ldots & 0 & 0 & 0 \\ 0 & 0 & R_0 & \tilde{D}_1 & \ldots & 0 & 0 & 0 \\ 0 & 0 & 0 & R_0 & \ldots & 0 & 0 & 0 \\ \ldots & \ldots & \ldots & \ldots & \ldots & \ldots & \ldots & \ldots \\ 0 & 0 & 0 & 0 & \ldots & R_0 & \tilde{D}_1 & 0 \\ 0 & 0 & 0 & 0 & \ldots & 0 & R_0 & \tilde{D}_1 \\ 0 & 0 & 0 & 0 & \ldots & 0 & 0 & R_A \end{pmatrix} \tag{3}$$

$$D_1^{(Z)} = \begin{pmatrix} 0 & \ldots & 0 & 0 & 0 \\ I_W \otimes C_t & \ldots & 0 & 0 & 0 \\ 0 & \ldots & 0 & 0 & 0 \\ \ldots & \ldots & \ldots & \ldots & \ldots \\ 0 & \ldots & I_W \otimes C_t & 0 & 0 \\ 0 & \ldots & 0 & I_W \otimes C_t & 0 \end{pmatrix}$$

*where matrices* $\tilde{D}_0$, $\tilde{D}_1 B_1$, $R_0$, $R_A$, $C_t$ *are defined as follows:*

$$\tilde{D}_0 = D_0 \otimes I_M$$
$$\tilde{D}_1 = D_1 \otimes I_M$$
$$B_1 = D_1 \otimes \left(\overline{\tau} \otimes \overline{\mathbf{1}}_M\right)$$
$$R_0 = D_0 \otimes I_M + I_W \otimes S - I_W \otimes C_t$$
$$R_A = (D_0 + D_1) \otimes I_M + I_W \otimes S - I_W \otimes C_t \tag{4}$$

$$C_t = \begin{bmatrix} \mu_{10} \\ \mu_{20} \\ \cdots \\ \mu_{M0} \end{bmatrix} \otimes \overline{\tau} = (-S\overline{\mathbf{1}}_M) \otimes \overline{\tau}$$

From now on $A \otimes B$ — is a Kronecker product of matrices A and B, and $I_K \in \mathbb{R}^{K \times K}$ — is an identity matrix of order $K$.

Thus, the given theorem states, that the output flow of the queuing system MAP/PH/1/N is again a MAP-flow.

Taking into account, that at each stage of the multi-stage system both a MAP flow and a cross traffic from the previous stage arrive, we can prove the following theorem:

Taking into account that at each station (phase) of the multi-stage system both a MAP flow from the previous station and a cross-traffic flow arrive, then the following theorem can be proved:

**Theorem 2.** *Composition of MAP-flows X and Y with transition rate matrices* $D_0^{(X)}$, $D_1^{(X)} \in \mathbb{R}^{M \times M}$ $D_0^{(Y)}$, $D_1^{(Y)} \in \mathbb{R}^{N \times N}$ *is a MAP-flow* $Z = X \otimes Y$, *which transition rate matrices* $D_0^{(Z)}$, $D_1^{(Z)} \in \mathbb{R}^{(MN) \times (MN)}$ *are defined as follows:*

$$D_0^{(Z)} = I_N \otimes D_0^{(X)} + D_0^{(Y)} \otimes I_M$$
$$D_1^{(Z)} = I_N \otimes D_1^{(X)} + D_1^{(Y)} \otimes I_M \tag{5}$$

Thus, the given theorem states, that a composition of two MAPs is again a MAP.

The formulated above theorems are used to calculate the performance characteristics of a multi-stage queuing system.

Further let's consider a method for computation of steady-state probability distribution. The stationary probabilities will allow to find the loss probability, the message arrival rates, the marginal distributions of queue lengths and other characteristics of the *MAP/PH/1/N* system.

The steady-state probability distribution $\overline{\theta} \in \mathbb{R}^L$ for the output MAP flow is the solution of the system of linear algebraic equations:

$$\begin{cases} \overline{\theta} D^{(o)} = \overline{\mathbf{0}}_L \\ \overline{\theta}\overline{\mathbf{1}}_L = 1 \end{cases} \tag{6}$$

Denote by $W$ a number of states of the input MAP-flow; let $M$ denote a number of nonabsorbing states of the Markov chain that defines the PH-distribution, and let $N$ denote the queue capacity of a server.

For the sake of convenience let as suppose that $\bar{\theta} = (\bar{\theta}_0 \ldots \bar{\theta}_{N+1})$, where each vector $\bar{\theta}_i \in \mathbb{R}^{WM}$ is a vector of components of distribution corresponding to a state $i$, i.e. when there $i$ customers in the system.

Let $\eta \in \mathbb{R}^{N+2}$ denote a probability distribution of the number of customers in the system. We can obtain the value of $\eta_n, n = \overline{0, N+1}$ from the vector $\bar{\theta}$:

$$\eta_n = \sum_i \bar{\theta}_{n,i} \tag{7}$$

Given matrices $D_0$, $D_1$, $D$ of the input MAP-flow we can get the steady-state distribution $\bar{\phi} \in \mathbb{R}^W$ of the input MAP-flow by solving the following linear algebraic system:

$$\begin{cases} \bar{\phi}D &= \bar{0}_W \\ \bar{\phi}1_W &= 1 \end{cases} \tag{8}$$

The obtained steady-state distribution allows to find the following characteristics:

– the average arrival rate of messages into the system $\lambda_{avg}$:

$$\lambda_{avg} = \bar{\phi}D_1\bar{1}_{W+1} \tag{9}$$

– the average number $N_{avg}$ of messages in the system, which is defined as the expected value of a random variable $n$:

$$N_{avg} = \sum_{n=0}^{N+1} \eta_n n \tag{10}$$

– and the loss probability $P_{loss}$ of a message arrived at the server:

$$P_{loss} = \bar{\theta}_{N+1}\frac{D_1}{\lambda_{avg}}\bar{1}_{W+1} \tag{11}$$

## 4   Calculation of the Performance Parameters of a Multi-stage System $MAP/PH/1/N \rightarrow \bullet/PH/1/N \rightarrow \cdots \rightarrow \bullet/PH/1/N$

In the previous section we described the construction of a MAP flow of customers arriving at an $i$-th stage and gave formulae for calculation of different characteristics of a $MAP/PH/1/N$ system.

The customer arriving to the system under study has to get sequentially the service at all stations of the tandem. However, due to lack of waiting area at the stations the customer can be lost at each of them. So, the quality of a service in the system is essentially defined be the probability of successful service of an arbitrary customer arriving at the tandem by all stations. Also, the probability of losses at different stations and subsystems of a tandem are important for performance evaluation of a tandem network and discovering and avoiding the so called bottlenecks in the network.

The algorithm 1 below (in a form of pseudocode) allows to use the obtained for $MAP/PH/1/N$ formulae to calculate the steady-state loss probabilities, arrival rates and average queue lengths for each station.

Let $K$ denote the number of stations, $\Psi_0 = MAP(A_0, A_1)$ – the flow of cross-taffic arriving at each station, $\Phi_i = MAP\left(B_0^{(i)}, B_1^{(i)}\right)$ – the disposal flow from $i$–th station, $\Psi_i = MAP\left(A_0^{(i)}, A_1^{(i)}\right)$ – arrival flow into the $i$-th station, $1 \leq i \leq K$. Let the service at each station follow the PH-distribution $\Omega = PH(\bar{\tau}, S)$.

On account of Theorem 2 we can state that:

$$\Psi_1 = \Psi_0$$
$$\Psi_i = \Phi_{i-1} \otimes \Psi_0, \, 2 \leq i \leq K \tag{12}$$

---

**Data:** $K$ — number of stations, $N$ — length size, $\Psi_0 = MAP(A_0, A_1)$ — cross traffic, $\Omega = PH(\bar{\tau}, S)$ — service time distribution

**Result:** $P_{loss}^{(i)}$ — probability of message loss at the $i$-th station,
$\lambda^{(i)}$ — average arrival rate at the $i$-th station,
$N_{avg}^{(i)}$ — mean number of messages at the $i$-th station

1  $i := 1$;
2  **while** $i \leq K$ **do**
3  $\quad$ **if** $i = 1$ **then**
4  $\quad\quad \mid \quad A_0^{(i)} := A_0, A_1^{(i)} := A_1$;
5  $\quad$ **else**
$\quad\quad\quad$ /* $\Psi_i = \Phi_{i-1} \otimes \Psi_0$ $\qquad\qquad\qquad\qquad\qquad\qquad$ */
6  $\quad\quad\quad$ calculate $A_0^{(i)}, A_1^{(i)}$ using (5) from $\Psi_0 = MAP(A_0, A_1)$
$\quad\quad\quad \Phi_i = MAP\left(B_0^{(i)}, B_1^{(i)}\right)$;
7  $\quad$ **end**
8  $\quad$ calculate $\Phi_i = MAP(B_0^{(i)}, B_1^{(i)})$ using (3);
9  $\quad$ using (1) calculate $\bar{\pi}^{(i)} = \left(\bar{\pi}_0^{(i)}, \ldots, \bar{\pi}_{N+1}^{(i)}\right)$ – steady-state probabilities of the flow $\Phi_i$;
10 $\quad$ using (1) calculate $\bar{\alpha}^{(i)}$ – steady-state probabilities of the flow $\Psi_i$;
11 $\quad$ calculate $\lambda^{(i)}$, using vector $\bar{\alpha}$, in accordance with (9);
12 $\quad$ calculate $N_{avg}^{(i)}$, using vector $\bar{\pi}$, in accordance with (10);
13 $\quad$ calculate $P_{loss}^{(i)}$, using vector $\bar{\pi}$, in accordance with (11);
14 $\quad$ $i := i + 1$;
15 **end**

**Algorithm 1.** Algorithm for calculating the parameters of a tandem system

According to the proposed algorithm, a software bundle was developed for calculation of the key performance characteristics of a wireless network. The main difficulty arising while calculating the parameters of the $MAP/PH/1/N \rightarrow \bullet/PH/1/N \rightarrow \cdots \rightarrow \bullet/PH/1/N$ system is the huge size of matrices describing the MAP-flows, arriving at an $i$-th station, $i = \overline{1,k}$. Their size grows exponentially, therefore a simulation model was developed for solving the high-dimension issues. The developed software allows to obtain an exact analytical solution for a tandem queue with a small number of stations and MAP with matrices of small dimension. This exact solution is used to tune up the simulation model.

## 5  Comparison of Results of Analytical Modeling and Simulation

Due to the enormously large matrices even in case of MAPs with 3 states and PH-distribution of service time with 2 phases, the exact solution was obtained for the following two cases:

(1) simple tandem system $M/M/1/N \rightarrow \bullet/M/1/N \rightarrow \cdots \rightarrow \bullet/M/1/N$ with Poisson cross-traffic
(2) system with a MAP input with 2 states $MAP/M/1/N \rightarrow \bullet/M/1/N \rightarrow \cdots \rightarrow \bullet/M/1/N$, where arrival rates were defined by matrices:

$$D_0 = \begin{pmatrix} -1 & 1 \\ 0 & -1 \end{pmatrix}, \ D_1 = \begin{pmatrix} 0 & 0 \\ 1 & 0 \end{pmatrix},$$

that corresponds to Erlang distribution with 2 phases and arrival rates equal to 1, service rates equal to 2, queue length $N = 2$, number of stations $K = 4$.

Parameters of the MAP were obtained from the compilation of real data, gathered on Moscow road traffic routes at different times of day.

The Fig. 2 shows the probability density function of time intervals between vehicles, obtained from the collected data and the probability density function of the MAP with following matrices:

$$D_0 = \begin{pmatrix} -0.85 & 0.85 & 0.0 \\ 0.0 & -1.1 & 0.2 \\ 0.0 & 0.5 & -4.0 \end{pmatrix}, \ D_1 = \begin{pmatrix} 0.0 & 0.0 & 0.0 \\ 0.9 & 0.0 & 0.0 \\ 3.5 & 0.0 & 0.0 \end{pmatrix}$$

The MAP fits the statistical data on the first two moments.

For both cases $P_{loss}$ (loss probability at each server) and $P_{busy}$ (probability that a server is busy) were found using both simulation and exact analytical calculation.

The comparison results are given in the Table 1.

It can be seen from the table that the results of analytical modelling and simulation have close agreement.

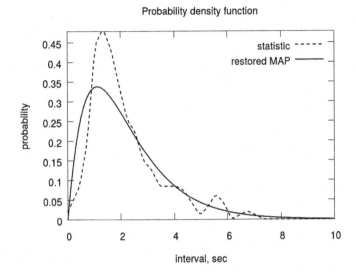

**Fig. 2.** Probability density functions of intervals (MAP and statistical distribution)

**Table 1.** Comparison of system characteristics: $P_{loss}^{sim}$ and $P_{loss}^{analytic}$ – stationary loss probabilities, obtained from simulation and analytical models; $P_{busy}^{sim}$ and $P_{busy}^{analytic}$ – stationary probabilities that the server is busy.

| Number of station | $P_{loss}^{sim}$ | $P_{loss}^{analytic}$ | $P_{busy}^{sim}$ | $P_{busy}^{analytic}$ |
|---|---|---|---|---|
| 1 | 0.0018 | 0.0019 | 0.2503 | 0.2495 |
| 2 | 0.0363 | 0.0365 | 0.4813 | 0.4813 |
| 3 | 0.1138 | 0.1139 | 0.6469 | 0.6480 |
| 4 | 0.1797 | 0.1815 | 0.7336 | 0.7350 |

## 6   Calculation of the Performance Characteristics of a Large-Scale Broadband Network

Unlike the analytical approach, simulation allows to obtain the results for a tandem system $MAP/PH/1/N \rightarrow \bullet/PH/1/N \rightarrow \cdots \rightarrow \bullet/PH/1/N$ with a cross traffic for the case of a large number of stages (stations). All the input flows were described by MAP, with matrices given above. The number of stations in the tandem queue equals $Q = 20$ and the maximum queue length $N = 10$. Such restrictions are met in real life when a distributed system operates on basis of very weak embedded platforms — the program components have buffers of limited capacity and the new messages are dropped when buffers overflow. In case of a small amount of available memory the size of buffers of some applications may be limited to 20 messages.

The service procedure was modeled with the use of exponential distributions with different intensities. We considered the following cases:

$$\mu \in \{2, 5, 10, 20, 100, 1000\},$$

starting from the case of high performance hardware and network ($\mu = 1000$) and proceeding to the case of a very weak hardware and low-performance network ($\mu = 2$). The last case appears to be relevant for the reason that under condition of poor communication channels between stations and in case of often downtimes, the average daily rate of service degrades significantly.

Figure 3 shows the average end-to-end delay for each station. The horizontal axis depicts the number of a station that performs the first transition of a message. The vertical axis depicts the stationary average time till this message leaves the system (when served at the last station). The figure shows that starting from $\mu = 100$ the delay doesn't exceed 200 ms. But if $\mu = 20$ it takes about 1.4 s to transfer a message from the most descend station.

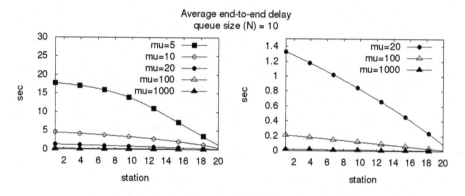

**Fig. 3.** Stationary average end-to-end delay

Figure 4 shows the average message loss ratio for each station. As it can be seen from this figure, there are almost no losses starting from $\mu = 20$. However, due to the random nature of traffic, it is preferable to have a higher performance in case of a large number of consequent stations.

Figure 5 shows the stationary non-delivery ratio of messages, sent by a specified station on their way to the last station. It is evident, that this value is a non-increasing function of a station number, and starting from $\mu = 20$ it is equal to zero nearly everywhere.

Figure 6 shows the average queue length for each station. For the reason that there additionally arrives a cross-traffic flow at each station, the arrival rate does not decrease as the station number increases. And as all servers have the same rate of service, the average queue length turns out to be a non-decreasing function of the station's number. As it can be seen from the figure, the higher is the service rate, the later the queue gets filed up. Particularly, in case $\mu = 20$ the queue of each station remains empty most of the time.

Finally, Figs. 7 and 8 illustrate the distributions of service delays at each station and distributions of time intervals between the outgoing messages at

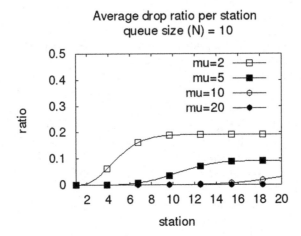

**Fig. 4.** Stationary loss ratio for each station

each station. The figures are given for different service rates and several stations. In particular, Fig. 8 displays that starting from a certain stage, which depends on the service rate, the output flows become almost indistinguishable.

It is evident, that at a low rate of service the system deteriorates very fast. It is connected mostly with the presence of cross-traffic: each consequent base station has not only to process the messages from the previous station, but the cross-traffic messages as well.

Particularly, the results show that for the high system performance it is necessary to pay attention to the quality of communication links between base stations. For a stable work the average rate of service should be not less than 20 messages per second. The rate of about 100 messages per second is desirable.

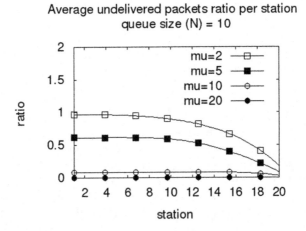

**Fig. 5.** Stationary non-delivery ratio

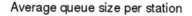

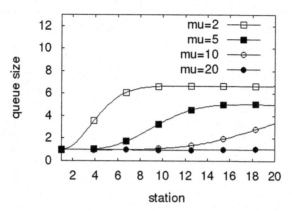

**Fig. 6.** Average queue length per station

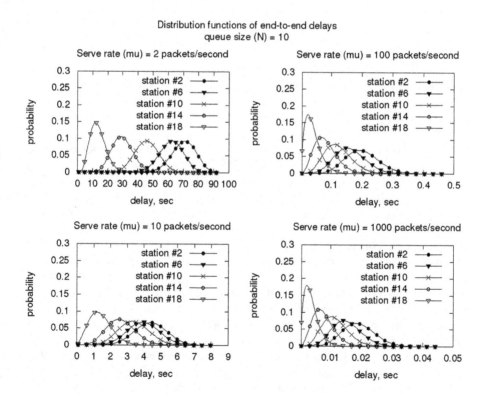

**Fig. 7.** Distribution density functions of end-to-end delays

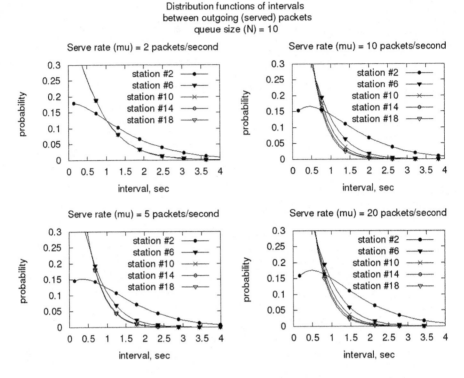

**Fig. 8.** Distribution density functions of time intervals between served messages

## 7 Summary

A novel approach was developed to evaluation of key characteristics of multistage (tandem) $MAP/PH/1/N \rightarrow \bullet/PH/1/N \rightarrow \cdots \rightarrow \bullet/PH/1/N$ queuing systems with a MAP input, a cross-traffic and PH-distributed service times.

Most of studies devoted to tandem queues are limited to the case of dual tandem queues, i.e. queues consisting of exactly two sequential servers with the stationary Poisson arrival process of customers. In our paper(s) we consider a tandem system, consisting of any finite number of stations with the MAP. The assumption that the arrival process of customers is defined by the MAP instead of stationary Poisson arrival process allows to adequately take into account the complexity and variance and correlated nature of inter-arrival times, typical for information flows in modern wireless communication networks.

We present an algorithm and a corresponding software package for exact calculation of the main performance characteristics of wireless networks of small dimension. For the large-scale case a simulation model was developed and the comparison of analytic and simulation results was carried out. The developed methods were successfully used in the process of design and implementation of a wireless broadband network along the city ring road of Kazan (M7 Volga route), Tatarstan, Russia.

**Acknowledgment.** This research was financially supported by the Ministry of Education and Science of the Russian Federation in the framework of the applied research project №14.613.21.0020 of 22.10.2014 (RFMEFI61314X0020).

# References

1. 802.11-2012 IEEE standard for information technology. Telecommunications, information exchange between local, metropolitan area networks. Specific Requirements Part 11: Wireless LAN Medium Access Control (MAC) and Physical Layer (PHY) Specifications – IEEE Std., March 2012
2. Vishnevsky, V.M., Portnoi, S.L., Shakhnovich, I.V.: WiMAX encyclopaedia. Way to 4G. Tekhnosfera, Moscow, p. 470 (2010) (in Russian)
3. Vishnevsky, V.M., Semenova, O.V.: Polling Systems: Theory and Applications for Broadband Wireless Networks, p. 317. Academic Publishing, London (2012)
4. Wu, Q., Zheng, J.: Performance modeling and analysis of IEEE 802.11 DCF based fair channel access for vehicle-to-roadside communication in a non-saturated state. Springer Wirel. Netw. **21**(1), 1–11 (2014)
5. Chakraborty, S., Nandi, S.: IEEE 802.11s mesh backbone for vehicular communication: fairness and throughput. IEEE Trans. Veh. Technol. **62**(5), 2193–2203 (2013)
6. Campolo, C., Cozzetti, H.A., Molinaro, A., Scopigno, R.: Augmenting vehicle-to-roadside connectivity in multi-channel vehicular ad hoc networks. J. Netw. Comput. Appl. **36**(5), 1275–1286 (2013)
7. Vishnevsky, V.M., Larionov, A.A., Semenova, O.V.: Performance evaluation of the high-speed wireless tandem network using centimeter and millimeter-wave channels. Probl. Upr. (4), 50–56 (2013). http://www.mathnet.ru/links/a66b2ecaada565991c942fd3cc5149cf/pu800.pdf
8. Vishnevsky, V., Semenova, O., Dudin, A., Klimenok, V.: Queueing model with gated service and adaptive vacations. In: IEEE International Conference on Communications Workshops, ICC Workshopps 2009, p. 5, June 2009
9. Klimenok, V., Dudin, A., Vishnevsky, V.: Tandem queueing system with correlated input and cross-traffic. In: Kwiecień, A., Gaj, P., Stera, P. (eds.) CN 2013. CCIS, vol. 370, pp. 416–425. Springer, Heidelberg (2013)
10. Vishnevsky, V.M., Dudin, A.N., Semenova, O.V., Klimenok, V.I.: Performance analysis of the BMAP/G/1 queue with gated servicing and adaptive vacations. Perform. Eval. **68**(5), 446–462 (2011)
11. Klimenok, V., Dudin, A., Vishnevsky, V.: On the stationary distribution of tandem queue consisting of a finite number of stations. In: Kwiecień, A., Gaj, P., Stera, P. (eds.) CN 2012. CCIS, vol. 291, pp. 383–392. Springer, Heidelberg (2012)
12. Gaidamaka, Y., Zaripova, E.: Comparison of polling disciplines when analyzing waiting time for signaling message processing at SIP-server. In: Dudin, A., et al. (eds.) ITMM 2015. CCIS 564 (Communications in Computer and Information Science), pp. 358–372. Springer International Publishing, Switzerland (2015)

# On the Waiting Time in the Discrete Cyclic–Waiting System of *Geo/G/*1 Type

Laszlo Lakatos[✉]

Eotvos Lorand University, Budapest, Hungary
lakatos1948@freemail.hu

**Abstract.** We continue to examine a discrete time queueing system where the service of a customer may start at the moment of arrival or at moments differing from it by the multiples of a given cycle time. We find the distribution and the mean value of waiting time in the case of general service time distribution.

**Keywords:** Geo/G/1 cyclic-waiting system · Optical signals

## 1 Introduction

Earlier we have considered a single-server queueing system where an entering customer might be accepted for service either at the moment of arrival or at moments differing from it by the multiples of a given cycle time $T$. Such problem was motivated by the transmission of optical signals: optical signals enter a node and they should be transmitted according to the FCFS rule. This information cannot be stored, if it cannot be served at once is sent to a delay line and returns to the node after having passed it. So, the signal can be transmitted from the node at the moment of its arrival or at moments that differ from it by the multiples of time necessary to pass the delay line. The original problem had been raised in connection with the landing of airplanes, later it appeared to be an exact model for the transmission of optical signals where because of the lack of optical RAM the fiber delay lines were used.

First this system was considered from the viewpoint of the number of present customers [1]. By using Koba's results [2] in [3] we investigated the distribution and characteristics of waiting time for the continuous time model. [4] solved this problem for the discrete time case if the service time had geometrical distribution. In the present paper we investigate the case of general discrete service time distribution.

## 2 Preliminaries and Theorem

We shortly repeat Koba's results [2] to find the waiting time distribution in the cyclic-waiting system.

© Springer International Publishing Switzerland 2016
V. Vishnevsky and D. Kozyrev (Eds.): DCCN 2015, CCIS 601, pp. 86–93, 2016.
DOI: 10.1007/978-3-319-30843-2_9

Let $t_n$ denote the time of arrival of the $n$th customer; its service will begin at the moment $t_n + T \cdot X_n$, where $T$ is the cycle time and $X_n$ is a nonnegative integer. Let $\xi_n = t_{n+1} - t_n$ and $\eta_n$ the service time of $n$th customer. Furthermore, let $X_n = i$, if

$$(k-1)T < iT + \eta_n - \xi_n \leq kT \qquad (k \geq 1),$$

then $X_{n+1} = k$, and if $iT + \eta_n - \xi_n \leq 0$, then $X_{n+1} = 0$. Hence, $X_n$ is a homogeneous Markov chain with transition probabilities $p_{ik}$, where

$$p_{ik} = P\{(k - i - 1)T < \eta_n - \xi_n \leq (k - i)T\}$$

if $k \geq 1$, and

$$p_{i0} = P\{\eta_n - \xi_n \leq -iT\}.$$

Introduce the notations

$$f_j = P\{(j-1)T < \eta_n - \xi_n \leq jT\}, \tag{1}$$

$$p_{ik} = f_{k-i} \quad \text{if} \quad k \geq 1, \quad p_{i0} = \sum_{j=-\infty}^{-i} f_j = \hat{f}_i. \tag{2}$$

The ergodic distribution of this chain satisfies the system of equations

$$p_j = \sum_{i=0}^{\infty} p_i p_{ij} \qquad (j \geq 0),$$

$$\sum_{j=0}^{\infty} p_j = 1.$$

Let us divide the cycle time $T$ into $n$ equal parts called slots. For each slot a new customer may enter with probability $r$, there is no entry with probability $1 - r$; the service time is $k$ slots with probability $q_k$. Denoting the interarrival time by $\xi$, the service time by $\eta$, their distributions are

$$P\{\xi = k\} = (1 - r)^{k-1} r, \qquad P\{\eta = k\} = q_k \quad (k \geq 1),$$

i.e. they have geometrical and general distributions, respectively.

Our result is formulated in the following

**Theorem 1.** *Let us consider the above described system and introduce a Markov chain whose states correspond to the waiting time (in the sense the waiting time is the number of actual state multiplied by $T$) at the arrival time of customers. The matrix of transition probabilities for this chain is*

$$
\begin{bmatrix}
\displaystyle\sum_{j=-\infty}^{0} f_j & f_1 & f_2 & f_3 & f_4 & \cdots \\
\displaystyle\sum_{j=-\infty}^{-1} f_j & f_0 & f_1 & f_2 & f_3 & \cdots \\
\displaystyle\sum_{j=-\infty}^{-2} f_j & f_{-1} & f_0 & f_1 & f_2 & \cdots \\
\displaystyle\sum_{j=-\infty}^{-3} f_j & f_{-2} & f_{-1} & f_0 & f_1 & \cdots \\
\vdots & \vdots & \vdots & \vdots & \vdots & \ddots
\end{bmatrix}
$$

*its elements are defined by (1) and (2). The generating function of the ergodic distribution is*

$$P(z) = \left[ 1 - \frac{\sum_{i=1}^{\infty} q_i \left\{ \left\lceil \frac{i}{n} \right\rceil - (1-r)^{i-1 \ (\mathrm{mod}\ n)} \frac{1-(1-r)^{\lceil \frac{i}{n} \rceil n}}{1-(1-r)^n} \right\}}{\frac{Q_1(1-r)^n}{(1-r)[1-(1-r)^n]}} \right]$$

$$\times \frac{\frac{Q_1}{1-r} - \frac{Q_1[1-(1-r)^n]}{1-r} \frac{z}{z-(1-r)^n}}{1 - F_+(z) - \frac{Q_1[1-(1-r)^n]}{1-r} \frac{z}{z-(1-r)^n}}, \tag{3}$$

*where*

$$F_+(z) = \sum_{j=1}^{\infty} f_j z^j, \qquad Q_1 = \sum_{j=1}^{\infty} q_j (1-r)^j;$$

*the condition of existence of ergodic distribution is*

$$\frac{\sum_{i=1}^{\infty} q_i \left\{ \left\lceil \frac{i}{n} \right\rceil - (1-r)^{i-1 \ (\mathrm{mod}\ n)} \frac{1-(1-r)^{\lceil \frac{i}{n} \rceil n}}{1-(1-r)^n} \right\}}{\frac{Q_1(1-r)^n}{(1-r)[1-(1-r)^n]}} < 1. \tag{4}$$

## 3    Proof of the Theorem

We find the transition probabilities $f_j$. We have

$$P\{\xi = k\} = (1-r)^{k-1} r, \qquad P\{\eta = k\} = q_k,$$

and we look for the distribution of $\eta - \xi$. Let $\eta - \xi = j > 0$. Then $\eta = \xi + j$, and the probability of this event is

$$\hat{q}_j = \sum_{i=1}^{\infty} (1-r)^{i-1} r q_{i+j} = \frac{r}{(1-r)^{j+1}} Q_{j+1},$$

where

$$Q_i = \sum_{k=i}^{\infty} q_k (1-r)^k.$$

If $\eta - \xi = -j \leq 0$ $(j \geq 0)$, then $\eta + j = \xi$, and we have

$$\hat{q}_{-j} = P\{\eta - \xi = -j\} = \sum_{i=1}^{\infty} q_i (1-r)^{i+j-1} r = \frac{r(1-r)^j}{1-r} Q_1.$$

The transition probabilities $f_j$ are obtained from these values, namely

$$f_j = \sum_{k=(j-1)n+1}^{jn} \hat{q}_k = \sum_{k=(j-1)n+1}^{jn} \frac{r}{(1-r)^{k+1}} \sum_{i=k+1}^{\infty} q_i(1-r)^i$$

for the positive jumps, and

$$f_{-j} = \sum_{k=jn}^{(j+1)n-1} \hat{q}_{-k} = \sum_{k=jn}^{(j+1)n-1} r(1-r)^{k-1}Q_1 = \frac{rQ_1}{1-r} \sum_{k=jn}^{(j+1)n-1} (1-r)^k$$

$$= \frac{Q_1[1-(1-r)^n]}{1-r}(1-r)^{jn}$$

for the nonpositive jumps. Later we show that in the case of positive jumps

$$f_j = \sum_{i=(j-1)n+2}^{jn} q_i \left[1 - (1-r)^{i-(j-1)-1}\right] + \sum_{i=jn+1}^{\infty} q_i[1-(1-n)^n](1-r)^{i-jn-1}.$$

Furthermore, let

$$\hat{f}_j = \sum_{i=-\infty}^{-j} f_i = \sum_{i=j}^{\infty} \frac{Q_1[1-(1-r)^n]}{1-r}(1-r)^{in} = \frac{Q_1}{1-r}(1-r)^{jn}.$$

By using the transition probabilities $f_j$, the ergodic probabilities are the solution of the system of equations

$$p_0 = p_0\hat{f}_0 + p_1\hat{f}_1 + p_2\hat{f}_2 + p_3\hat{f}_3 + \dots$$
$$p_1 = p_0f_1 + p_1f_0 + p_2f_{-1} + p_3f_{-2} + \dots$$
$$p_2 = p_0f_2 + p_1f_1 + p_2f_0 + p_3f_{-1} + \dots$$
$$\vdots$$

Multiplying the $j$-th equation by $z^j$, summing up from zero to infinity, for the generating function $\sum_{j=0}^{\infty} p_j z^j$ we have (as in [1])

$$P(z) = P(z)F_+(z) + \sum_{j=1}^{\infty} p_j z^j \sum_{i=0}^{j-1} f_{-i}z^{-i} + \sum_{j=0}^{\infty} p_j\hat{f}_j, \tag{5}$$

where $F_+(z)$ is the generating function of positive jumps, because of its complexity its determination is given later.

In this expression

$$\sum_{i=0}^{j-1} f_{-i}z^{-i} = \sum_{i=0}^{j-1} \frac{Q_1[1-(1-r)^n]}{1-r}(1-r)^{in}z^{-i}$$

$$= \frac{Q_1[1-(1-r)^n]}{1-r} \frac{1 - \left(\frac{(1-r)^n}{z}\right)^j}{1 - \frac{(1-r)^n}{z}},$$

$$\sum_{j=1}^{\infty} p_j z^j \sum_{i=0}^{j-1} f_{-i} z^{-i} = \sum_{j=1}^{\infty} p_j z^j \frac{Q_1[1-(1-r)^n]}{1-r} \frac{1-\left(\frac{(1-r)^n}{z}\right)^j}{1-\frac{(1-r)^n}{z}}$$

$$= \frac{Q_1[1-(1-r)^n]}{1-r} \frac{z}{z-(1-r)^n} [P(z)-P((1-r)^n)],$$

$$\sum_{j=0}^{\infty} p_j \hat{f}_j = \sum_{j=0}^{\infty} p_j \frac{Q_1}{1-r}(1-r)^{jn} = \frac{Q_1}{1-r} P((1-r)^n).$$

Substituting these expressions, (5) yields

$$P(z)\left[1 - F_+(z) - \frac{Q_1[1-(1-r)^n]}{1-r} \frac{z}{z-(1-r)^n}\right]$$

$$= P((1-r)^n)\left[\frac{Q_1}{1-r} - \frac{Q_1[1-(1-r)^n]}{1-r} \frac{z}{z-(1-r)^n}\right].$$

The value of $P((1-r)^n)$ can be found from the condition $P(1)=1$. By using l'Hospital's rule we get

$$P((1-r)^n) = 1 - F'_+(1)\frac{(1-r)[1-(1-r)^n]}{Q_1(1-r)^n},$$

and the generating function takes on the form

$$P(z) = \left[1 - F'_+(1)\frac{(1-r)[1-(1-r)^n]}{Q_1(1-r)^n}\right]$$

$$\times \frac{\dfrac{Q_1}{1-r} - \dfrac{Q_1[1-(1-r)^n]}{1-r}\dfrac{z}{z-(1-r)^n}}{1 - F_+(z) - \dfrac{Q_1[1-(1-r)^n]}{1-r}\dfrac{z}{z-(1-r)^n}}. \qquad (6)$$

From this generating function

$$p_0 = \left[1 - F'_+(1)\frac{(1-r)[1-(1-r)^n]}{Q_1(1-r)^n}\right]\frac{Q_1}{1-r}.$$

It is positive if

$$F'_+(1)\frac{(1-r)[1-(1-r)^n]}{Q_1(1-r)^n} < 1, \qquad (7)$$

which serves as stability condition.

## 4    The Generating Function of Positive Jumps

As we have mentioned the probability of $\eta - \xi = j \ (j \geq 1)$ is

$$\hat{q}_j = \frac{r}{(1-r)^{j+1}} \sum_{i=j+1}^{\infty} q_i(1-r)^i = r \sum_{i=j+1}^{\infty} q_i(1-r)^{i-j-1}.$$

By using $\hat{q}_j$ we compute the transition probabilities $f_j$. We have

$$f_1 = \sum_{i=1}^{n} \hat{q}_i = \hat{q}_1 + \hat{q}_2 + \ldots + \hat{q}_n,$$

this sum is represented in the table (for the sake of simplicity we omit the factor $r$ in all elements)

$$
\begin{array}{cccccc}
q_2 & q_3(1-r) & \ldots & q_n(1-r)^{n-2} & q_{n+1}(1-r)^{n-1} & q_{n+2}(1-r)^n & \ldots \\
 & q_3 & & \ldots \; q_n(1-r)^{n-3} & q_{n+1}(1-r)^{n-2} & q_{n+2}(1-r)^{n-1} & \ldots \\
 & & \vdots & \vdots & & \vdots & \\
 & & q_n & & q_{n+1}(1-r) & q_{n+2}(1-r)^2 & \ldots \\
 & & & & q_{n+1} & q_{n+2}(1-r) & \ldots
\end{array}
$$

Consequently,

$$
\begin{aligned}
f_1 ={}& rq_2 + rq_3[1+(1-r)] + rq_4[1+(1-r)+(1-r)^2] + \ldots + rq_n[1+\ldots+(1-r)^{n-2}] \\
& + rq_{n+1}[1+(1-r)+\ldots+(1-r)^{n-1}] + rq_{n+2}[(1-r)+(1-r)^2+\ldots+(1-r)^n] \\
& + rq_{n+3}[(1-r)^2+(1-r)^3+\ldots+(1-r)^{n+1}] + \ldots \\
={}& \sum_{i=2}^{n} q_i[1-(1-r)^{i-1}] + \sum_{i=n+1}^{\infty} q_i[1-(1-r)^n](1-r)^{i-n-1}.
\end{aligned}
$$

On a similar way one gets

$$f_2 = \sum_{i=n+2}^{2n} q_i \left[1-(1-r)^{i-n-1}\right] + \sum_{i=2n+1}^{\infty} q_i \left[1-(1-r)^n\right](1-r)^{i-2n-1},$$

and in the general case

$$f_k = \sum_{i=(k-1)n+2}^{kn} q_i \left[1-(1-r)^{i-(k-1)n-1}\right] + \sum_{i=kn+1}^{\infty} q_i \left[1-(1-r)^n\right](1-r)^{i-kn-1}.$$

After some arithmetics the corresponding generating function will be

$$
\begin{aligned}
F_+(z) = \sum_{k=1}^{\infty} z^k \Bigg\{ & \sum_{i=(k-1)n+2}^{kn} q_i + \frac{1}{(1-r)^{kn+1}} \sum_{i=kn+1}^{\infty} q_i(1-r)^i \\
& - \frac{1}{(1-r)^{(k-1)n+1}} \sum_{i=(k-1)n+2}^{\infty} q_i(1-r)^i \Bigg\}.
\end{aligned}
$$

Its derivative at $z = 1$ is

$$F'_+(1) = \sum_{i=2}^{n} q_i + 2 \sum_{i=n+2}^{2n} q_i + 3 \sum_{i=2n+2}^{3n} + \ldots + k \sum_{i=(k-1)n+2}^{kn} q_i + \ldots$$

$$+ \sum_{k=1}^{\infty} k q_{(k-1)n+1} - \sum_{k=1}^{\infty} \frac{1}{(1-r)^{(k-1)n+1}} \sum_{i=(k-1)n+1}^{\infty} q_i (1-r)^i$$

$$= \sum_{k=1}^{\infty} k \sum_{i=(k-1)n+1}^{kn} q_i - \sum_{k=1}^{\infty} \frac{1}{(1-r)^{(k-1)n+1}} \sum_{i=(k-1)n+1}^{\infty} q_i (1-r)^i.$$

The second term of this expression is

$$\sum_{k=1}^{\infty} \sum_{i=(k-1)n+1}^{\infty} q_i (1-r)^{i-(k-1)n-1},$$

or

$$q_1 \; q_2(1-r) \ldots q_n(1-r)^{n-1} \; q_{n+1}(1-r)^n \ldots q_{2n+1}(1-r)^{2n} \ldots$$
$$q_{n+1} \qquad \ldots q_{2n+1}(1-r)^n \ldots$$
$$q_{2n+1} \qquad \ldots$$

etc.

From each $n$ columns one can factor out

$$q_i(1-r)^{i-1 \;(\mathrm{mod}\; n)},$$

there remain the powers of $(1-r)^n$, their sums are of the form $\frac{1-(1-r)^{jn}}{1-(1-r)^n}$, i.e. in the first $n$ columns $\frac{1-(1-r)^n}{1-(1-r)^n}$, in the second $n$ columns $\frac{1-(1-r)^{2n}}{1-(1-r)^n}$, etc. The resulting sum will be

$$\sum_{i=1}^{\infty} q_i(1-r)^{i-1 \;(\mathrm{mod}\; n)} \frac{1-(1-r)^{\lceil \frac{i}{n} \rceil n}}{1-(1-r)^n},$$

and

$$F'_+(1) = \sum_{k=1}^{\infty} k \sum_{i=(k-1)n+1}^{kn} q_i - \sum_{i=1}^{\infty} q_i(1-r)^{i-1 \;(\mathrm{mod}\; n)} \frac{1-(1-r)^{\lceil \frac{i}{n} \rceil n}}{1-(1-r)^n},$$

which can be written in the form

$$F'_+(1) = \sum_{i=1}^{\infty} q_i \left\{ \left\lceil \frac{i}{n} \right\rceil - (1-r)^{i-1 \;(\mathrm{mod}\; n)} \frac{1-(1-r)^{\lceil \frac{i}{n} \rceil n}}{1-(1-r)^n} \right\}.$$

The substitution of this value into (6) leads to the expression (3), the condition of ergodicity (7) gives (4).

**Remark 1.** By using the generating function one can compute the mean value of waiting time (measured in cycles). In our case it is equal to

$$\overline{C} = P'(1) = \frac{F''_+(1) - \dfrac{2Q_1(1-r)^n}{(1-r)[1-(1-r)^n]}}{2\left\{\dfrac{Q_1(1-r)^n}{(1-r)[1-(1-r)^n]} - F'_+(1)\right\}} - \frac{1}{1-(1-r)^n},$$

where

$$F''_+(1) = \sum_{k=2}^{\infty} k(k-1) \sum_{i=(k-1)n+1}^{kn} q_i - \sum_{k=1}^{\infty} \frac{2k}{(1-r)^{kn+1}} \sum_{i=kn+1}^{\infty} q_i(1-r)^i.$$

**Remark 2.** One can check that in the case of geometrical service time distribution $q_i = q^{i-1}(1-q)$ the above formulas lead to the results of [4].

# References

1. Lakatos, L., Szeidl, L., Telek, M.: Introduction to Queueing Systems with Telecommunication Applications. Springer, USA (2013)
2. Koba, E.V.: On a GI/G/1 queueing system with repetition of requests for service and FCFS service discipline. Dopovidi NAN Ukrainy, no. 6, pp. 101–103 (2000) (in Russian)
3. Lakatos, L., Efrosinin, D.: Some aspects of waiting time in cyclic-waiting systems. Commun. Comput. Inf. Sci. **356**, 115–121 (2013)
4. Lakatos, L., Efrosinin, D.: A discrete time probability model for the waiting time of optical signals. Commun. Comput. Inf. Sci. **279**, 114–123 (2014)

# Quickest Multidecision Abrupt Change Detection with Some Applications to Network Monitoring

Igor Nikiforov[✉]

Université de Technologie de Troyes, UTT/ICD/LM2S, UMR 6281, CNRS,
12, rue Marie Curie, CS 42060, 10004 Troyes Cedex, France
nikiforov@utt.fr

**Abstract.** The quickest change detection/isolation (multidecision) problem is of importance for a variety of applications. Efficient statistical decision tools are needed for detecting and isolating abrupt changes in the properties of stochastic signals and dynamical systems, ranging from on-line fault diagnosis in complex technical systems (like networks) to detection/classification in radar, infrared, and sonar signal processing.

**Keywords:** Sequential change detection/isolation · Multidecision problems · Anomaly detection · Network monitoring

## 1 Introduction

The quickest change detection/isolation (multidecision) problem is of importance for a variety of applications. Efficient statistical decision tools are needed for detecting and isolating abrupt changes in the properties of stochastic signals and dynamical systems, ranging from on-line fault diagnosis in complex technical systems (like networks) to detection/classification in radar, infrared, and sonar signal processing. The early on-line fault diagnosis (detection/isolation) in industrial processes (SCADA systems) helps in preventing these processes from more catastrophic failures.

The quickest multidecision detection/isolation problem is the generalization of the quickest changepoint detection problem to the case of $K - 1$ post-change hypotheses. It is necessary to detect the change in distribution as soon as possible and indicate which hypothesis is true after a change occurs. Both the rate of false alarms and the misidentification (misisolation) rate should be controlled by given levels.

## 2 Problem Statement

Let $X_1, X_2, \ldots$ denote the series of observations, and let $\nu$ be the serial number of the *last pre-change* observation. In the case of multiple hypothesis, there are several possible post-change hypotheses $\mathcal{H}_j$, $j = 1, 2, \ldots, K - 1$. Let $\mathbb{P}_k^j$ and $\mathbb{E}_k^j$

© Springer International Publishing Switzerland 2016
V. Vishnevsky and D. Kozyrev (Eds.): DCCN 2015, CCIS 601, pp. 94–101, 2016.
DOI: 10.1007/978-3-319-30843-2_10

denote the probability measure and the expectation when $\nu = k$ and $\mathcal{H}_j$ is the true post-change hypothesis, and let $\mathbb{P}_\infty$ and $\mathbb{E}_\infty = \mathbb{E}_0^0$ denote the same when $\nu = \infty$, i.e., there is no change. Let (see [1] for details)

$$\mathbb{C}_\gamma = \left\{ \delta = (T, d) : \min_{0 \le \ell \le K-1} \min_{1 \le j \ne \ell \le K-1} \mathbb{E}_0^\ell \left( \inf_{r \ge 1} \{T_r : d_r = j\} \right) \ge \gamma \right\}, \quad (1)$$

where $T$ is the stopping time, $d$ is the final decision (the number of post-change hypotheses) and the event $\{d_r = j\}$ denotes the first false alarm of the $j$-th type, be the class of detection and isolation procedures for which the average run length (ARL) to false alarm and false isolation is at least $\gamma > 1$. In the case of detection–isolation procedures, the risk associated with the detection delay is defined analogously to Lorden's worst-worst-case and it is given by [1]

$$\mathsf{ESADD}(\delta) = \max_{1 \le j \le K-1} \sup_{0 \le \nu < \infty} \left\{ \operatorname{esssup} \mathbb{E}_\nu^j [(T - \nu)^+ | \mathcal{F}_\nu] \right\}. \quad (2)$$

Hence, the minimax optimization problem seeks to

Find $\delta_{\mathrm{opt}} \in \mathbb{C}_\gamma$ such that $\mathsf{ESADD}(\delta_{\mathrm{opt}}) = \inf_{\delta \in \mathbb{C}_\gamma} \mathsf{ESADD}(\delta)$ for every $\gamma > 1$, (3)

where $\mathbb{C}_\gamma$ is the class of detection and isolation procedures with the lower bound $\gamma$ on the ARL to false alarm and false isolation defined in (1).

Another minimax approach to change detection and isolation is as follows [2,3]; unlike the definition of the class $\mathbb{C}_\gamma$ in (1), where we fixed *a priori* the changepoint $\nu = 0$ in the definition of false isolation to simplify theoretical analysis, the false isolation rate is now expressed by the maximal probability of false isolation $\sup_{\nu \ge 0} \mathbb{P}_\nu^\ell(d = j \ne \ell | T > \nu)$. As usual, we measure the level of false alarms by the ARL to false alarm $\mathbb{E}_\infty T$. Hence, define the class

$$\mathbb{C}_{\gamma,\beta} = \left\{ \delta = (T, d) : \mathbb{E}_\infty T \ge \gamma, \max_{1 \le \ell \le K-1} \max_{1 \le j \ne \ell \le K-1} \sup_{\nu \ge 0} \mathbb{P}_\nu^\ell(d = j | T > \nu) \le \beta \right\}. \quad (4)$$

Sometimes Lorden's worst-worst-case ADD is too conservative, especially for recursive change detection and isolation procedures, and another measure of the detection speed, namely the maximal conditional average delay to detection $\mathsf{SADD}(T) = \sup_\nu \mathbb{E}_\nu(T - \nu | T > \nu)$, is better suited for practical purposes. In the case of change detection and isolation, the SADD is given by

$$\mathsf{SADD}(\delta) = \max_{1 \le j \le K-1} \sup_{0 \le \nu < \infty} \mathbb{E}_\nu^j(T - \nu | T > \nu). \quad (5)$$

We require that the $\mathsf{SADD}(\delta)$ should be *as small as possible* subject to the constraints on the ARL to false alarm and the maximum probability of false isolation. Therefore, this version of the minimax optimization problem seeks to

Find $\delta_{\mathrm{opt}} \in \mathbb{C}_{\gamma,\beta}$ such that $\mathsf{SADD}(\delta_{\mathrm{opt}}) = \inf_{\delta \in \mathbb{C}_{\gamma,\beta}} \mathsf{SADD}(\delta)$

for every $\gamma > 1$ and $\beta \in (0, 1)$. (6)

A detailed description of the developed theory and some practical examples can be found in the recently published book [4].

## 3    Efficient Procedures of Quickest Change Detection/isolation

*Asymptotic Theory.* In this paragraph we recall a lower bound for the worst mean detection/isolation delay over the class $\mathbb{C}_\gamma$ of sequential change detection/isolation tests proposed in [1]. First, we start with a technical result on sequential multiple hypotheses tests and then we give an asymptotic lower bound for $\mathrm{ESADD}(\delta)$.

**Lemma 1.** *Let* $(X_k)_{k \geq 1}$ *be a sequence of i.i.d. random variables. Let* $\mathcal{H}_0, \ldots, \mathcal{H}_{K-1}$ *be* $K \geq 2$ *hypotheses, where* $\mathcal{H}_i$ *is the hypothesis that* $X$ *has density* $f_i$ *with respect to some probability measure* $\mu$, *for* $i = 0, \ldots, K-1$ *and assume the inequality*

$$0 < \rho_{ij} \stackrel{def.}{=} \int f_i \log \frac{f_i}{f_j} d\mu < \infty, \quad 0 \leq i \neq j \leq K-1,$$

*to be true.*

*Let* $\mathbb{E}_i(N)$ *be the average sample number (ASN) in a sequential test* $(N, \delta)$ *which chooses one of the* $K$ *hypotheses subject to a* $K \times K$ *error matrix* $A = [a_{ij}]$, *where* $a_{ij} = \mathbb{P}_i(\text{accepting } \mathcal{H}_j)$, $i, j = 0, \ldots, K-1$.

Let us reparameterize the matrix $A$ in the following manner :

$$\begin{pmatrix} 1 - \sum_{\ell=1}^{K-1} \alpha_\ell & \alpha_1 & \cdots & \alpha_{K-1} \\ \gamma_1 & 1 - \sum_{\ell=2}^{K-1} \beta_{1,\ell} - \gamma_1 \cdots & & \beta_{1,K-1} \\ \gamma_2 & \beta_{2,1} & \cdots & \beta_{2,K-1} \\ \cdots & \cdots & \cdots & \cdots \\ \gamma_i & \beta_{i,1} & \cdots & \beta_{i,K-1} \\ \cdots & \cdots & \cdots & \cdots \\ \gamma_{K-1} & \beta_{K-1,1} & \cdots 1 - \sum_{\ell=1}^{K-2} \beta_{K-1,l} - \gamma_{K-1} \end{pmatrix}.$$

Then a lower bound for the ASN $\mathbb{E}_i(N)$ is given by the following formula :

$$\mathbb{E}_i(N) \geq \max \left\{ \frac{(1 - \tilde{\gamma}_i) \ln \left( \sum_{\ell=1}^{K-1} \alpha_l \right)^{-1} - \log 2}{\rho_{i0}}, \right.$$

$$\left. \max_{1 \leq j \neq i \leq K-1} \left( \frac{(1 - \tilde{\gamma}_i) \ln \beta_{ji}^{-1} - \log 2}{\rho_{ij}} \right) \right\}$$

for $i = 1, \ldots, K - 1$, where

$$\tilde{\gamma}_i = \gamma_i + \sum_{\ell=1, \ell \neq i}^{K-1} \beta_{i,\ell}.$$

**Theorem 1.** *Let* $(Y_k)_{k \geq 1}$ *be an independent random sequence observed sequentially :*

$$\mathcal{L}(Y_k) = \begin{cases} P_0 & \text{if } k \leq \nu \\ P_\ell & \text{if } k \geq \nu + 1 \end{cases}, \quad \nu = 0, 1, 2, \ldots, \quad \text{for } 1 \leq \ell \leq K - 1$$

*The distribution* $P_\ell$ *has density* $f_\ell$, $\ell = 0, \ldots, K - 1$. *An asymptotic lower bound for* $\mathsf{ESADD}(\delta)$, *which extends the result of Lorden [5] to multiple hypotheses case, is:*

$$\mathsf{ESADD}(T; \gamma) \gtrsim \frac{\log \gamma}{\rho^*} \quad \text{as } \gamma \to \infty,$$

*where*

$$\rho^* \stackrel{\text{def.}}{=} \min_{1 \leq \ell \leq K-1} \ \min_{0 \leq j \neq \ell \leq K-1} \rho_{\ell,j} \quad \text{and} \quad 0 < \rho_{\ell,j} \stackrel{\text{def.}}{=} \mathbb{E}_1^l \left( \log \frac{f_l(Y_i)}{f_j(Y_i)} \right) < \infty$$

*is the K-L information.*

*Generalized CUSUM Test.* The generalized CUSUM (non recursive) test asymptotically attains the above mentioned lower bound [1]. Let us introduce the following stopping time and final decision

$$\tilde{N} = \min\{\tilde{N}^1, \ldots, \tilde{N}^{K-1}\}; \quad \tilde{d} = \mathrm{argmin}\{\tilde{N}^1, \ldots, \tilde{N}^{K-1}\}$$

of the detection/isolation algorithm. The stopping time $\tilde{N}^\ell$ is responsible for the detection of hypothesis $\mathcal{H}_\ell$:

$$\tilde{N}^\ell = \inf_{k \geq 1} \tilde{N}^\ell(k), \ \tilde{N}^\ell(k) = \inf \left\{ n \geq k : \min_{0 \leq j \neq \ell \leq K-1} S_k^n(\ell, j) \geq h \right\}$$

$$\tilde{N}^\ell = \inf \left\{ n \geq 1 : \max_{1 \leq k \leq n} \min_{0 \leq j \neq \ell \leq K-1} S_k^n(\ell, j) \geq h \right\}, \quad S_k^n(\ell, j) = \sum_{i=k}^n \log \frac{f_\ell(Y_i)}{f_j(Y_i)}.$$

The generalized matrix recursive CUSUM test, which also attains the asymptotic lower bound, has been considered in [6,7]. Let us introduce the following stopping time and final decision

$$\widehat{N}_r = \min\{\widehat{N}^1, \ldots, \widehat{N}^{K-1}\}; \quad \tilde{d}_r = \mathrm{argmin}\{\widehat{N}^1, \ldots, \widehat{N}^{K-1}\}$$

of the detection/isolation algorithm. The stopping time $\widehat{N}^\ell$ is responsible for the detection of hypothesis $\mathcal{H}_l$ :

$$\widehat{N}^\ell = \inf \left\{ n \geq 1 : \min_{0 \leq k \neq j \leq K-1} Q_n(\ell, j) \geq h \right\},$$

$$Q_n(\ell, j) = (Q_{n-1}(\ell, j) + Z_n(\ell, j))^+, \quad Z_n(\ell, j) = \log \frac{f_\ell(Y_n)}{f_j(Y_n)}$$

For some safety critical applications, a more tractable criterion consists in minimizing the maximum detection/isolation delay:

$$\mathsf{SADD}(\delta) = \max_{1 \le j \le K-1} \sup_{0 \le \nu < \infty} \mathbb{E}_\nu^j(T - \nu | T > \nu). \tag{7}$$

subject to :

$$\mathbb{C}_{\gamma,\beta} = \left\{ \delta \;:\; \mathbb{E}_\infty T \ge \gamma, \; \max_{1 \le \ell \le K-1} \max_{1 \le j \ne \ell \le K-1} \sup_{\nu \ge 0} \mathbb{P}_\nu^\ell(d = j | T > \nu) \le \beta \right\}.$$

for $1 \le \ell, j \ne \ell \le K - 1$. An asymptotic lower bound in this case is given by the following theorem [3].

**Theorem 2.** *Let* $(Y_k)_{k \ge 1}$ *be an independent random sequence observed sequentially:*

$$\mathcal{L}(Y_k) = \begin{cases} P_0 \; if \, k \le \nu \\ P_\ell \; if \, k \ge \nu + 1 \end{cases}, \quad \nu = 0, 1, 2, \ldots, \; for \; 1 \le \ell \le K - 1$$

*Then*

$$\mathsf{SADD}(N; \gamma, \beta) \gtrsim \max \left\{ \frac{\log \gamma}{\rho_{\mathsf{d}}^*}, \frac{\log \beta^{-1}}{\rho_{\mathsf{i}}^*} \right\} \quad as \; \min\{\gamma, \beta^{-1}\} \to \infty,$$

*where* $\rho_{\mathsf{d}}^* = \min_{1 \le j \le K-1} \rho_{j,0}$ *and* $\rho_{\mathsf{i}}^* = \min_{1 \le \ell \le K-1} \min_{1 \le j \ne \ell \le K-1} \rho_{\ell,j}$.

*Vector Recursive CUSUM Test.* If $\gamma \to \infty$, $\beta \to 0$ and $\log \gamma \ge \log \beta^{-1}(1 + o(1))$, then the above mentioned lower bound can be realized by using the following recursive change detection/isolation algorithm [2] :

$$N_r = \min_{1 \le \ell \le K-1} \{N_r(\ell)\}, \quad d_r = \arg \min_{1 \le \ell \le K-1} \{N_r(\ell)\},$$

where $N_r(\ell) = \inf \{n \ge 1 : \min_{0 \le j \ne \ell \le K-1} [S_n(\ell, j) - h_{\ell,j}] \ge 0\}$,

$$S_n(\ell, j) = g_n(\ell, 0) - g_n(j, 0), \quad g_n(\ell, 0) = (g_{n-1}(\ell, 0) + Z_n(\ell, 0))^+,$$

with $Z_n(\ell, 0) = \log \frac{f_\ell(Y_n)}{f_0(Y_n)}$, $g_0(\ell, 0) = 0$ for every $1 \le \ell \le K - 1$ and $g_n(0, 0) \equiv 0$,

$$h_{\ell,j} = \begin{cases} h_d \; if \; 1 \le \ell \le K - 1 \; and \; j = 0 \\ h_i \; if \; 1 \le j, \ell \le K - 1 \; and \; j \ne \ell \end{cases}.$$

## 4   Applications to Network Monitoring

In this section the above mentioned theoretical results are illustrated by application of the proposed detection/isolation procedures to the problem of network monitoring.

Let us consider a network composed of $r$ nodes and $n$ mono-directional links, where $y_\ell$ denotes the volume of traffic on the link $\ell$ at discrete time $k$ (see details in [8,9]). For the sake of simplicity, the subscript $k$ denoting the time is omitted now. Let $x_{i,j}$ be the Origin-Destination (OD) traffic demand from node $i$ to node $j$ at time $k$. The traffic matrix $X = \{x_{i,j}\}$ is reordered in the lexicographical order as a column vector $X = [(x_{(1)}, \dots, x_{(m)})]^T$, where $m = r^2$ is the number of OD flows.

Let us define an $n \times m$ routing matrix $A = [a_{\ell,k}]$ where $0 \leq a_{\ell,k} \leq 1$ represents the fraction of OD flow $k$ volume that is routed through link $\ell$. This leads to the linear model

$$Y = AX,$$

where $Y = (y_1, \dots, y_n)^T$ is the Simple Network Management Protocol (SNMP) measurements. Without loss of generality, the known matrix $A$ is assumed to be of full row rank, i.e., $\operatorname{rank} A = n$.

The problem consists in detecting and isolating a significant volume anomaly in an OD flow $x_{i,j}$ by using only SNMP measurements $y_1, \dots, y_n$. In fact, the main problem with the SNMP measurements is that $n \ll m$. To overcome this difficulty a parsimonious linear model of non-anomalous traffic has been developed in the following papers [10–17].

The derivation of this model includes two steps: (i) description of the ambient traffic by using a spatial stationary model and (ii) linear approximation of the model by using piecewise polynomial splines.

The idea of the spline model is that the non-anomalous (ambient) traffic at each time $k$ can be represented by using a known family of basis functions superimposed with unknown coefficients, i.e., it is assumed that

$$X_k \approx B\mu_k, \quad k = 1, 2, \dots,$$

where the $m \times q$ matrix $B$ is assumed to be known and $\mu_t \in \mathbb{R}^q$ is a vector of unknown coefficients such that $q < n$. Finally, it is assumed that the model residuals together with the natural variability of the OD flows follow a Gaussian distribution, which leads to the following equation:

$$X_k = B\mu_k + \xi_k \tag{8}$$

where $\xi_k \sim \mathcal{N}(0, \Sigma)$ is Gaussian noise, with the $m \times m$ diagonal covariance matrix $\Sigma = \operatorname{diag}(\sigma_1^2, \dots, \sigma_m^2)$. The advantages of the detection algorithm based on a parametric model of ambient traffic and its comparison to a non-parametric approach are discussed in [11,14], (see also [18] for PCA based approach). Hence, the link load measurement model is given by the following linear equation :

$$Y_k = AB\mu_k + A\xi_k = H\mu_k + \zeta_k + [\theta_\ell], \tag{9}$$

where $Y_k = (y_1, \dots, y_n)_k^T$ and $\zeta_k \sim \mathcal{N}(0, A\Sigma A^T)$. Without any loss of generality, the resulting matrix $H = AB$ is assumed to be of full column rank. Typically, when an anomaly occurs on OD flow $\ell$ at time $\nu + 1$ (change-point), the vector $\theta_\ell$ has the form $\theta_\ell = \varepsilon a(\ell)$, where $a(\ell)$ is the $\ell$-th normalized column of $A$ and $\varepsilon$ is the intensity of the anomaly. The goal is to detect/isolate the presence of an anomalous vector $\theta_\ell$, which cannot be explained by the ambient traffic model $X_k \approx B\mu_k$.

Therefore, after the de-correlation transformation, the change detection/isolation problem is based on the following model with nuisance parameter $X_k$ :

$$Y_k = HX_k + \xi_k + \theta(k, \nu), \quad \xi_k \sim \mathcal{N}(0, \sigma^2 I_n), \quad k = 1, 2, \ldots, \tag{10}$$

where $H$ is a full rank matrices of size $n \times q$, $n > q$, and $\theta(k, \nu)$ is a change occurring at time $\nu + 1$, namely :

$$\theta(k, \nu) = \begin{cases} 0 & \text{if } k \neq \nu \\ \theta_\ell & \text{if } k \geq \nu + 1 \end{cases}, \quad 1 \leq \ell \leq K - 1.$$

This problem is invariant under the group $G = \{Y \to g(Y) = Y + HX\}$ (see details in [19]). The invariant test is based on maximal invariant statistics. The solution is the projection of $Y$ on the orthogonal complement $R(H)^\perp$ of the column space $R(H)$ of the matrix $H$. The parity vector $Z = WY$ is a maximal invariant to the group $G$.

$$WH = 0, \quad W^T W = P_H = I_r - H(H^T H)^- H^T, \quad WW^T = I_{n-q}.$$

Transformation by $W$ removes the interference of the nuisance parameter $X$

$$Z = WY = W\xi \, (+W\theta).$$

Hence, the sequential change detection/isolation problem can be re-written as

$$Z_k = WY_k = W\xi_k + W\theta(k, \nu), \quad \xi_k \sim \mathcal{N}(0, \sigma^2 I_{n-q}), \quad k = 1, 2, \ldots.$$

**Theorem 3.** *Let $(Y_k)_{k \geq 1}$ be the output of the model given by (10) observed sequentially. Then the generalized CUSUM or matrix recursive CUSUM tests attain the lower bound corresponding to the minimax setup :*

$$\mathsf{ESADD}(N; \gamma) \gtrsim \frac{\log \gamma}{\overline{\rho^*}} \text{ as } \gamma \to \infty, \quad \overline{\rho^*} \overset{def.}{=} \inf_{X^\ell, X^j} \min_{1 \leq \ell \leq K-1} \min_{0 \leq j \neq \ell \leq K-1} \rho_{\ell, j}(X^\ell, X^j)$$

*where $X^\ell$ (resp. $X^j$) corresponds to the hypothesis $\mathcal{H}_\ell$ (resp. $\mathcal{H}_j$). The vector recursive CUSUM test attains the lower bound*

$$\mathsf{SADD}(N; \gamma, \beta) \gtrsim \max\left\{ \frac{\log \gamma}{\overline{\rho^*}_d}, \frac{\log \beta^{-1}}{\overline{\rho^*}_i} \right\} \text{ as } \gamma \to \infty, \ \beta \to 0, \log \gamma \geq \log \beta^{-1}(1 + o(1)),$$

*where*

$$\overline{\rho^*}_d = \inf_{X^j, X^0} \min_{1 \leq j \leq K-1} \rho_{j,0}(X^j, X^0) \text{ and } \overline{\rho^*}_i = \inf_{X^\ell, X^j} \min_{1 \leq \ell \leq K-1} \min_{1 \leq j \neq \ell \leq K-1} \rho_{\ell, j}(X^\ell, X^j).$$

**Acknowledgement.** This work was partially supported by the French National Research Agency (ANR) through ANR CSOSG Program (Project ANR-11-SECU-0005).

# References

1. Nikiforov, I.V.: A generalized change detection problem. IEEE Trans. Inf. Theor. **41**(1), 171–187 (1995)
2. Nikiforov, I.V.: A simple recursive algorithm for diagnosis of abrupt changes in random signals. IEEE Trans. Inf. Theor. **46**(7), 2740–2746 (2000)
3. Nikiforov, I.V.: A lower bound for the detection/isolation delay in a class of sequential tests. IEEE Trans. Inf. Theor. **49**(11), 3037–3047 (2003)
4. Tartakovsky, A., Nikiforov, I., Basseville, M.: Sequential Analysis: Hypothesis Testing and Changepoint Detection. CRC Press, Taylor & Francis Group, Boca Raton (2015)
5. Lorden, G.: Procedures for reacting to a change in distribution. Annals Math. Stat. **42**, 1897–1908 (1971)
6. Oskiper, T., Poor, H.V.: Online activity detection in a multiuser environment using the matrix CUSUM algorithm. IEEE Trans. Inf. Theor. **48**(2), 477–493 (2002)
7. Tartakovsky, A.G.: Multidecision quickest change-point detection: previous achievements and open problems. Sequential Anal. **27**, 201–231 (2008)
8. Lakhina, A. et al.: Diagnosing network-wide traffic anomalies. In: SIGCOMM (2004)
9. Zhang, Y. et al.: Network anomography. In: IMC 2005 (2005)
10. Fillatre, L., Nikiforov, I., Vaton, S. Sequential Non-Bayesian Change Detection-Isolation and Its Application to the Network Traffic Flows Anomaly Detection. In: Proceedings of the 56th Session of ISI, Lisboa, 22–29 August 2007, pp. 1–4 (special session)
11. Casas, P., Fillatre, L., Vaton, S., Nikiforov, I.: Volume anomaly detection in data networks: an optimal detection algorithm vs. the PCA approach. In: Valadas, R., Salvador, P. (eds.) FITraMEn 2008. LNCS, vol. 5464, pp. 96–113. Springer, Heidelberg (2009)
12. Fillatre, L., Nikiforov, I., Vaton, S., Casas, P.: Network traffic flows anomaly detection and isolation. In: 4th edition of the International Workshop on Applied Probability, IWAP 2008, 7–10 July 2008, Compiègne, p. 1–6 (invited paper)
13. Fillatre, L., Nikiforov, I., Casas, P., Vaton, S.: Optimal volume anomaly detection in network traffic flows. In: 16th European Signal Processing Conference (EUSIPCO 2008), Lausanne, p. 5, 25–29 August 2008
14. Casas, P., Fillatre, L., Vaton, S., Nikiforov, I.: Volume anomaly detection in data networks: an optimal detection algorithm vs. the PCA approach. In: Valadas, R., Salvador, P. (eds.) FITraMEn 2008. LNCS, vol. 5464, pp. 96–113. Springer, Heidelberg (2009)
15. Casas, P., Vaton, S., Fillatre, L., Nikiforov, I.V.: Optimal volume anomaly detection and isolation in large-scale IP networks using coarse-grained measurements. Comput. Netw. **54**, 1750–1766 (2010)
16. Casas, P., Fillatre, L., Vaton, S., Nikiforov, I.: Reactive robust routing: anomaly localization and routing reconfiguration for dynamic networks. J. Netw. Syst. Manage. **19**(1), 58–83 (2010)
17. Fillatre, L., Nikiforov, I.: Asymptotically uniformly minimax detection and isolation in network monitoring. IEEE Trans. Signal Process. **60**(7), 3357–3371 (2012)
18. Ringberg, H. et al.: Sensitivy of PCA for traffic anomaly detection. In: SIGMETRICS (2007)
19. Fouladirad, M., Nikiforov, I.: Optimal statistical fault detection with nuisance parameters. Automatica **41**(7), 1157–1171 (2005)

# An M/M/1 Queue with $n$ Undesired Services and a Desired Service

A. Krishnamoorthy[1], A.S. Manjunath[2], and V.M. Vishnevsky[3]([✉])

[1] Department of Mathematics, Cochin University of Science and Technology,
Cochin 682022, India
achyuthacusat@gmail.com
[2] Department of Mathematics, Government College Kottayam,
Kottayam 686013, India
manjunadem@gmail.com
[3] V. A. Trapeznikov Institute of Control Sciences, Russian Academy of Sciences,
65 Profsoyuznaya Street, Moscow 117997, Russia
vishn@inbox.ru

**Abstract.** We study an M/M/1 queue with a set of services in undesired states and a desired state. Arrivals are according to a Poisson process with mean rate $\lambda$. An arrival enters into desired service with probability $\theta$, otherwise goes to one of the undesired services with complementary probability depending on the environmental factors. When the service time in undesired states exceeds a threshold random variable, the customer is pushed out of the system. Otherwise the customer, after recognizing that the current service is not the right one moves to desired service state. We discuss the case of $n$ undesired services and a desired service.

**Keywords:** Undesired/desired service states · Random threshold clock

## 1 Introduction

In the analysis of classic queueing models it is assumed that the server is completely aware of the exact service requirement of the customer (see for example Gross et al. [4]). It is also true that the customer knows the type of service he needs. Thus there is no conflict on the service provided to the customer. However,

A. Krishnamoorthy—Research is supported by the Department of Science and Technology, Government of India under grant INT/RUS/RMES/P-3/2014, dated 15/4/2015.
A.S. Manjunath—Research is supported by the University Grants Commission, Government of India, under Faculty Development Programme (Grant No. F.FIP/12thPlan/KLMG003TF05).
V.M. Vishnevsky—Research is supported by the Ministry of Education and Science of the Russian Federation under grant №14.613.21.0020, dated 22/10/2014 (RFMEFI61314X0020).

V. Vishnevsky and D. Kozyrev (Eds.): DCCN 2015, CCIS 601, pp. 102–110, 2016.
DOI: 10.1007/978-3-319-30843-2_11

there are several real life situations where the server and/customer are (is) not knowledgeable about the service requirement. This is especially the case where several types of services are provided. As a concrete example we have vehicles for repair at service stations; patients consulting physicians for diagnosis and medication. If the right service required is not identified and instead the diagnosis turned out to be false the result could be disastrous. A wrong diagnosis and consequent service provided may sometimes turn out to be even fatal/may result in the equipment getting service, rendered totally unusable. It is this type of problem that we analyze in this note.

We start with a simple situation. More complex situations will be analyzed in a follow up paper. All underlying distributions assumed in this paper are exponential. Even with this simple assumption we will soon notice that the phase type distribution [3] gets generated in a sufficiently general setting. We have a system providing $n + 1$ different services. Service time requirement for any type during any visit is exponentially distributed with parameter depending both the current type and the type to which service proceeds immediately after the present. An initial probability vector for admitting to the type of service is assumed. For clarity in exposition we call $n$ types of services 'undesired' and the remaining one, $(n + 1)^{th}$ type as the 'desired'. For different customers the 'undesired' and 'desired' types could be different. Further there could be more than one type in the desired class. These considerations will be taken up in an extension of the present work. Thus for this paper our assumption is that $n$ services are 'undesired' (with sometimes serious consequences) and the remaining one is the desired. However, this is hard to identify. Customers arrive according to a Poisson process of rate $\lambda$ to a single server counter. At the time when taken for service the service requirement is correctly diagnosed with probability $\theta$; with complementary probability $(1 - \theta)$ the identification goes wrong. As a consequence of correct diagnosis (prob. $\theta$), service immediately starts with duration having exponential distribution with parameter $\mu$. The customer leaves the system after service. However, if initially the customer is admitted to the group of undesired service (with probability $p_i$, it is diagnosed as requiring type $i$ service requirement, $i = 1, 2, \ldots, n$), it may stay in this class, moving from one undesired to another undesired, until finally all turn out to be failure and the customer turns out to be unfit for further service (desired or undesired). It may also happen that at some stage of service in undesired class, the service provider identifies that the customer is being served in the undesired set of services and so immediately takes him to the desired service stage. The customer gets required service here and leaves. But then how long is it possible to stay in service in the undesired set of states? We assume that a random clock (RC) with exponentially distributed duration starts ticking the moment a customer starts getting his service in a state in the undesired class. If correct diagnosis is made of the desired service during its sojourn in the undesired set of states before this random clock realizes, then the customer is immediately transferred for service in the desired state. On completion of service, assumed exponentially distributed with parameter $\mu$, the customer leaves the system. On the other hand if the RC realizes before the customer's service need is correctly diagnosed, then no further service is provided to that customer since it is rendered useless as a consequence of services in the undesired set of states. The service system pays a heavy penalty for this. In case correct diagnosis is done right at the beginning, then the customer pays for service at the time when he leaves. On the other hand if the customer initially stays in service in the undesired class and is able to quit it before RC realization, then on service completion in the desired state, that customer is provided a small compensation (for the undesired service he was forced upon).

The readers are referred to earlier two papers: Madan [1] introduced a service system with one main service and a second optional service in which customers are served by the same server in the main and optional phases; the service time in first phase being arbitrarily distributed and that in the second phase having exponential distribution. Medhi [2] extended the above model to the case where the second optional service too has arbitrary distribution, for which he establishes the Pollaczek-Khintchine formula. Our model is to some extent a dual of those discussed in Madan and Medhi. However, what we call the 'optional' services in the present context are not 'really not optional', but something inadvertent. They are rather provided due to wrong identification.

The immediate application of the problem described above could be found in communication on highways: Emergency messages are to be passed on within the shortest time possible. Such messages generated in a time slot should all be transmitted within that time slot; else they are bound to have lost their significance. This part is taken care of by the RC. At the time when the message is generated if the server is found to be busy then the message passes through several contention windows (CWs); each time the final slot of the contention window is reached, the message checks for the server availability. If found to be busy, the message goes to the next level of the CW.

In Sect. 2, the mathematical model is described. This section also provides the steady state analysis and some performance measures, including the expected service time of a customer.

**Notations:**

- CTMC: Continuous time Markov chain
- LIQBD: Level independent quasi-birth and death process
- $\mathbf{e}$: Column vector of $1's$ with appropriate order
- $\mathbf{x}_i$: Probability that number of customers in the system is $i$
- $\mathbf{x}_{.j}$: Probability that the server is in the $j^{th}(1 \leq j \leq n+1)$ phase of service.

## 2　Mathematical Formulation

**Assumptions.** The assumptions leading to the formulation of the mathematical model are

- Arrival of customers to the system is from a Poisson stream with mean arrival rate $\lambda(> 0)$.

- The probability that a customer selected to provided desired service is $\theta$.
- A customer falls in the $i^{th}$ undesired phase with probability $p_i$ so that $p_1 + p_2 + \cdots + p_n = 1$.
- A single server is providing both undesired and desired service. Service times are exponentially distributed with mean service rate $\mu$ for desired service and $\mu_{ij}$ for the rate of transition to $j^{th}$ state of undesired service after completing service in $i^{th}$ state of the $n-$ undesired services, $i \in \{1, 2, 3, \ldots, n\}$.
- A threshold clock is set which follows exponential distribution with mean rate $\alpha$ so that the customer is pushed out of the system if the clock expires before service completion in undesired service.

Let $N(t)$ be the number of customers in the system and $S(t)$ the service phase at time $t$, which is $i, 1 \le i \le n$ if server is at the $i^{th}$ undesired service and $n+1$ if at desired service. Then $\{(N(t), S(t)), t \ge 0\}$ is a CTMC with state space

$$\Omega = \{0\} \cup \{(i,j), i \in Z^+, j = 1, 2, \ldots, n, n+1\}$$

The infinitesimal generator $Q$ of this CTMC is a LIQBD where

$$Q = \begin{pmatrix} B_{00} & B_{01} & & & \\ B_{10} & B_1 & B_0 & & \\ & B_2 & B_1 & B_0 & \\ & & \ddots & \ddots & \ddots \end{pmatrix}$$

with, $B_{00} = [-\lambda]$, $B_{01} = \begin{bmatrix} \lambda(1-\theta)p_1 & \lambda(1-\theta)p_2 & \cdots & \lambda(1-\theta)p_n & \lambda\theta \end{bmatrix}$ $B_{10} = \begin{bmatrix} \alpha & \alpha & \cdots & \alpha & \mu \end{bmatrix}^T$, $B_0 = \lambda I_{n+1}$

$$B_2 = \begin{pmatrix} \alpha(1-\theta)p_1 & \alpha(1-\theta)p_2 & \cdots & \alpha(1-\theta)p_n & \alpha\theta \\ \alpha(1-\theta)p_1 & \alpha(1-\theta)p_2 & \cdots & \alpha(1-\theta)p_n & \alpha\theta \\ \cdots\cdots & \cdots\cdots & \cdots & \cdots\cdots & \cdots \\ \ddots\ddots & \ddots\ddots & \ddots & \ddots\ddots & \ddots \\ \alpha(1-\theta)p_1 & \alpha(1-\theta)p_2 & \cdots & \alpha(1-\theta)p_n & \alpha\theta \\ \mu(1-\theta)p_1 & \mu(1-\theta)p_2 & \cdots & \mu(1-\theta)p_n & \mu\theta \end{pmatrix}$$

$$B_1 = \begin{pmatrix} -(\mu_{11}+\alpha+\lambda) & \mu_{12} & \mu_{13} & \cdots & \mu_{1n} & \mu_{1(n+1)} \\ \mu_{21} & -(\mu_{22}+\alpha+\lambda) & \mu_{23} & \cdots & \mu_{2n} & \mu_{2(n+1)} \\ \vdots & \vdots & & \ddots & & \vdots \\ \mu_{n1} & \mu_{n2} & -(\mu_{nn}+\alpha+\lambda) & \cdots & \mu_{nn} & \mu_{n(n+1)} \\ 0 & 0 & \cdots & & 0 & -(\lambda+\mu) \end{pmatrix}$$

Here,

$$\mu_{ii} = \sum_{\substack{j=1 \\ j \ne i}}^{n+1} \mu_{ij}, \ 1 \le i \le n.$$

## 2.1   Steady State Analysis

We proceed with the steady-state analysis of the queueing system under study. Naturally we have to look for the condition for stability.

**Stability Condition.** We consider $B = B_0 + B_1 + B_2$ for determining the stability condition for the original system where, $B$ is given by

$$B(i,j) = \begin{cases} \alpha(1-\theta)p_i - (\mu_{ii} + \alpha), & j = i, 1 \le i \le n \\ \alpha(1-\theta)p_j + \mu_{ij}, & i \ne j, 1 \le i, j \le n \\ \mu(1-\theta)p_j, & i = n+1, 1 \le j \le n \\ \alpha\theta + \mu_{i(n+1)}, & j = n+1, 1 \le i \le n \\ \mu(\theta-1), & i = j = n+1 \end{cases}$$

Let $\boldsymbol{\Pi} = (\pi_1, \pi_2, \ldots, \pi_n, \pi_{n+1})$ be the steady-state probability distribution of the Markov chain corresponding to the generator $B$. Then $\pi_i's$ are computed below.

We have $\boldsymbol{\Pi}B = 0$ and $\boldsymbol{\Pi}\mathbf{e} = 1$, the first one being a system of $n$ homogeneous linear equations. Replace the last equation in this system by the second equation resulting in a new system of $n + 1$ equations given by $C\boldsymbol{\Pi} = C'$, where

$$C = [c_{ij}] = \begin{cases} \alpha(1-\theta)p_i - (\mu_{ii} + \alpha), & j = i, 1 \le i \le n \\ \alpha(1-\theta)p_i + \mu_{ji}, & i \ne j, 1 \le i, j \le n \\ \mu(1-\theta)p_i, & j = n+1, 1 \le i \le n \\ 1, & i = n+1, 1 \le j \le n+1 \end{cases}$$

and

$$C'(i,j) = \begin{cases} 1, & i = n+1, j = 1 \\ 0, & \text{elsewhere} \end{cases}$$

Define

$$d_{nj} = -\frac{c_{nj}}{c_{n,(n+1)}}; \quad 1 \le j \le n$$

and for $1 \le k \le (n-1)$,

$$d_{(n-k)j} = -\frac{c_{(n-k),j} + \sum_{m=1}^{k} U_{mj}}{c_{(n-k),(n-k+1)} + \sum_{m=1}^{k} U_{m(n-k+1)}}; \quad 1 \le j \le n$$

For $1 \le m \le k$,

$$U_{mj} = \sum_{\substack{r_l = n-k+m+2-l \\ 1 \le l \le m \\ r_0 = n+2, \, r_{m+1} = j}}^{r_{l-1}-1} c_{(n-k)r_1} \prod_{j=1}^{m} d_{r_j-1, \, r_{j+1}}$$

Setting

$$V_m = \sum_{\substack{k_j = m \\ 1 \leq j \leq m \\ k_0 = n}}^{k_{j-1}} \prod_{j=1}^{m-1} d_{(k_j - j + 1)(k_{j+1} - j + 1)} d_{(k_m - m + 1)1}$$

we get

$$\pi_1 = \left(1 + \sum_{m=1}^{n} V_m\right)^{-1}$$

and

$$\pi_{j+1} = \sum_{k=1}^{j} d_{jk}\, \pi_k, \quad \text{for } 1 \leq j \leq n.$$

The LIQBD description of the model indicates that the queueing system is stable if and only if

$$\Pi B_0 \mathbf{e} < \Pi B_2 \mathbf{e}.$$

This gives the stability condition as

**Lemma 1.** *The system is stable if and only if*

$$\lambda < \alpha(1 - \pi_{n+1}) + \mu\pi_{n+1} \tag{1}$$

*Proof.* One can calculate left (decrease by one in the number of customers) and right (increase by one in the number of customers) drifts for stability, left drift rate should be larger than right drift rate. One may also proceed heuristically to arrive at the above result.

**Steady-State Probability Vector.** Assuming that Eq. (1) is satisfied, we briefly out line the computation of the steady state probability of the system. Let $\mathbf{x}$ denote the steady-state probability vector of the generator $Q$. Then

$$\mathbf{x}Q = 0, \qquad \mathbf{x}\mathbf{e} = 1 \tag{2}$$

Assuming that the stability condition (1) holds and partitioning $\mathbf{x}$ as $\mathbf{x} = (\mathbf{x}_0, \mathbf{x}_1, \mathbf{x}_2, \ldots)$ we obtain

$$\mathbf{x}_n = \mathbf{x}_1 R^{n-1}, n \geq 2 \tag{3}$$

where $R$ is the minimal non negative solution to the matrix quadratic equation

$$R^2 A_2 + R A_1 + A_0 = 0 \tag{4}$$

The two boundary equations involving $\mathbf{x}_0$ are

$$\mathbf{x}_0 A_{00} + \mathbf{x}_1 A_{10} = 0, \tag{5}$$

$$\mathbf{x}_0 A_{01} + \mathbf{x}_1 [A_1 + R A_2] = 0 \tag{6}$$

These together with the normalizing condition in (2) gives

$$x_1 = x_0 V \text{ where } V = -A_{01}[A_1 + RA_2]^{-1} \tag{7}$$

$$x_0[1 + V(I - R)^{-1}e] = 1 \tag{8}$$

Now we look at a few of the system performance measures.

– Expected number of customers in the system

$$L = \sum_{i=1}^{\infty} i\, x_i e$$

– Expected waiting time in the system

$$W = \frac{L}{\lambda}$$

**Expected Service Time of a Customer.** We consider a Markov chain with state space $\{1, 2, \ldots, n, n+1\} \bigcup \{\Delta\}$, where $k(1 \le k \le n)$ is the phase of undesired service, $n+1$ means the desired service and $\{\Delta\}$ denotes the absorbing state which represents the realization of the threshold clock or completion of service from desired service. The infinitesimal generator of this CTMC is

$$W_s = \begin{bmatrix} T & T^0 \\ \mathbf{0} & 0 \end{bmatrix}$$

where

$$T = \begin{pmatrix} -(\mu_{11} + \alpha) & \mu_{12} & \mu_{13} & \cdots & \mu_{1(n+1)} \\ \mu_{21} & -(\mu_{22} + \alpha) & \mu_{23} & \cdots & \mu_{2(n+1)} \\ \vdots & \vdots & \ddots & \ddots & \vdots \\ \mu_{n1} & \mu_{n2} & -(\mu_{nn} + \alpha) & \cdots & \mu_{n(n+1)} \\ 0 & 0 & & \cdots & -\mu \end{pmatrix}, T^0 = \begin{pmatrix} \alpha \\ \alpha \\ \vdots \\ \alpha \\ \mu \end{pmatrix}$$

Now, the service time of a customer is the time until absorption of the Markov chain. The distribution of $W_s$ is Phase type with initial probability vector

$$\alpha = \left( (1 - \theta)p_1 \ (1 - \theta)p_2 \ \cdots \ (1 - \theta)p_n \ \theta \right).$$

Then the expected time a customer spend in service is $-\alpha T^{-1}e$.

**Numerical Illustration.** Figure 1: Illustrates the expected service time in the system with variation in the values of the underlying parameters. It is assumed that $\lambda = 3, \mu = 6, n = 3, p = [0.3 \ 0.3 \ 0.4], [\mu_{ij}] = \begin{bmatrix} -9 & 2 & 3 & 4 \\ 2 & -7 & 3 & 2 \\ 1 & 2 & 3 & -6 \end{bmatrix}$.

Figure 2: Illustrates the status (probability for the server to be in a particular state of service or idle state) for different values of $\lambda$. It is assumed that $\alpha = 4, \mu = 6, \theta = 0.7, n = 3, p = [0.3 \ 0.3 \ 0.4], [\mu_{ij}] = \begin{bmatrix} -12 & 3 & 4 & 5 \\ 4 & -15 & 5 & 6 \\ 3 & 4 & 3 & -10 \end{bmatrix}$.

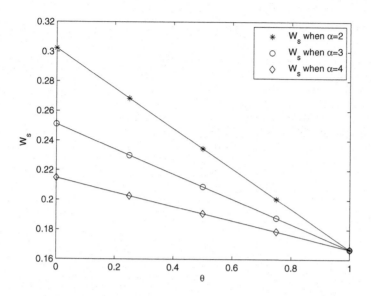

**Fig. 1.** Expected service time of a customer

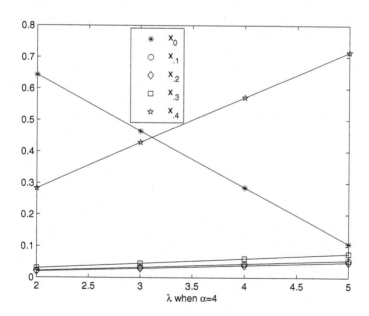

**Fig. 2.** Status of the server

# References

1. Madan, K.C.: An M/G/1 queue with second optional service. Queueing Syst. **34**(1–4), 37–46 (2000)
2. Medhi, J.: A single server poisson input queue with a second optional channel. Queueing Syst. **42**, 239–242 (2002). http://link.springer.com/article/10.1023%2FA%3A1020519830116
3. Neuts, M.F.: Matrix-Geometric Solutions in Stochastic Models. Johns Hopkins University Press, Baltimore (1981)
4. Gross, D., et al.: Observation of Strains: Fundamentals of Queueing Theory. John Wiley and Sons, Inc., Hoboken (2011)

# Swarm of Public Unmanned Aerial Vehicles as a Queuing Network

R. Kirichek$^{(\boxtimes)}$, A. Paramonov, and A. Koucheryavy

The Bonch-Bruevich Saint-Petersburg State University of Telecommunications,
Saint Petersburg, Russia
kirichek@sut.ru, alex-in-spb@yandex.ru, akouch@mail.ru

**Abstract.** Unmanned aerial vehicles which are used to build flying ubiquitous sensor networks are viewed as a queuing system and their swarm — as a queuing network. It is proved that a sufficiently large number of UAVs swarm can be considered as a network of Jackson. The distribution of the lengths of the shortest paths for the UAVs swarms with a cube and a sphere is determined.

**Keywords:** Public flying ubiquitous sensor network · Unmanned Aerial Vehicle · The queuing system · The queuing network · The length of the shortest path

## 1 Introduction

One of the most attractive areas of the networks and communication systems has recently been Flying Ad Hoc Networks (FANET) [1–3]. Initially used mainly for military purposes, UAVs are currently used in civilian applications [4,5]. By analogy with the division of terrestrial in the Ad Hoc network [6,7] and ubiquitous or wireless sensor networks [8,9] in the field of Ad Hoc networks there were flying ubiquitous sensor network FUSN [10]. Widespread public unmanned aerial vehicles and related networking features FUSN enable to identify a new class of public communications networks FUSN-P (Public) [11]. One of the main features of FUSN-P is that the UAV is operated usually by nonprofessional users, so that it requires the simplest handling of them during operation. For this purpose, in [11] in the FUSN-P it was proposed to use the UAV flight for the data collection from the sensor fields on a given route. Simultaneous use of multiple UAVs leads both to creation of a swarm and to the possibility of considering it as a swarm of the queuing network. Notable works of UAV swarms as a part of FANET usually pursued the target of cooperation the UAV opportunities for solving military tasks, for search of the target, etc. [12–14]. We believe that the wide spread of public unmanned aerial vehicles enables to consider a separate UAV as a queuing system [10] and a swarm as a queuing network.

© Springer International Publishing Switzerland 2016
V. Vishnevsky and D. Kozyrev (Eds.): DCCN 2015, CCIS 601, pp. 111–120, 2016.
DOI: 10.1007/978-3-319-30843-2_12

## 2   UAV as a Queuing System

Let sensory nodes FUSN-P which are considered, for example, for the head [15], are located on the UAV, which perform the flight of the sensor field territory (terrestrial network USN) and collect the data from the terrestrial-based sensor nodes. While servicing a plurality of nodes, the UAV can be seen as a queuing system, the input of which receives the entity (terrestrial sensor nodes in the service area) which can expect the service within the time of their stay in the area of accessibility. The entities (nodes) that have not been serviced during this time are denial of the service. The flow rate is dependent on the radius of the service area, the density of nodes and the speed of the UAV. To serve the UAV terrestrial sensor assembly some time is spent and the node should be in the area of accessibility during the period of service (Fig. 1).

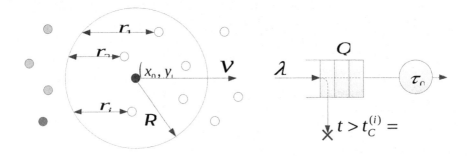

**Fig. 1.** UAV as a queuing system.

If the coordinates of the terrestrial nodes are accidental, the entry system receives the random flow of the entities. The properties of this flow are determined by the properties of the sensor field (publishing sites on the surface), the radius of the service drones and its speed. We will make the following assumptions:

- the sensory field is a Poisson field;
- the UAV is believed to move in a straight line at a constant velocity v;
- the zone service is a circle with a radius R.

Define the distribution function for the incoming flow entities. For this purpose we will examine the service area of the UAV at time 0 and at time t. During t the entities (nodes) which are found in the area that is defined by a shift of the UAV service area for time t will go in the system. According to the properties of the Poisson field, the probability of presence of n points (nodes) in a certain area is determined by the Poisson distribution and depends only on the field area. The probability of presence of z entities (nodes) in the field S is

$$p_z = \frac{a^z}{z!}e^{-a} \tag{1}$$

where a = p * S; p - is the number of points (nodes) in a unit area; S - is the field area.

$$p_z(t) = \frac{(p \cdot S(t))^z}{z!} e^{-p \cdot S(t)} \tag{2}$$

The field area can be defined as

$$S(t) = 2R \cdot vt \tag{3}$$

The flow rate, i.e. the average number of entities per unit of time is equal to

$$\gamma = p \cdot 2R \cdot vt \tag{4}$$

The distribution of the time interval between the entities We will consider the random variable T as the time interval between two successive events in the stream and will find its distribution function.

$$F(t) = P(T < t) \tag{5}$$

Then the probability that z entities will go to the time section of the length t is

$$P(T \geq t) = 1 - F(t) \tag{6}$$

Therefore, the probability can be calculated by the formula

$$P(T \geq t) = p_0(t) = e^{-p \cdot 2 \cdot R \cdot vt} \tag{7}$$

Considering this fact, the distribution function of the time interval between the entities is

$$F(t) = 1 - e^{-p \cdot 2 \cdot R \cdot vt} \tag{8}$$

Thus, the elementary flow will enter the system, the time intervals between the entities, which are distributed exponentially with a mean.

$$\bar{a} = \frac{1}{\rho \cdot 2R \cdot v} \tag{9}$$

## 3   Swarm of UAV-P as a Queuing Network

Taking into consideration the above mentioned facts, the flow of entities (messages), which arrives at the node of each of the UAV has the properties of a simple entity flow. Beside the flow of messages from a particular terrestrial sensor field, the viewed nodes receive the traffic flows from other nodes on the network.

Further we will assume that the output flow of messages from $i$ node with probability $r_{ij}$ is an input to the node $j$. With probability $1 - \sum_{j=1}^{n} r_{ij}$ the entities will leave the node i and will be sent to the external environment, i.e., to the gateway, Fig. 2.

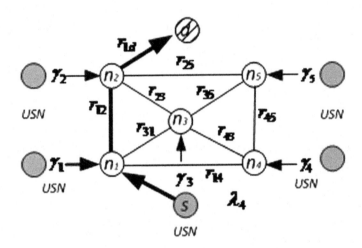

**Fig. 2.** Model of data delivery route between the source (s) and receiver (t).

In the general case, the service time of the messages on the route segment $t$ consists of two main components: the time of sending the message on channel $\tau$ and the time-out state of the channel readiness$\psi$, which are generally random.

Changing of the channel status is a random process that occurs under the influence of many independent factors (events), such as the entry and stepping out of communication range due to the random deviations from the desired path of movement, the effect of interference from transmitters located on the other elements of the system and others. It is expected that with a sufficiently large number of independent events the channel readiness intervals will have the distribution which is close to exponential distribution, therefore, the state of waiting time readiness $\psi$ will also have the similar distribution.

If the time distribution of the message sending through the communication channel $\tau$ is close to an exponential one, then the assumption of exponential distribution of service time $t$ is quite possible.

If we strengthen the above mentioned conditions of the network by the assumption of exponential service time of messages in the nodes, these conditions will coincide with the conditions of the network Jackson [16].

$$\overline{T} = \sum_{j=1}^{M} \frac{\lambda_j}{\gamma} T_j, \tag{10}$$

where M - is the number of channels in the network; n - is the number of network nodes; $T_j$ - is the delay in the j-th channel; $\gamma = \sum_{i=1}^{n} \gamma_i$ - is the total traffic network; $\lambda$ - is the total traffic served in the j-th channel.- is the delay in the j-th channel;

The value

$$T_j = \frac{1}{\mu_j - \lambda_j}, \tag{11}$$

where $\mu_j = \frac{1}{t_j}$ - is the service rate in the j-th channel.

Delivery time for a particular route network $\theta_k$ can be estimated by using the properties of the Jackson network. It is known that each node of the network can be considered as independent QS M/M/1, and the whole route — as a series of independent QS M/M/1, Fig. 3.

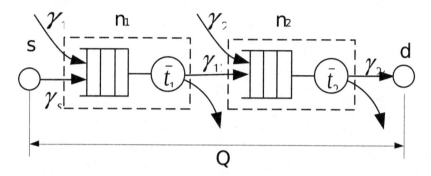

**Fig. 3.** Model of data delivery route between the source (s) and receiver (t).

The distribution function of time to deliver a message in this system can be described by Erlang distribution.

In case of equality of all $\lambda_i = \lambda$, and $\mu_i = \mu$ with the average value $m \cdot t$, which is the average time to deliver a message on the route $\theta_k = m_k \cdot t$ where $m_k$ is the number of channels in the k-route (Fig. 4).

$$S(x, m) = \frac{m \cdot \mu \cdot (m \cdot \mu \cdot x)^{m-1}}{(m-1)!} e^{-m \cdot \mu \cdot x} \qquad (12)$$

The order m, in this case, corresponds to the number of transits (hops), assuming that the message transmission (service) for each of them is equal.

The more accurate approximation of the viewed network as the Jackson network is, the more n there are and the nearer service time of distribution blitz to exponential distribution is. With a relatively small number of n nodes and a small number of routes network the properties can significantly differ from the properties of the Jackson network. In this case, the route pattern can be described as a multiphase system G/G/1. Getting of the distribution function of delivery time, in this case, can be very difficult. However, an approximate estimate of the average delivery time to the j channel route is possible, as it is shown in [16]

$$\tilde{T}_j \approx \frac{\rho_j \cdot \bar{t}_j}{2(1 - \rho_j)} \left( \frac{\sigma_{a_j}^2 + \sigma_{t_j}^2}{\bar{t}_j^2} \right) \left( \frac{\bar{t}_j^2 + \sigma_{t_j}^2}{a_j^2 + \sigma_{t_j}^2} \right) \qquad (13)$$

Where $p_j = \lambda_j \bar{t}_j$; $\sigma_{a_j}^2$ - dispersion of intervals between messages; $\sigma_{t_j}^2$ - dispersion of service time in j channel; $\bar{t}_j$ - Service time in j channel; $a_j = \frac{1}{\lambda_j}$ - the mean value of the interval between messages in j channel.

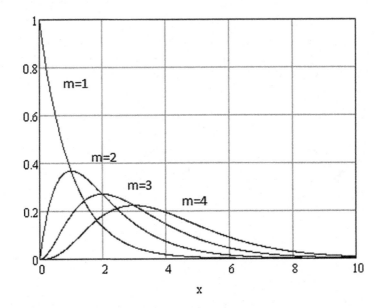

**Fig. 4.** The probability density of the delivery time on the route of m length = 1,2,3,4 hops.

Then the delivery time on the route will be equal to

$$\theta_k = \sum_{j=1}^{m_k} \widetilde{T}_j \tag{14}$$

where $m_k$ is the number of channels in the k route.

It should be noted that the more accurate the estimate of the mean delivery time on the route 4 and 5 is, the higher the intensity of the message flow is $\lambda_j$ and the smaller relation between the service time in the channel and time of messages receipt on the input of each channel.

It is obvious that one of the determining factors of the delivery time is the number of "hops" (channels) in the route m. This number depends on the used methods of routing. It is logical to assume that the route of the minimum length (with a minimum number of "hops") is chosen. Figure 5 shows the implementation of the random distribution of nodes in the space which is defined by a cube $200 \times 200 \times 200$ m (a) and by an equal volume of a sphere (b).

Figure 6 shows the distribution of the lengths of the shortest paths in the network which is formed by nodes that are arranged in a cube, with a communication node radius of 50 m.

This distribution was obtained by simulation. The shortest route was chosen by the criterion of a minimum number of hops. The average path length was 4.47 hop. For comparison, the same figure shows a Poisson distribution with a mean of 4.47. The connectivity probability was 0.98.

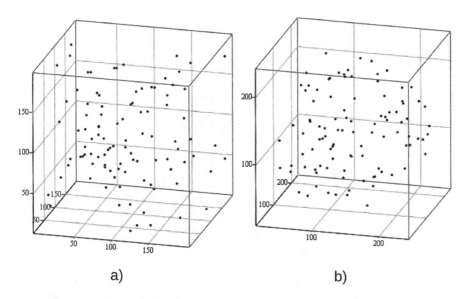

**Fig. 5.** Random placement of 100 nodes in the cube $200 \times 200 \times 200\,\mathrm{m}$ (a) and in the sphere of equal volume (b) 14.

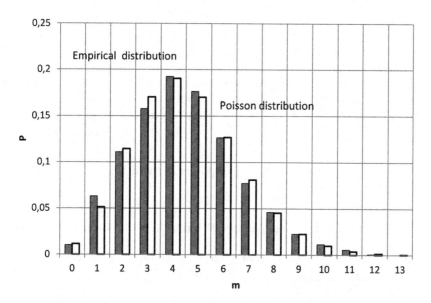

**Fig. 6.** Distribution of the lengths of the shortest paths in the network of 100 nodes in the cube $200 \times 200 \times 200$ m.

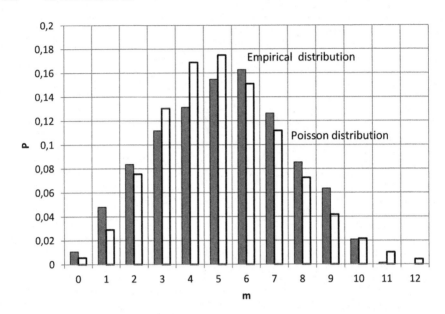

**Fig. 7.** The distribution of the lengths of the shortest paths in the network of 100 nodes in an equal volume area.

Figure 7 shows the distribution of the lengths of the shortest paths in the network with a random arrangement of 100 nodes in the area with a communication network node radius of 50 m.

The average path length was 5.18 hop. For comparison, the same figure also shows a Poisson distribution with a mean of 5.18. In this case, the network connectivity was 0.94.

The connectivity probability can be defined as the probability of falling into the sphere of a given radius of at least one node.

Out of the properties of the Poisson field, this probability is

$$P_{\geq 1} = 1 - e^{-a} \tag{15}$$

Where a - is the expected number of points in the field.

$$a = V \cdot \rho \tag{16}$$

where $V = \frac{4}{3}\pi \cdot x^3$ is the sphere volume of radius x; $p$ - is the sphere volume of radius x;

Then the dependence of the probability density and connectivity of the network node communication radius is equal to

$$P = 1 - e^{-\frac{4}{3}\pi \cdot x^3 \cdot \rho} \tag{17}$$

For the simulated network, starting from (17), it is equal to 0.999. The values of connectivity which are obtained from the simulation results are within the error due to the finite size of the sample. It should be noted that the expression (17) gives the probability of connectivity for the unlimited Poisson field. In this case, the field is limited with a certain volume. In the case of restrictions, "edge effect" takes place which considers that the probability of connectivity for the nodes near the border is less than for the nodes that are closer to the center of considered limiting volume of the figure. This is obvious when considering a node which is located strictly at the boundary field.

The adjacent to it node can be located within the area. If the boundary is the plane, the extent to which communication with the neighboring node is possible is less than half for the site located near the center of the examined area (if the communication range is smaller than the area of the node). In this regard, it should be expected that the assessment of the connection probability (17) is the upper bound. Also the closer to the probability the value of the connected network will be (17), the larger the ratio of bounding shape to its surface area is. It is obvious that by increasing of the geometric dimensions the ratio will increase. As it is seen from the given figures in the case of considering the limited space of a cube, the length of the shortest path is well described by a Poisson distribution. In the case when the space is limited by a sphere, the distribution of the lengths of the shortest paths differs from the Poisson distribution to a greater extent. The average lengths of the shortest path (in the race) in the cases of cube and sphere are expected to vary.

## 4   Conclusion

1. While the organization of interaction with UAVs USN nodes to collect data under the certain conditions, the network connections between the UAV can be seen as a queuing network.
2. When a sufficiently large number of nodes which are located on the UAV model, the network Jackson can be used. In this case, the delivery time of data between the sources and receiver will obey the law of Erlang.
3. With a relatively small number of UAV to estimate the time of the data delivery it is possible to use familiar approximate estimates for systems G/G/1.
4. The number of "hops" in the shortest route between the nodes of the UAV is distributed according to the law which is close to the Poisson law that enables to estimate the length of the routes and the delay of the data delivery.

**Acknowledgments.** The reported study was supported by RFBR, research project No15 07-09431a "Development of the principles of construction and methods of self-organization for Flying Ubiquitous Sensor Networks".

## References

1. Bekmezci, I., Sahingoz, O.K., Temel, S.: Flying Ad-Hoc networks: a survey. Ad Hoc Netw. **11**(3), 1254–1270 (2013)

2. Sahingoz, O.K.: Networking model in flying Ad Hoc networks (FANETs): concepts and challenges. J. Intell. Robot. Syst. **74**(1–2), 513–527 (2014)

3. Singh, S.K.: A comprehensive survey on fanet: challenges and advancements. Int. J. Comput. Sci. Inf. Technol. **6**(3), 2010–2013 (2015)

4. Rosário, D., Zhao, Z., Braun, T., Cerqueira, E., Santos, A.: A comparative analysis of beaconless opportunistic routing protocols for video dissemination over flying Ad-hoc networks. In: Balandin, S., Andreev, S., Koucheryavy, Y. (eds.) NEW2AN/ruSMART 2014. LNCS, vol. 8638, pp. 253–265. Springer, Heidelberg (2014)

5. de Freitas, E.P., Heimfarth, T., Netto, I.F., Lino, C.E., Pereira, C.E., Ferreira, A.M., Wagner, F.R., Larsson, T.: UAV relay network to support WSN connectivity. In: Proceedings of the ICUMT (2010)

6. Vinel, A., Vishnevsky, V., Koucheryavy, Y.: A simple analytical model for the periodic broadcasting in vehicular Ad-Hoc networks. In: IEEE Globecom Workshops GLOBECOM 2008 (2008)

7. Jakubiak, J., Koucheryavy, Y.: State of the art and research challenges for VANETs. In: 5th IEEE Consumer Communications and Networking Conference, CCNC 2008 (2008)

8. Akyildiz, I.F., Vuran, M.C., Akan, O.B., Su, W.: Wireless sensor networks: a survey revisited. Comput. Netw. J. **38**(4), 393–422 (2005)

9. Koucheryavy, A., Salim, A.: prediction-based clustering algorithm for mobile wireless sensor networks. In: Proceedings of the International Conference on Advanced Communication Technology: ICACT 2010, Phoenix Park, Korea (2010)

10. Kirichek, R., Paramonov, A., Koucheryavy, A.: Flying ubiquitous sensor networks as a quening system. In: Proceedings of the International Conference on Advanced Communication Technology: ICACT 2015, Phoenix Park, Korea (2015)

11. Koucheryavy, A., Vladyko, A., Kirichek, R.: State of the art and research challenges for public flying ubiquitous sensor networks. In: Balandin, S., Andreev, S., Koucheryavy, Y. (eds.) NEW2AN/ruSMART 2015. LNCS, vol. 9247, pp. 299–308. Springer, Heidelberg (2015)

12. Altshuler, Y., Yanovsky, V., Wagner, I., Bruckstein, A.: The cooperative hunters-efficient cooperative search for smart targets using UAV swarms. In: Second International Conference on Informatics in Control, Automation and Robotics (ICINCO), the First International Workshop on Multi-Agent Robotic Systems (MARS) (2005)

13. McCunea, R.R., Madey, G.R.: Swarm control of UAVs for cooperative hunting with DDDAS. In: International Conference on Computational Science, ICCS (2013)

14. Madey, G., Blake, M., Poellabauer, C., Lu, H., McCune, R., Wei, Y.: Applying DDDAS principles to command, control and mission planning for UAV swarms. Procedia Comput. Sci. **9**, 1177–1186 (2012)

15. Futahi, A., Koucheryavy, A., Paramonov, A., Prokopiev, A.: Ubiquitous sensor networks in the heterogeneous LTE network. In: Proceedings of the International Conference on Advanced Communication Technology: ICACT 2015, Phoenix Park, Korea (2015)

16. Kleinrock, L.: Queueing Systems. Wiley, New York (1975). Vol. 1 Theory

# Approach to the Analysis of Probability Measures of Cloud Computing Systems with Dynamic Scaling

Yuliya Gaidamaka, Eduard Sopin$^{(\boxtimes)}$, and Margarita Talanova

Peoples' Friendship University of Russia,
Miklukho-Maklaya street 6, 117198 Moscow, Russia
{ygaidamaka,esopin}@sci.pfu.edu.ru, matalanova@gmail.com

**Abstract.** Cloud computing paradigm proved itself to be a good approach for efficient high-performance computational infrastructure design. Besides performance, one of the most important measures for cloud service providers is energy efficiency of the system. For better energy efficiency, in light load case additional virtual machines are switched off, and switched on again when load increases. Dynamic scaling increases performance of the system, but leads to increase in service costs. To avoid ping-pong effect in dynamic scaling, hysteresis approach is applied. Considering number of virtual machines in modern cloud platforms, special attention should be paid to efficiency of computing algorithm. In this work, we describe behavior of cloud computing system in terms of queuing system with threshold-based hysteretic control of active servers number and noninstantaneous server activation. We use state elimination technique to develop efficient algorithm for stationary probabilities calculations and estimation of its performance measures.

**Keywords:** Cloud computing · Hysteretic load control · Dynamic scaling · State elimination method

## 1 Introduction

Cloud computing grants to a user computing resources and infrastructure in the form of Internet service [1]. Cloud computing systems are applied to data storage and processing, to distributed computing for scientific and business solutions. Modern cloud systems are usually designed scalable so that the system can cope with high load and may reduce energy consumption during the periods of low load. One of the ways for system scalability realization is dynamic activation-deactivation of servers and virtual machines [2–4]. Analysis of such systems can be performed using queuing systems with threshold based control of active servers number [5,6]. But the main problem of cloud systems modeling

This work was supported in part by the Russian Foundation for Basic Research, projects No. 14-07-00090, 15-07-03051, 15-07-03608.

© Springer International Publishing Switzerland 2016
V. Vishnevsky and D. Kozyrev (Eds.): DCCN 2015, CCIS 601, pp. 121–131, 2016.
DOI: 10.1007/978-3-319-30843-2_13

in terms of teletraffic theory is high computing complexity of derived algorithms [7–9]. In [5], a queuing system with hysteretic dynamic scaling and noninstantaneous additional server activation is applied for video-on-demand service, stationary characteristics of system are obtained by means of matrix-geometric methods which are not applicable for cloud computing because of extremely large amount of servers in modern cloud platforms. In [3,4], cloud computing system with dynamic scaling was described in terms of queuing system with hysteretic control of active servers number, and for a case of three servers, the effective computing algorithm for stationary probability distribution was developed. Complexity of received algorithm is linear because of state elimination technique [11] used for queuing system analysis. In [10], the same approach was applied for any number of servers in the system. The weak side of proposed in [10] model is extremely large state space, which has quadratic dependence on number of servers. Moreover, the majority of states has a negligible effect on overall system behavior. Therefore, in this article, we consider a simplified model with reduced state space and perform its analysis using the same state elimination approach. Finally, in numerical analysis section we provide a comparative analysis of the initial and simplified models.

## 2   Simplified Model Description

For performance measures evaluation we describe cloud computing system dynamic adding and removing additional servers in terms of queuing system with $K$ servers and hysteretic control of active servers number (Fig. 1). An arriving customer enter the system if the total amount of customers is less than system capacity $R$, otherwise the customer is considered lost. Customers arrive according to the Poisson process with rate $\lambda$. We consider servers to be uniform, the service times are exponentially distributed with rate $\mu$.

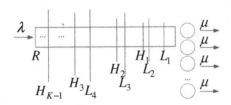

**Fig. 1.** Queuing model with thresholds $H$ and $L$.

In the empty system, there is only one active server. Server activation/deactivation procedures are governed by number of customers in the system upper threshold vector $H = (H_1, H_2, ..., H_{K-1})$, $H_1 < H_2 < ... < H_{K-1}$ and lower threshold vector $L = (L_1, L_2, ..., L_{K-1})$, $L_1 < L_2 < ... < L_{K-1}$ where $L_{k+1} < H_k$, $k = \overline{1, K-2}$ and $L_k < H_k$, $k = \overline{1, K-1}$. Customers are served in FCFS (First Come First Served) order. System functions as follows:

– if there is $H_k$ customers in the system, arriving customer causes an additional $(k+1)$th server activation procedure, server activation time is not instant and is exponentially distributed with rate $\alpha$;
– if there is $L_k$ customers, then on departure of a customer either server shuts down in case of no running server activation procedure or one of server activation procedures stops.

For the simplification of initial model [10] we assume that in case of two running server activation procedures, third server activation procedure is completed immediately on start. Thus, it is impossible to have 3 running activation procedures at once. Provided assumption significantly reduces state space of the system for big values of $K$.

Functioning of system is described by Markov process with a set of states

$$
S = \left\{ (k,i,n) \left|
\begin{array}{ll}
0 \le n \le H_1 & \text{for } k = 1, i = 1, \\
L_1 \le n \le H_2 & \text{for } k = 2, i = 1,2 \\
L_{k-1} \le n \le H_k & \text{for } k = \overline{3, K-1}, i = \overline{k-2, k}, \\
L_{K-1} \le n \le R & \text{for } k = K, i = \overline{K-2, K},
\end{array}
\right. \right\}
$$

where $k$ is necessary number of servers; $i$ is number of the active servers; $n$ is number of customers in the system.

For the chosen arrangement of threshold values, Fig. 2 depicts state transition diagram for the system with $K = 5$ servers. Let us denote state levels as $(k,i)$.

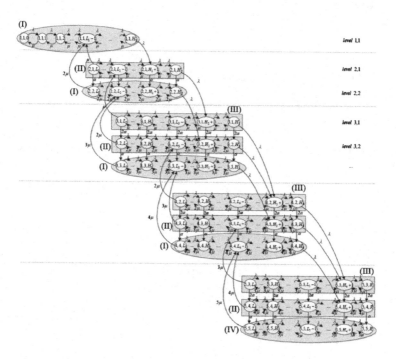

**Fig. 2.** State transition diagram for the system with $K = 5$ devices.

# 3   Stationary Probability Distribution Calculation Algorithm

For calculation of stationary probabilities, the method of state elimination is used. We introduce several auxiliary probabilities and intensity, similar to [4]:

- $a_{k,i,n}$ is probability that starting from state $(k,i,n)$ the system will pass to $(k,i,n-1)$ earlier, than to states on levels $k+1,i$ or $k,i+1$,
- $a_{k,i}^*$ is probability that starting from state $(k,i,L_k)$ the system will pass to $(k,i,L_k-1)$ earlier, than to state $(k,i+1,L_k-1)$,
- $\Lambda_{k,i,n}$ is intensity of direct transition from states of level $k,i-1$ to a state $(k,i,n)$ considering that states $(k,i,n+1)$, $(k,i,n+2)$, ... are eliminated,
- $A_{k,i,n}$ is probability that starting from state $(k,i,H_{k-1}+1)$ the system will pass to $(k,i,n)$ earlier, than to states of levels $k+1,i$ or $k,i+1$,
- $A_{k,i,n}^*$ is probability that starting from state $(k,i,H_{k-1}+1)$ the system will pass to $(k,i,L_k-1)$ earlier, than to state $k,i+1,L_k-1$,
- $C_{k,i}$ is probability that starting from a state $(k+1,i,H_k+1)$ the system will pass to $(k,i,L_k-1)$ earlier, than to states of levels $k,i+1$.

For these auxiliary probabilities and intensities recurrent relations (2–8), (11–24), (26–32), (34–38) hold true.

Stationary probabilities for states $(k,i,n)$ for $k = \overline{1,K-1}, i = k$ are calculated by formula (1).

$$\pi_{k,i,n} = \frac{\Lambda_{k,i,n} + \lambda\pi_{k,i,n-1}}{i\mu + \lambda(1 - a_{k,i,n+1})}, L_{k-1} \le n \le H_k, \tag{1}$$

where threshold value $L_0 := 0$, initial probabilities are

$$\begin{cases} \pi_{k,i,L_{k-1}-1} := 1 \text{ for } k = 1, \\ \pi_{k,i,L_{k-1}-1} := 0 \text{ for } k = \overline{2,K-1}. \end{cases}$$

The auxiliary probabilities and intensities used in (1) are calculated using recurrent formulas (2–8).

$$a_{k,i,H_k+1} := 0; \tag{2}$$

$$a_{k,i,n} = \frac{i\mu}{i\mu + \lambda(1 - a_{k,i,n+1})}, L_k + 1 \le n \le H_k; \tag{3}$$

$$a_{k,i,n+1} := 1, L_{k-1} \le n \le L_k - 1, \tag{4}$$

$$\Lambda_{k,i,H_k+1} := 0; \tag{5}$$

$$\Lambda_{k,1,n} := 0, L_{k-1} \le n \le H_k; \tag{6}$$

$$\Lambda_{k,i,n} = \alpha\pi_{k,i-1,n} + a_{k,i,n+1}\Lambda_{k,i,n+1}, L_k \le n \le H_k, \tag{7}$$

$$\Lambda_{k,i,n} = \alpha\sum_{v=n}^{H_k}\pi_{k,i-1,v} + \lambda(1 - C_{k,i-1})\pi_{k,i-1,H_k}, L_{k-1} \le n \le L_k - 1, \tag{8}$$

Stationary probabilities for states $(k, i, n)$ for $k = \overline{2, K-1}, i = k - 1$ are calculated by formulas (9) and (10).

$$\pi_{k,i,n} = \frac{\Lambda_{k,i,n} + \lambda \pi_{k,i,n-1} + \lambda A_{k,i,n} \pi_{k-1,i,H_{k-1}}}{i\mu + \alpha + \lambda(1 - a_{k,i,n+1})}, L_{k-1} \leq n \leq L_k - 2, L_k \leq n \leq H_k, \quad (9)$$

$$\pi_{k,i,n} = \frac{\Lambda_{k,i,n} + \lambda \pi_{k,i,n-1} + \lambda A_{k,i}^* \pi_{k-1,i,H_{k-1}}}{i\mu + \alpha + \lambda(1 - a_{k,i}^*)}, n = L_k - 1, \quad (10)$$

where initial probabilities are $\pi_{k,i,L_{k-1}-1} := 0$.

The auxiliary probabilities and intensity used in formulas (9) and (10) are calculated using recurrent relations (11–24).

$$a_{k,i,H_k+1} := 0; \quad (11)$$

$$a_{k,i,n} = \frac{i\mu}{i\mu + \alpha + \lambda(1 - a_{k,i,n+1})}, L_{k-1} + 1 \leq n \leq L_k - 2, L_k \leq n \leq H_k; \quad (12)$$

$$a_{k,i,L_k-1} = \frac{i\mu}{i\mu + \alpha + \lambda(1 - a_{k,i}^*)}; \quad (13)$$

$$a_{k,i}^* = a_{k,i,L_k} + (1 - a_{k,i,L_k})\frac{\lambda C_{k,i}}{\lambda + \alpha}, \quad (14)$$

$$\Lambda_{k,i,H_k+1} := 0; \quad (15)$$

$$\Lambda_{k,1,n} := 0, L_{k-1} \leq n \leq H_k; \quad (16)$$

$$\Lambda_{k,i,n} = 2\alpha \pi_{k,i-1,n} + a_{k,i,n+1} \Lambda_{k,i,n+1}, L_k \leq n \leq H_k; \quad (17)$$

$$\Lambda_{k,i,n} = 2\alpha \sum_{v=n}^{H_k} \pi_{k,i-1,v} + \lambda(1 - C_{k,i-1})\pi_{k,i-1,H_k}, L_{k-1} \leq n \leq L_k - 1 \quad (18)$$

$$C_{k,i} = \frac{i\mu}{i\mu + \alpha + \lambda(1 - a_{k+1,i,L_k+1})} \prod_{v=L_k+1}^{H_k+1} a_{k+1,i,v}, \quad (19)$$

$$A_{k,i,H_{k-1}+1} := 1; \quad (20)$$

$$A_{k,i,n} := 0, H_{k-1} + 2 \leq n \leq H_k; \quad (21)$$

$$A_{k,i,n} = \prod_{v=n+1}^{H_{k-1}+1} a_{k,i,v}, L_k - 1 \leq n \leq H_{k-1}; \quad (22)$$

$$A_{k,i}^* = A_{k,i,L_k-1} + (1 - A_{k,i,L_k-1})\frac{\lambda C_{k,i}}{\lambda + \alpha}; \quad (23)$$

$$A_{k,i,n} = A_{k,i}^* \prod_{v=n+1}^{L_k-1} a_{k,i,v}, L_{k-1} \leq n \leq L_k - 2. \quad (24)$$

Stationary probabilities for states $(k, i, n)$, $k = \overline{3, K}$, $i = k - 2$ are calculated by formula (25).

$$\pi_{k,i,n} = \frac{\lambda \pi_{k,i,n-1} + \lambda A_{k,i,n}(\pi_{k-1,i,H_{k-1}} + \pi_{k-1,i-1,H_{k-1}})}{i\mu + 2\alpha + \lambda(1 - a_{k,i,n+1})}, L_{k-1} \leq n \leq H_k, \quad (25)$$

where threshold value $H_K := R$ probability $\pi_{2,0,H_2} := 0$, initial probabilities are $\pi_{k,i,L_{k-1}-1} := 0$.

The auxiliary probabilities and intensity used in formula (25) are calculated by recurrent relations (26–32).

$$a_{k,i,H_k+1} := 0, k \leq K - 1; \quad (26)$$

$$a_{K,i,R+1} := 1, \quad (27)$$

$$a_{k,i,n} = \frac{i\mu}{i\mu + 2\alpha + \lambda(1 - a_{k,i,n+1})}, L_{k-1} + 1 \leq n \leq H_k, \quad (28)$$

$$A_{k,i,H_{k-1}+1} := 1; \quad (29)$$

$$A_{k,i,n} := 0, H_{k-1} + 2 \leq n \leq H_k; \quad (30)$$

$$A_{k,i,n} = \prod_{v=n+1}^{H_{k-1}+1} a_{k,i,v}, L_{k-1} \leq n \leq H_{k-1}. \quad (31)$$

$$C_{k,i} = \frac{i\mu}{i\mu + 2\alpha + \lambda(1 - a_{k+1,i+1,L_k+1})} \prod_{v=L_k+1}^{H_k+1} a_{k+1,i+1,v}, \quad (32)$$

Stationary probabilities for states $(k, i, n)$ for $k = K, i = \{k, k - 1\}$ are calculated on a formula (33).

$$\pi_{k,i,n} = \frac{\Lambda_{k,i,n} + \lambda \pi_{k,i,n-1} + \lambda A_{K,i,n} \pi_{K-1,i,H_{K-1}}}{i\mu + (K - i)\alpha + \lambda(1 - aK, i, n + 1)}, L_{k-1} \leq n \leq R, \quad (33)$$

where initial probabilities are $\pi_{k,i,L_{k-1}-1} := 0$.

The auxiliary probabilities and intensities used in formula (33) are calculated by recurrent relations (34–38).

$$a_{K,i,R+1} := 1, \quad (34)$$

$$a_{K,i,n} = \frac{i\mu}{i\mu + (k - i)\alpha + \lambda(1 - a_{k,i,n+1})}, L_{K-1} + 1 \leq n \leq R, \quad (35)$$

$$a_{K,K,n} := 0, L_{K-1} + 1 \leq n \leq R, \quad (36)$$

$$A_{K,K-1,n} = \prod_{v=n+1}^{H_{K-1}+1} a_{K,K-1,v}, L_{k-1} \leq n \leq H_{k-1}. \quad (37)$$

$$\Lambda_{k,i,n} = \alpha \sum_{v=n}^{R} \pi_{k,i-1,v}, L_{k-1} \leq n \leq R, \quad (38)$$

To recieve stationary probabilities let $\widetilde{\pi}_{1,1,0} = 1$. Then, using formulas (1), (9), (10), (25), (33) we will calculate nonnormalized probabilities $\widetilde{\pi}_{k,i,n}$, $(k, i, n) \in S$ and the normalizing constant $G = \sum_{(k,i,n) \in S} \widetilde{\pi}_{k,i,n}$. Then stationary probabilities are $\pi_{k,i,n} = \frac{\widetilde{\pi}_{k,i,n}}{G}$, $(k, i, n) \in S$.

## 4    Stationary Characteristics of Queueing System

Knowing stationary probability distribution of the system, it is possible to estimate some probabilistic and time characteristics of the considered system, formulas are given below.

Average number of customers in the system:

$$N = \sum_{(k,i,n) \in S} n \pi_{k,i,n}. \tag{39}$$

Average number of required servers:

$$MD = \sum_{k=1}^{K} k \sum_{i=1}^{K} \sum_{n=1}^{R} \pi_{k,i,n}. \tag{40}$$

Average number of busy servers:

$$MF = \sum_{i=1}^{K} i \sum_{k=1}^{K} \sum_{n=1}^{R} \pi_{k,i,n}. \tag{41}$$

Blocking probability:

$$\pi = \sum_{i=1}^{K} \pi_{K,i,R}. \tag{42}$$

Average sojourn time:

$$T = \frac{N}{\lambda(1 - \pi)}. \tag{43}$$

Note that three last characteristics correspond to performance measures of cloud computing system with dynamic scaling, namely to average number of the activated virtual machines, blocking probability of user request and average response time of the system. The average number of the activated virtual machines $MF$ can serve for assessment of energy efficiency of cloud computing system at known energy consumption on maintenance of the virtual computer in an active state.

## 5    Numerical Analysis

In this section we perform a comparative analysis of initial system, proposed in [10] and simplified model, described above. We compare two main performance characteristics for both models, i.e. mean response time and blocking probability. We provide calculations for the system with $K = 20$ servers, serving rate $\mu = 1$ and two different arrival rate intensities $\lambda = \{12, 18\}$ for various values of server switching rate $\alpha$, which is key parameter that makes behavior of the systems differ from each other.

We used two threshold arrangements for numerical analysis:

$$\begin{cases} H_i = 25i, & i = \overline{1, K-1}, R = 500; \\ L_i = H_{i-1} - 1, & i = \overline{2, K-1}, L_1 = 12; \end{cases} \tag{44}$$

and

$$\begin{cases} H_i = 50i, & i = \overline{1, K-1}, R = 1000; \\ L_i = H_{i-1} - 1, & i = \overline{2, K-1}, L_1 = 12. \end{cases} \tag{45}$$

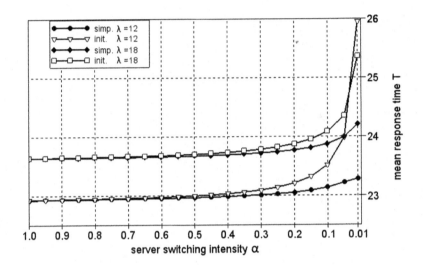

**Fig. 3.** Mean response time comparison for thresholds arrangement (44)

Figures 3 and 4 show mean response time of both models for thresholds arrangements (44) and (45) respectively. As it is seen of Fig. 3, simplified and initial model have almost the same response time for $\frac{\mu}{\alpha} \approx 1$. With the growth of $\frac{\mu}{\alpha}$ ratio to 10, relative error of the simplified model increases to approximately 2 % for low load and 1 % for high load case and reaches 10 % and 5 % when $\frac{\mu}{\alpha} = 100$ for light and heavy loads respectively. Note that simplified model show lower mean response time because with low additional server activation rate we assumed activation to complete immediately in some cases.

With wider inter-threshold range arrangement (45), relative error of the simplified significantly decreases (Fig. 4). It becomes nearly 0.2 % in $\frac{\mu}{\alpha} = 10$ case and 3 % in $\frac{\mu}{\alpha} = 100$ case. ,

Figures 5 and 6 show blocking probability of both models for thresholds arrangement 44 and 45 respectively. Figures show that in light load case, blocking probability relative error of the simplified model is extremely high even for $\frac{\mu}{\alpha} = 5$, but it may be considered feasible in heavy load case. The reason of that is more frequent immediate server activation, which increases system utilization. However, in heavy load cases system shows almost maximum performance already, and additional factors do not have significant effect.

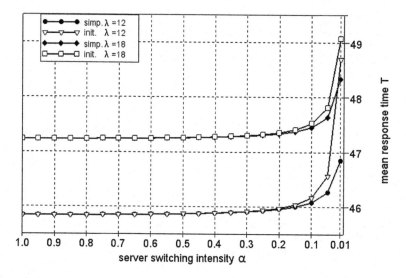

**Fig. 4.** Mean response time comparison for thresholds arrangement (45)

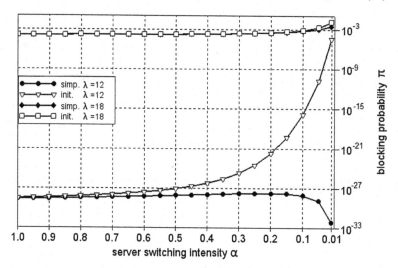

**Fig. 5.** Blocking probability comparison for thresholds arrangement (44)

## 6    Conclusion

In this paper, we proposed the simplified model for performance analysis of cloud computing system with dynamic server activation. For the considered model, we developed an efficient stationary probabilities computing algorithm based on state elimination technique. Simplificaiton of the model significantly decreases state space complexity of corresponding Markov process and, consequently, computing complexity of the algorithm.

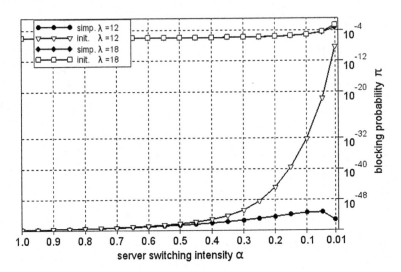

**Fig. 6.** Blocking probability comparison for thresholds arrangement (45)

Our analysis showed that the simplified system gives good precision of the mean response time for a variety of system and load parameters. Moreover, increasing distance between corresponding lower and upper thresholds, improves the degree of calculation accuracy. However, blocking probability precision can be considered feasible only in heavy load cases.

Another way to increase computation accuracy of the simplified model is to increase maximum simultaneously running server activation procedures to three or four and it will be done in our further research.

**Acknowledgments.** This work was supported in part by the Russian Foundation for Basic Research, projects No. 14-07-00090, 15-07-03051, 15-07-03608.

# References

1. ETSI Cloud Standards Coordination. Final Report 2013, ver. 1.0. http://www.etsi.org/images/files/Events/2013/2013_CSC_Delivery_WS/CSC-Final_report-013-CSC_Final_report_v1_0_PDF_format-.PDF
2. Goswami, V., Patra, S.S., Mund, G.B.: Performance analysis of cloud with queue-dependent virtual machines. In: 1st International Conference on Recent Advances in Information Technology, Dhanbad, India, pp. 357–362. IEEE Press (2012)
3. Pechinkin, A., Gaidamaka, Y., Sopin, E., Talanova, M.: Performance analysis of cloud computing systems with dynamic scaling. In: IX International Annual Scientific and Practical Conference "Modern Information Technologies and IT Education", pp. 395–406. INTUIT.RU, Moscow (2014)

4. Shorgin, S.Y., Pechinkin, A.V., Samouylov, K.E., Gaidamaka, Y.V., Gudkova, I.A., Sopin, E.S.: Threshold-based queuing system for performance analysis of cloud computing system with dynamic scaling. In: 12th International Conference of Numerical Analysis and Applied Mathematics ICNAAM-2014, pp. 1–3. AIP Publishing, USA (2015)
5. Golubchik, L., Lui, J.C.S.: Bounding of performance measures for threshold-based queuing systems: theory and application to dynamic resource management in video-on-demand servers. IEEE Trans. Comput. **51**(3), 353–372 (2002)
6. Gaidamaka, Y.V., Pechinkin, A.V., Razumchik, R.V., Samuylov, A.K., Samouylov, K.E., Sokolov, I.A., Sopin, E.S., Shorgin, S.Y.: The distribution of the return time from the set of overload states to the set of normal load states in a system M|M|1—<L,H><H,R> with hysteretic load control. Inform. Appl. **7**(4), 20–33 (2013)
7. Basharin, G.P., Gaidamaka, Y.V., Samouylov, K.E.: Mathematical theory of tele-traffic and its application to the analysis of multiservice communication of next generation networks. Autom. Control Comput. Sci. **47**(2), 62–69 (2013)
8. Mokrov, E.V., Samouylov, K.E.: Modeling of cloud computing as a queuing system with batch arrivals. T-Comm — Telecommun. Transp. **7**(11), 139–141 (2013)
9. Mokrov, E.V., Chukarin, A.V.: Performance analysis of cloud computing system with live migration. T-Comm — Telecommun. Transp. **8**(8), 64–67 (2014)
10. Gaidamaka, Y., Sopin, E.S., Talanova, M.O.: Performance measures analysis of cloud computing systems with hysteretic control. T-Comm — Telecommun. Transp. **9**(9), 54–60 (2015)
11. Bocharov, P.P., D'Apice, C., Pechinkin, A.V., Salerno, S.: Queueing Theory. VSP Publishing, Ultrecht, Boston (2004)

# Synthesis of Noise-Like Signal Based on Ateb-Functions

Ivanna Dronyuk$^{(\boxtimes)}$, Maria Nazarkevych, and Olha Fedevych

Lviv Polytechnic National University, Bandera street 28a, Lviv 79015, Ukraine
ivanna.droniuk@gmail.com, mar.nazarkevych@gmail.com,
olha.fedevych@gmail.com
http://www.lp.edu.ua/ikni

**Abstract.** The mathematical model of the noise signal based on Ateb-functions was proposed. Connectivity of Ateb-functions with the curve of superellipse was shown. Algorythm for synthesis Gaussian white noise signal based on the periodical Ateb-functions is created. Investigation for statistical characteristics of generated signals using MATLAB 7.0 environment is presented. Application the proposed method of noise-like signals synthesis to protection information during data transfer in computer networks, in particular networks with CDMA technology is discussed.

**Keywords:** Noise-like signal · Ateb-functions · Gaussian white noise · Superellipse

## 1 Introduction

When it comes to protecting data in computer networks, there is often a need to generate some noise signals. This is especially true for wireless communication systems, including networks which use Code Division Multiple Access (CDMA) technology [1]. CDMA technology is based on a SST (DH-SS Direct Sequence Spread Spectrum) transfer technology, when information is as if smeared on a wide range of frequencies. Harmonic oscillations often play the role of information carrier. In book [2] the author describes a classical approche to describe additive, white and Gaussian models of noise. The noise is present in all communication systems and it was an obstacle for the detection of signals and has to be eliminated by the filters. With the further development of telecommunication technologies, the information transmission methods based on noise signals were emerged [3]. During the building of modern confidential communication systems, the noise signals transformed from the obstacles to the main signal carriers.

However, during the last decade, people have begun to use noise signals as the carrier of main information transmitted by connection channels. Noise signals have several advantages in terms of information security both in communication cable channels and in radio channels. Data transferring in noise signal can be a good and effective alternative for the methods of cryptographic protection.

V. Vishnevsky and D. Kozyrev (Eds.): DCCN 2015, CCIS 601, pp. 132–140, 2016.
DOI: 10.1007/978-3-319-30843-2_14

It is known [1], that the carrier oscillation, formed on the basis of noise signal, enables effective recognition of signals by form. Noise signals also has another advantage - it gives a possibility to provide transmitting protection on the level of physical channel, which is especially important during the construction of protected multi-access communication systems. Noise-like signals is a type of noise signals, where the role of carrier can be played by signal constructions implemented on the basis of harmonic oscillations.

Oscillations that arise in nonlinear systems with multiple degrees of freedom generalize harmonic oscillations and are described by the relevant differential equations. Mathematically, these oscillations can be described by using Ateb-functions. This is the idea behind this paper: to apply the method of construction of noise-like signal based on harmonic oscillations to generalize oscillations, described by Ateb-functions, and to build noise-like oscillations based on Ateb-functions. As noise-like signals are used in CDMA technology, among others, the proposed approach is relevant and can have a wide practical use to protect the transmission of information in computer networks. The proposed method can be used for noise signal synthesis in both hardware and software domains. The purpose of this research is to develop mathematical model based on Ateb-functions for the problem of synthesis of discrete noise signal with given spectral and autocorrelation properties.

## 2    Construction of the Mathematical Model

At first, the necessary notations and formulas related to Ateb-functions, further required to build noise signal, shall be introduced. The details concerning Ateb-functions properties can be found in [4]. Within this article, we shall introduce notation $sa(n, m, \omega)$ for Ateb-sine and $ca(m, n, \omega)$ for Ateb-cosine. It is known [4] that Ateb-functions satisfy the identity

$$ca^{m+1}(m, n, \omega) + sa^{n+1}(n, m, \omega) = 1. \tag{1}$$

Let $\Gamma(x)$ denotes the Gamma function. The period $\Pi(m, n)$ of Ateb-functions shall be calculated on the following formula

$$\Pi(m, n) = \frac{\Gamma\left(\frac{1}{n+1}\right) \Gamma\left(\frac{1}{m+1}\right)}{\Gamma\left(\frac{1}{n+1} + \frac{1}{m+1}\right).} \tag{2}$$

In the case when $n = m = 1$, we get $ca(1, 1, \omega) = \cos(\omega)$, $sa(1, 1, \omega) = \sin(\omega)$ that satisfies Eq. (1), namely $\cos(\omega)^2 + \sin(\omega)^2 = 1$. Thus, Ateb-functions are generalization of usual trigonometric functions.

Now the relationship between the considered Ateb-functions and plane algebraic Lame's curve, which is also called generalized superellipse, shall be shown. Identity (1), satisfied by periodic Ateb-functions, can be represented in a graphical way in the form of superellipse. Let us consider the formula of generalized superellipse as [5]

$$\left|\frac{x}{a}\right|^p + \left|\frac{y}{b}\right|^q = 1, \quad where \quad p, \, q > 0. \tag{3}$$

For simplicity, let us to assume $a = b = 1$, and $p, q$ are related to the parameters of Ateb-functions as follow $p = m + 1, q = n + 1$. The substitution $x = ca(n, m, \omega), y = sa(m, n, \omega)$ into Eq. (3) of a generalized supcrellipse, transforms (3) to the identity (1). Thus it shows that the main Ateb-functions identity (1) is the representation of a superellipse Eq. (3).

Now it is turning to the description of the method of forming a noise signal based on the Ateb-functions. Taking into consideration characteristic features of real noise signal, the real noise signal $s(t)$ - is a set of simultaneously existing power oscillations of frequency, phase, and amplitude, which are random. The spectrum of noise signals covers a wide band of frequencies. If this spectrum is uniform at all frequencies from 0 to $\infty$, then such noise is called white. In practice, this noise cannot be obtained, but for any hardware, whose bandwidth is in many times less than noise signal spectrum, the noise can be considered as white. The power of noise signal used is determined by the bandwidth of the device on the input of which it comes. If a controlled noise signal with duration T is radiated, and acceptance is going through a concerted filter or a correlation circuit, then during the output of correlator with the coincidence of the time of received and reference signals ($\tau = 0$) the energy would be released

$$E = \int_0^T s(t)^2 dt. \tag{4}$$

In conditions that this energy is also random variable with average mean $E_0$ and dispersion $\sigma^2$, fluctuation of energy of carrier oscillation can be seen as interference, which is distorting the signal. If the noise is limited by $\Delta f$ stripe and has a large base B, then the relative standard deviation of value is equal [6]

$$[h] \frac{\sigma^2}{E_0} \approx \frac{1}{\Delta f T}. \tag{5}$$

In order to minimize distortion of information transferred by noise, it is appropriate to use signal with a large base. Then, based on (4), the results of measurements of noise energy will become average during the time which in several times exceeds the correlation time $\tau$ of the investigated signal. White noise is an ideal noise signal that has an infinite spectrum and correlation function in the form of a delta function. In practice, white noise cannot be obtained, because equipment with unlimited pass band is needed for its generation and processing. For this reason in real systems noise with limited stripe $\Delta f$ is used. Work [6] considers the method of construction of noise-like signals based on harmonic oscillations. As Ateb-functions can generate a wider range of fluctuations that are generalizations of harmonious and are periodic, the present work generalizes the proposed in [4] method based on Ateb-functions. The main advantage

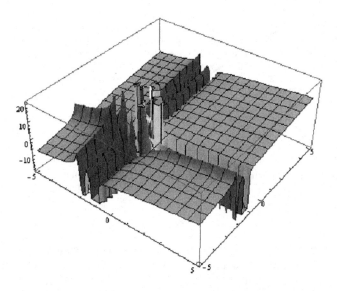

**Fig. 1.** Dependence of period $\Pi(m,n)$ from the parameters $n$ and $m$, which are belonging to the interval $[-5;5]$.

of the proposed approach lies in variable period of Ateb-functions defined by the formula (2). Depending on the parameters $n$ and $m$ of Ateb-functions, this allows choose such time period that best suits the task of hiding information based on the generated noise-like signal. Figure 1 shows the dependence of the period $\Pi(m,n)$ on the parameters $n$ and $m$ and by changing the parameters of the latter in the range of $-5$ to 5. Calculations were carried out by using the Eq. (2) in the computing environment Wolfram Mathematika 7.0. Figure 1 shows that the value of the period mostly increases with the increase of arguments, however, it also has a number of local extremes and some asymptotes for the negative values of the arguments. This allows you to select the required value of period depending on the width of the channel. When building a noise-like signal with a view to make it proximal to the white noise, a number of the following requirements shall be taken into account: the signal must be broadband, i.e. base signal $B = FT \gg 1$, where $T$ - the duration of the signal; $F$ - bandwidth of frequencies of signal.

## 3 Algorythm for Synthesis of White Noise Signal Based on the Periodical Ateb-Functions

Then the noise-like signal can be written in the form:

$$u(m,n,t) = U(t)v(t)sa(m,n,2\Pi(n,m)s_0 t + \phi(t)), \qquad (6)$$

where $sa(m,n,t)$ - Ateb-sine with parameters $m$, $n$, $U(t)$ - an amplitude and $\phi(t)$ - a phase mutually independent random functions, which are changing slowly

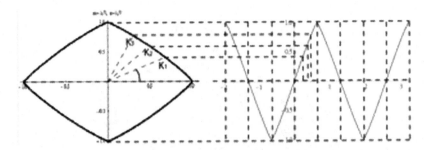

**Fig. 2.** Algorythm of forming a discrete noise sygnal based on $sa(\frac{1}{7}, \frac{1}{7}, t)$ (on the left hand superellipse is presented, on the right hand - the graph of Ateb-sine), dashed lines correspond to discrete signal points.

in comparison with Ateb-sine; $s_0$ - center frequency of spectrum of the noise; $v(t)$ normalized pseudo-random sequence of numbers with a given distribution law.

Let us consider in details the benefits of Eq. (6), using the properties of Ateb-functions. It can be seen (Fig. 1) that the period of function $ca(3,3)$ is by an order bigger than the period of the function $ca(0.05, 0.05)$. We shall estimate the speed of growth and decay of Ateb-sine function depending on the parameters. It is known [7] that the derivative of Ateb-sine is given in the formula

$$\frac{dsa(n, m, t)}{dt} = \frac{2}{n+1} ca^m(m, n, t) \tag{7}$$

Ateb-sine function figure looks similar to cosine and is presented on Fig. 2. Whereas $|ca(\bullet)| \leq 1$, function $sa(m, n, t)$ grows more slowly with the increase of parameters n and m. This property is also demonstrated on Fig. 2. With the help of parameters n and m, the abovementioned properties allow influence the spectrum of noise-similar signal $u(m, n, t)$ generated on the Eq. (6).

Method of noise signal formation based on Ateb-sinus is shown on Fig. 2. Method will be applied for forming continuous or discrete signals. This figure shows a system of rectangular coordinates with the starting point $K$ with coordinate $(1; 0)$ when initial phase in (6) $\phi(t) = 0$, which moves counterclockwise with a constant speed by superellipse (point $K_1, K_2, K_3$ for discrete case), and noise signal $u(m, n, t)$ is formed by the formula (6). Graphs of synthesized signals with differential base B are shown on Figs. 3, 4. These and next figures are created using MatLab 7.0 computation system. In order to improve noise capacity and to investigate the statistical characteristics of the generated signals, normalized autocorrelation functions and spectrums were discussed. White noise is a random signal with a constant power spectral density. A special type is the Gaussian white noise (GWN), which has a Gaussian distribution amplitude. It is a mathematical approach for describing a real noise signals. According to these mathematical descriptions, generated by (6) broadband noise-like signals belong to the Gaussian white noise, which sometimes is named as noise-like signals. Statistical characteristics GWN: a mean is equal to zero, and variance is a finite.

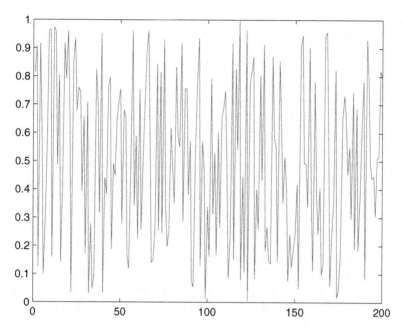

**Fig. 3.** The noise, generated on the basis of the Eq. (6), signal base $B = 200$ and Ateb parameters $n = \frac{1}{7}, m = \frac{1}{7}$.

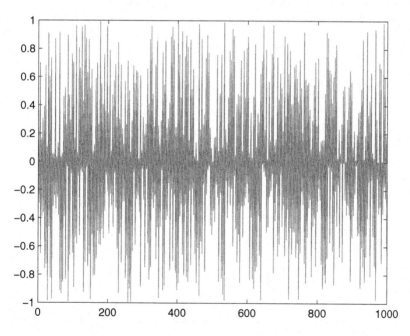

**Fig. 4.** The noise, generated on the basis of the Eq. (6), signal base $B = 1000$ and Ateb parameters $n = \frac{1}{20}, m = \frac{1}{20}$.

# 4    Investigation of Statistical Characteristics of a Generated Noise Signals

It is interesting to investigate some characteristic of noise signals, which was generated on the basis of Ateb-functions. Using the expression (6) and imitation modeling method [4], the noise sequence $u(m, n, t)$ were formed. It is a investigation autocorrelation function of these signals and their spectrums using MatLab 7.0 computation system. Analysis of the spectrums of all investigated implementations of signals shows a uniformity of spectrums of signals in the band $\Delta f$, i.e. their proximity to the white noise in this band. Spectrums of signals look similarly. Autocorrelation function (ACF) of the noise signal is shown on Fig. 5 with paramaters $B = 200, n = \frac{1}{7}, m = \frac{1}{7}$, and on Fig. 6 with signal base $B = 1000$ and Ateb parameters $n = \frac{1}{20}, m = \frac{1}{20}$. ACF in all investigated cases has one big peak near zero and tends rapidly to zero after that. This peak is presented on Fig. 6 too, but it is visually merges with vertical axis.

At the heart of CDMA is situated the SST (DH-SS Direct Sequence Spread Spectrum) transfer technology, where information is "smeared" on a wide range of frequencies. The sequence of information bits is multiplied by a pseudorandom sequence of short pulses. Then, signal in a wider frequency range with much smaller intensity is received. To decode this sequence, you should know pseudorandom sequence, which was used during the transfer. This coding mechanism provides protection of signal from eavesdropping. It is need to know pseudo-randomness sequence-key. Wideband of signal allows renewing a signal, especially if interferences are narrowband. Similarly signal is protected and from temporary disappearance on separate frequencies (fading). For CDMA

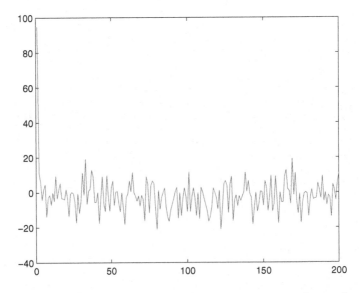

**Fig. 5.** ACF of the generated noise signal $B = 200, n = \frac{1}{7}, m = \frac{1}{7}$.

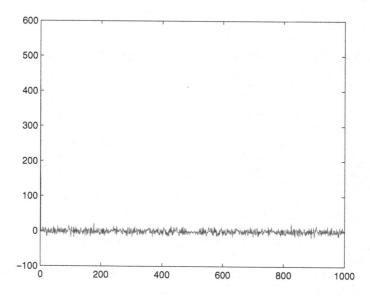

**Fig. 6.** ACF of the generated noise signal $B = 1000, n = \frac{1}{20}, m = \frac{1}{20}$.

technology can be used wideband signals [1], in particular noise-like signals. So, the proposed method of synthesis of noise-like signals can be used to protect an information during data transfer, as well as in other cases, where Gaussian noise signals with a large information capacity are required.

## 5    Conclusion

The mathematical model of the noise signal based on Ateb-functions was proposed. The connection between Ateb-functions and the curve of superellipse was shown. The implementation of the method of synthesizing continuous or discrete signal structures with given spectral and autocorrelation characteristics based on Ateb-functions was presented. The required properties of formed noise implementation are provided by varying the parameters of Ateb-functions. The application of noise signals for the purposes of information security in modern communication systems, including the use of CDMA technology was discussed.

In future investigation it will be interesting to change amplitude, frequency and phase of signal are changed in accordance with the distribution law of random or pseudorandom numbers. It will be also interesting to investigate a comparative analysis of implementations of noise signals based on different distribution laws.

# References

1. Ipatov, V.P.: Spread Spectrum and CDMA: Principles and Applications. John Wiley and Sons, Ltd. (2005). ISBN: 0-470-09178-9
2. Sklar, B.: Digital Communications: Fundamentals and Applications, 2nd edn. Prentice Hall P T R, New Jersey (2001)
3. Rizzo, C., Brookson, C.: Security for ICT - the Work of ETSI, ETSI White Paper No. 1, Fourth edition, January 2012
4. Grytsyk, V.V., Dronyuk, I.M., Nazarkevych, M.A.: Information technologies of document protection by Ateb-functions means. Part 1: Building an Ateb-functions database to protect documents/Problems of control and informatics, **2**, 139–152 (2009). (In Russian)
5. Sokolov, D.D.: Lame curve. In: Hazewinkel, M. (ed.) Encyclopedia of Mathematics. Springer (2001). ISBN 978-1-55608-010-4
6. Korchinskiy, V.V.: Method of modeling of noise signals for systems of transmission confidential information/Vestnik NTU "HPI" - Kharkiv, **38**(1011), 99–104 (2013). (In Russian)
7. Senyk, P.M.: About Ateb-functions/Rep. USSR Academy of Sciences, Ser. A, No.1, pp. 23–27 (1968). (In Ukrainian)

# Approach to Estimation of Performance Measures for SIP Server Model with Batch Arrivals

Yurii Orlov[1], Yuliya Gaidamaka[2], and Elvira Zaripova[2(✉)]

[1] Keldysh Institute of Applied Mathematics Russian Academy of Sciences,
Miusskaya sq. 4, 125047 Moscow, Russia
ov31509f@yandex.ru
[2] Peoples' Friendship University of Russia, Miklukho-Maklaya street 6,
117198 Moscow, Russia
{ygaidamaka,ezarip}@sci.pfu.edu.ru
http://api.sci.pfu.edu.ru

**Abstract.** In this paper an approach to analysis of dependence of Session Initiation Protocol server model with batch arrivals performance measures on batch size distribution is considered. Proposed approach employs non-parametric methods of statistical analysis. It is shown that there is statistical reliable dependence of performance measures, taken for signaling traffic analysis, on distance between distributions in definite norm. On the basis of proposed analysis elasticity coefficients were evaluated depending on distance between batch size distributions. This approach enables to get correction factors for estimation of these parameters in case distribution functions differ from uniform.

**Keywords:** Optimization · SIP mathematical model · Distribution function · Norm · Performance measure · Queuing system · Parameter sensibility · Sample · Batch arrivals

## 1 Introduction

Developing telecommunication services are successfully provided via IP-based Multimedia Subsystem (IMS), where Session Initiation Protocol (SIP) is the main signaling protocol [1]. Signaling traffic load is auxiliary for communication nodes and is used for providing communications services to users. Signaling messages have so-called *life time*. When life time is over, information becomes not actual, signaling messages are retransmitted and overload the SIP server that processes them. Among the other problems with Session Initiation Protocol the overloading problem was stated by IETF in [2], and solution requirements that address the problems were formulated in [3]. Providing telecommunication

This work was supported in part by the Russian Foundation for Basic Research, projects No. 15-07-03051, 15-07-03608.

services of required quality results in necessity of detailed research of communication system structure, statistic data analysis, implementation of processing algorithm for different types of signaling messages in order to increase number of successfully initiated sessions and decrease of service and sojourn time at a SIP server [6]. Many papers from the IETF documents [3–5] to a number of scientific papers [11–16, 18–24] were devoted to developing and investigating mechanisms of SIP-server overload control. For example, signaling messages service process was presented as asymmetric polling system [9], as queueing model with threshold control [10]. Sometimes you can see research even with unreliable servers [17].

In this paper were chosen the following parameters for the investigation: average queue length and average waiting time. The first one enables estimation of SIP server's buffer capacity during the busy hour, the second when summed with service time enables estimation of SIP server sojourn time and is comparable to the message life time.

Process of signaling messages arrival and processing at a SIP server is performed via a single-server queue with batch arrivals and vacations. The vacation models the time interval when server processes messages that differ from signaling ones. In a model the batch arrival of customers corresponds to the simultaneous requests of the group of online subscribers.

In [6–8] there is an estimation of average value of queue length and average waiting time for general distribution of batch size. In [8] the analysis for the four batch size distribution functions has been done: Zipf, geometric, logarithmic and uniform distribution. In [8] a statistical dependence of investigating parameters on batch size distribution in case of Poisson process has been analyzed. In contrast with the papers where an investigation was performed only for geometrical distribution, in [8] it is recommended to use a uniform distribution of batch size that essentially simplifies formulas for calculations. In this paper an approach to performance measures analysis is proposed, sensitivity of model parameters to batch size distribution was estimated.

Distribution function variation was evaluated in [8] under different norms. Those with highest reliability of estimated statistical dependence were chosen. It turned out that analysis of numeric evaluation of SIP server performance parameters sensitivity to batch size distribution variation can be carried out with high reliability (0.99).

Since only sampling distribution functions of parameters are available for preliminary analysis, so we want to evaluate model parameters with statistical fluctuation of sampling distribution function when the distribution is not converge to any general population because of non-stationary behavior. We denote, that the batch size is a measured on practice, whereas average waiting time and average queue length depend on messages service time and processor vacation time. Distribution functions of the last two parameters are assumed to be known, because they depend on the SIP server hardware implementation. However, confidence interval for average queue length may not be got from corresponding empirical distribution, because we do not know this distribution function.

We can evaluate these parameters by means of the simplified models only. In [8] the sensitivity analysis for one of models was carried out. Analysis was later applied to empirical distribution function of batch size.

## 2    SIP Server Model as a Queue with Batch Arrivals and Vacations

A mathematical model of signaling messages processing at a SIP server is investigated as a queuing system with batch arrivals and vacations. According to Basharin-Kendall notation this system is denoted as $M^{[X]}|G|1|\infty$. Let suppose that a batch of customers arrives according to Poisson process with rate $\lambda$. Customer service time is a random variable with a distribution function $B(t)$, where $b_1$ is the mean value and $b_2$ is the finite second moment. If there are no customers in the queue, the server goes for a vacation. The vacation time is a random variable with a distribution function $V(t)$ with finite first and second moments $v_1$ and $v_2$.

Let $f(k)$ be probability that batch size is equal to $k \geq 1$. We denote the corresponding distribution function $F(k)$. For the queuing system $M^{[X]}|G|1|\infty$ with vacations we can find average queue length and average waiting time depending on offered load $\rho$ which is estimated as follows:

$$\rho = \lambda b_1 l^{(1)}, \tag{1}$$

where $l^{(1)}$ is average batch size with distribution function $F(k)$.

In [6] a generating function was obtained for the queue length distribution for a single server model with vacations. The result is expressed as follows:

$$P(z) = \frac{1-\rho}{\lambda v_1} \cdot \frac{1-z}{1-L(z)} \cdot \frac{1-\phi(\lambda, z)}{\beta(\lambda, z) - z}, \tag{2}$$

where $L(z)$ is a generating function for batch size with the distribution function $F(k)$: $L(z) = \sum\limits_{k=0}^{\infty} f(k)z^k$. Other functions mentioned in (2) are expressed with the following equations: $v_1 = \int\limits_0^{\infty} t\, dV(t)$, $\phi(\lambda, z) = \int\limits_0^{\infty} e^{-\lambda t(1-L(z))}\, dV(t)$, and $\beta(\lambda, z) = \int\limits_0^{\infty} e^{-\lambda t(1-L(z))}\, dB(t)$. According to [8] average queue length depending on load $\rho$ is obtained by the generating function (2) as follows:

$$N = \lim_{z \to 1} P'(z) = \frac{v_2}{2v_1 b_1}\rho + \frac{l^{(2)} - l^{(1)}}{2l^{(1)}}\frac{\rho}{1-\rho} + \frac{b_2}{2b_1^2}\frac{\rho^2}{1-\rho}. \tag{3}$$

Taking into account the notation $v_s = \int\limits_0^{\infty} t^s dV(t)$, $b_s = \int\limits_0^{\infty} t^s dB(t)$ and $l^{(s)} = \sum\limits_k k^s f(k)$, $s = 1; 2$, the average waiting time is obtained as follows:

$$\tau = \frac{Nb_1}{\rho} = \frac{v_2}{2v_1} + \frac{l^{(2)} - l^{(1)}}{2l^{(1)}}\frac{b_1}{1-\rho} + \frac{b_2}{2b_1}\frac{\rho}{1-\rho}. \tag{4}$$

You can see the average queue length (Fig. 1) and average waiting time (Fig. 2) for 4 batch size distributions for exponential service time and vacation time with the following initial data $b_1 = 15$ ms – average service time, $v_1 = 15l_1$ ms – average vacation time, $l_1 = 3$ – average batch size. This is the starting point of the reseaches. The plots (Figs. 1 and 2) have similar behavior. We investigate the sensitivity of the model parameters to probability variation of batch size.

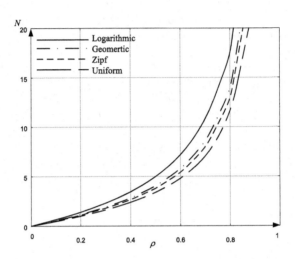

**Fig. 1.** Average queue length depending on load.

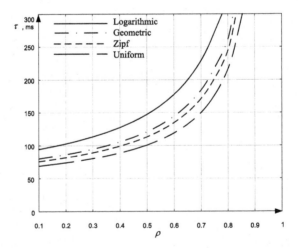

**Fig. 2.** Average waiting time depending on load.

## 3    Sensitivity of the Model Parameters to Probability Variation

As a base of our analysis we use 4 batch size distributions with following distribution functions (Table 1), where $f_i(k)$ is the probability, that batch size is equal $k$. We assume, that maximum batch size is equal to 8. The distribution functions can be found by formula $F_i(n) = \sum_{k=1}^{n} f_i(k)$.

We consider 5 types of norms and define the distance between two DFs by the formulas from the Table 2, where the distance from $i$ to $j$ distribution functions is $\rho_{ij}^{norm}$. The first step of our investigation was to compare distances between DFs in several norms.

Table 3 shows the distances between batch size DFs in considered norms. For example, the first row shows the distance between Logarithmic and Geometric batch size distributions, the last row – the distance between Zipf and Uniform distributions. Statistics show that the distances are dependent from each other. So, for the next step of our investigation we can use any appropriate norm.

**Table 1.** Distribution functions

| Logarithmic DF | $f_1(k) = \frac{1}{Z_1} \frac{(0,85)^k}{k}$, | $Z_1 = \sum_{k=1}^{8} \frac{(0,85)^k}{k}$ |
|---|---|---|
| Geometric DF | $f_2(k) = \frac{1}{Z_2}(0,67)^k$, | $Z_2 = \sum_{k=1}^{8}(0,67)^k$ |
| Zipf DF | $f_3(k) = \frac{1}{Z_3 k}$, | $Z_3 = \sum_{k=1}^{8} 1/k$ |
| Uniform DF | $f_4(k) = 1/8$ | |

**Table 2.** Definition of distance in norms

| $C$ norm | $\rho_{ij}^C = \max\limits_{k} |f_i(k) - f_j(k)|$ |
|---|---|
| $L1$ norm | $\rho_{ij}^{L1} = \sum_{k=1}^{8} |f_i(k) - f_j(k)|$ |
| Hellinger distance | $\rho_{ij}^{He} = \sum_{k=1}^{8} \left(\sqrt{f_i(k)} - \sqrt{f_j(k)}\right)^2$ |
| Kullback-Leibler distance | $\rho_{ij}^{KL} = \sum_{k=1}^{8} f_i(k) \ln\left(f_i(k)/f_j(k)\right)$ |
| Supplement for total area $S$ | $\rho_{ij}^{S} = 1 - \sum_{k=1}^{8} \min\left(f_i(k), f_j(k)\right)$ |

**Table 3.** Distances between batch size distributions

| Comparison | $C$ | $L1$ | He | KL | $S$ |
|---|---|---|---|---|---|
| Logarithmic, Geometric | 0,13 | 0,27 | 0,02 | 0,04 | 0,134 |
| Logarithmic, Zipf | 0,11 | 0,26 | 0,03 | 0,05 | 0,129 |
| Logarithmic, Uniform | 0,35 | 0,86 | 0,26 | 0,52 | 0,431 |
| Geometric, Zipf | 0,05 | 0,18 | 0,01 | 0,03 | 0,090 |
| Geometric, Uniform | 0,22 | 0,71 | 0,18 | 0,33 | 0,354 |
| Zipf, Uniform | 0,24 | 0,60 | 0,13 | 0,26 | 0,302 |

Let find out how to change the values $N$ and $\tau$ in formulas (3)–(4), if probability variation of $f(k)$ is low. Let a new distribution can be expressed as $\tilde{f}(k) = f(k) + \varepsilon(k)f(k)$, moreover, under the same norming the equality $\sum_k \varepsilon(k)f(k) = 0$ is fulfilled. Let introduce the variable $E = \sup_k |\varepsilon(k)|$. Then we get the following estimation of distance between distributions in $L1$ norm:

$$\varepsilon = \left\| f - \tilde{f} \right\|_{L1} = \sum_k \left| f(k) - \tilde{f}(k) \right| = \sum_k |\varepsilon(k)|\, f(k) \leq E. \tag{5}$$

Let use (5) to get estimation of some differentiable function variation depending on average batch size in case $E \ll 1$:

$$\left| \delta l^{(1)} \right| = \left| \sum_k k \left( f(k) - \tilde{f}(k) \right) \right| < \sum_k k \left| f(k) - \tilde{f}(k) \right| = \sum_k k \, |\varepsilon(k)|\, f(k) \leq E l^{(1)};$$

$$\left| \delta u \left( l^{(1)} \right) \right| = \left| \frac{\partial u \left( l^{(1)} \right)}{\partial \left( l^{(1)} \right)} \right| \cdot \left| \delta l^{(1)} \right| \leq E u \left| \frac{\partial \ln u}{\partial \ln l} \right| \Big|_{l=l^{(1)}} + o\left( E \right). \tag{6}$$

Logarithmic derivative of function with respect to parameter's logarithm is called function sensitivity to parameter variation. Then, from (6) comes that variation of some function from the mean value, obtained due to distribution variation, at linear approximation on $E$ does not exceed this function multiplied by the supremum of distribution density variation and by modulus of specified sensitivity. In our case estimations of average queue length (3) and average waiting time (4), that are linear for $E$, are expressed as follows:

$$|\delta N| \leq \frac{E\rho}{2} \cdot \left| \frac{v_2}{v_1 b_1} + \frac{l^{(2)} - l^{(1)}}{l^{(1)}} \frac{b_1}{(1-\rho)^2} + \frac{b_2}{b_1^2} (2-\rho) \frac{\rho}{(1-\rho)^2} \right|;$$

$$|\delta \tau| \leq \frac{E\rho}{2 (1-\rho)^2} \cdot \left| \frac{l^{(2)} - l^{(1)}}{l^{(1)}} b_1 + \frac{b_2}{b_1} \right|. \tag{7}$$

However, theoretical estimations (7) do not possess adequate accuracy as they appear to be too excessive. Despite this fact they cannot be improved within of functions to be chosen for empirical distribution functions. Unimprovability comes out of existence of a variation when $\varepsilon(k) = \{0; \pm E\}$. Inadequacy of accuracy comes out of condition (5): as far as $\varepsilon \leq E$, then for heavily nonuniform distributions the variation norm in the form of $\sup_k |\varepsilon(k)|$ is too crude estimate, since for such kind of distributions the distance between distributions can be significantly less than $E$. That is why in [4] an analysis of sensitivity of these parameters based on the numerical results for various functions $f(k)$ was carried out. It was found out that four types of norms – L1 and C for $F(k)$, L1 for $f(k)$ and similar to them the forth norm that is a supplement for total area $S$ to the unity of two densities – determine rather exactly the variation of values $N$ and $\tau$ under variation of densities.

Let us denote variation of $f(k)$ by $\delta f$, which is denominated in norm C for distribution function. We denote by $\delta N(\rho)$ variation of average queue length and by $\delta \tau(\rho)$ variation of average waiting time for a given value of load $\rho$. Data analysis showed that under variation of uniform distribution, which determined on the interval from 1 to the maximum batch size (in our case maximum is equal to 8), there is a relationship between $\delta N(\rho)$ and $\delta f$, also $\delta \tau(\rho)$ and $\delta f$ with determination 0.99:

$$\left| \frac{\delta N(\rho)}{N(\rho)} \right| = 0,2\,(1 - \ln \rho)\,(\delta f)^{0,27};$$

$$\left| \frac{\delta \tau(\rho)}{\tau(\rho)} \right| = \frac{0,106 + 0,041 \ln \rho}{\rho}\,(\delta f)^{0,27}; \tag{8}$$

$$\rho \in [0,1;\ 0,9].$$

## 4   Algorithm for Average Queue Length Calculation

Practical application of the described method is the algorithm for average queue length calculation.

<u>Step 1.</u> Choose the basic predicted batch size distribution of $f(k)$ on the interval $1 \leq k \leq M$, for example, uniform distribution: $f(k) = 1/8$.

<u>Step 2.</u> Calculate the average queue length $N_0$ for the basic distribution:

$$N_0 = \frac{v_2}{2v_1 b_1}\rho + \frac{M-1}{3}\frac{\rho}{1-\rho} + \frac{b_2}{2b_1^2}\frac{\rho^2}{1-\rho}, \tag{9}$$

For basic distribution function Eq. (3) is considered as null approximation.

<u>Step 3.</u> Get the empirical batch size distribution function $F(k)$ from measurements. Let consider that empirical distribution, being investigated, is stationary, $F(k)$ is its distribution function.

<u>Step 4.</u> Calculate the distance between empirical and uniform distributions:

$$\delta f = \sup_k |F(k) - k/M|. \tag{10}$$

<u>Step 5.</u> Substituting the result of (10) in (8) we get estimation of average queue length that corresponds to the following empirical distribution:

$$N/N_0 \approx 1 + 0,2\,(1 - \ln \rho)\,(\delta f)^{0,27},\ \rho \in [0,1;\ 0,9] \tag{11}$$

Estimation for average waiting time variation is expressed in the same way.

Equation (11) is computationally much simpler than calculation of the generating function in accordance with (2), where $L(z)$ is calculated through empirical distribution of $F(k)$. That is, firstly, rather difficult and, secondly, leads to calculation errors that may exceed approximation inaccuracy for (11).

# 5  Average Queue Length for Geometric Batch Size Distribution

For example, let consider that an empirical distribution of $f_n(k)$ is taken from a general population $f(k)$ that has geometric distribution with a parameter $q$ on the interval $1 \leq k \leq M$. That means $f(k) = \frac{1-q}{1-q^M} q^{k-1}$.

We show you algorithm for average queue length calculation with geometric distribution function.

Step 1. Basic batch size distribution:
  $f(k) = 1/8$, $1 \leq k \leq M$ - uniform distribution.

Step 2. Average queue length $N_0$ for the uniform distribution:

$$N_0(\rho), \quad \rho \in [0,1; \ 0,9].$$

Step 3. Empirical batch size distribution is geometric distribution with following parameters: $q = 0,67$, $M = 8$ [8].

$$F(k) = \frac{1-q}{1-q^M} q^{k-1}, \quad 1 \leq k \leq M.$$

Step 4. Distance between empirical and basic distributions:

$$\delta f = 0,35.$$

Step 5. Estimation of average queue length:

$$N \approx 0,14 \, (1 - \ln \rho) \, N_0(\rho), \quad \rho \in [0,1; \ 0,9].$$

# 6  Conclusion

This paper presents the approach to estimation of the performance measures for SIP server model with batch arrivals and vacations depending on the batch size distribution. Investigation of this particular dependence was motivated by the fact that the batch size distribution is not known as a general population and, moreover, cannot be recognized as far as empirical evaluations of this population are non-stationary. That is why approximate evaluation methods, that are not associated with a specified functional class of mentioned distributions, are of great significance and actuality. Therefore, the method that considers coefficients of the model parameters sensitivity to adjustment of the distance between distribution functions seems to be efficient among nonparametric techniques. This method may be used for non-stationery distributions when non-stationary behavior is interpreted as definite variation of some basic distribution (for example, uniform). This approach enables to circumvent technical difficulty coming from absence of convergence theorem both for probability and the norm for random variables being investigated.

Proposed approach to evaluation of performance measures of a Session Initiation Protocol server model and given analysis of parameters sensitivity leads to recommendations for engineers to use simple formulas for preliminary evaluation of presence signaling messages service.

**Acknowledgment.** This work was supported in part by the Russian Foundation for Basic Research, projects No. 15-07-03051, 15-07-03608.

We thank Professor Konstantin Samouylov from Peoples' Friendship University of Russia for comments that greatly improved the paper.

# References

1. Rosenberg, J., et al.: IETF RFC 3261 SIP: Session Initiation Protocol (2002). http://www.ietf.org/rfc/rfc3261.txt
2. Rosenberg, J.: IETF RFC 5390 Requirements for Management of Overload in the Session Initiation Protocol (2008). https://tools.ietf.org/html/rfc5390
3. Hilt, V., Noel, E., Shen, C., Abdelal, A.: IETF RFC 6357 Design Considerations for Session Initiation Protocol (SIP) Overload, Control (2011). https://tools.ietf.org/html/rfc6357
4. Gurbani, V., Hilt, V., Schulzrinne, H.: IETF RFC 7339 Session Initiation Protocol (SIP) Overload Control (LBOC) (2014). https://tools.ietf.org/html/rfc7339
5. Noel, E., Williams, P.M.: IETF RFC 7415 Session Initiation Protocol (SIP) Rate Control (RBOC) (2015). https://tools.ietf.org/html/rfc7415
6. Samouylov, K.E., Sopin, E.S.: On analysis of M[X]|G|1|r queuing system. Bulletin of Peoples' Friendship University of Russia. Series Mathematics, Information Sciences, Physics, no. 1, pp. 91–97 (2011)
7. Gaidamaka, Y., Pechinkin, A., Razumchik, R., Samouylov, K., Sopin, E.: Analysis of M|G|1|R queue with batch arrivals and two hysteretic overload control policies. Int. J. Appl. Math. Comput. Sci. **24**(3), 519–534 (2014)
8. Gaidamaka, Y.V., Zaripova, E.R., Orlov, Y.N.: Analysis of the impact the batch size distribution on parameters of the SIP-server queueing model with batch arrivals. KIAM Preprint no. 27, pp. 1–16, Moscow (2015). http://library.keldysh.ru/preprint.asp?id=2015-27
9. Gaidamaka, Y., Zaripova, E.: Comparison of polling disciplines when analyzing waiting time for signaling message processing at SIP-server. In: Dudin, A., et al. (eds.) ITMM 2015. CCIS, vol. 564, pp. 358–372. Springer International Publishing, Switzerland (2015)
10. Gaidamaka, Y.V.: Model with threshold control for analyzing a server with an SIP protocol in the overload mode. Autom. Control Comput. Sci. **47**(4), 211–218 (2013)
11. Abaev, P., Gaidamaka, Y., Samouylov, K.E.: Modeling of hysteretic signaling load control in next generation networks. In: Andreev, S., Balandin, S., Koucheryavy, Y. (eds.) NEW2AN/ruSMART 2012. LNCS, vol. 7469, pp. 440–452. Springer, Heidelberg (2012)
12. Abaev, P., Gaidamaka, Y., Samouylov, K.E.: Queuing model for loss-based overload control in a SIP server using a hysteretic technique. In: Andreev, S., Balandin, S., Koucheryavy, Y. (eds.) NEW2AN/ruSMART 2012. LNCS, vol. 7469, pp. 371–378. Springer, Heidelberg (2012)

13. Gaidamaka, Y., Pechinkin, A., Razumchik, R., Samouylov, K., Sopin, E.: Analysis of M|G|1|R queue with batch arrivals and two hysteretic overload control policies. Int. J. Appl. Math. Comput. Sci. **24**(3), 519–534 (2014)

14. Abaev, P., Gaidamaka, Y., Pechinkin, A., Razumchik, R., Samouylov, K., Shorgin, S.: Hysteretic control technique for overload problem solution in network of SIP servers. Comput. Inform. **33**(1), 218–236 (2014)

15. Shorgin, S., Samouylov, K., Gaidamaka, Y., Etezov, S.: Polling system with threshold control for modeling of SIP server under overload. In: Advances in Intelligent Systems and Computing, vol. 240, pp. 97–107 (2014)

16. Abhayawardhana, V.S., Babbage, R.: A traffic model for the IP multimedia subsystem (IMS). In: Proceedings of 65th Vehicular Technology Conference, pp. 783–787 (2007)

17. Vishnevsky, V., Kozyrev, D., Semenova, O.: Redundant queuing system with unreliable servers. In: 2014 6th International Congress on Ultra Modern Telecommunications and Control Systems and Workshops (ICUMT), pp. 283–286 (2014)

18. Abaev, P.O., Gaidamaka, Y.V., Pechinkin, A.V., Razumchik, R.V., Shorgin, S.Y.: Simulation of overload control in SIP server networks. In: Proceedings of the 26th European Conference on Modelling and Simulation, ECMS 2012, Koblenz, Germany, pp. 533–539 (2012)

19. Abaev, P.O., Pechinkin, A.V., Razumchik, R.V.: On analytical model for optimal SIP server hop-by-hop overload control. In: Proceedings of the 4th International Congress on Ultra Modern Telecommunications and Control Systems (IEEE ICUMT-2012), Saint-Petersburg, Russia, pp. 299–304 (2012)

20. Abaev, P., Gaidamaka, Y., Samouylov, K., Shorgin, S.: Design and software architecture of SIP server for overload control simulation. In: Proceedings of the 27th European Conference on Modelling and Simulation (ECMS 2013), 27–30 May, Aalesund, Norway, pp. 580–586 (2013)

21. Samouylov, K., Abaev, P., Gaidamaka, Y., Pechinkin, A., Razumchik, R.: Analytical modelling and simulation for performance evaluation of SIP server with histeretic overload control. In: Proceedings of the 28th European Conference on Modelling and Simulation (ECMS 2014), 27–30 May, Brescia, Italy, pp. 603–609 (2014)

22. Gaidamaka, Y., Pechinkin, A., Razumchik, R.: Time-related stationary characteristics in queueing system with constant service time under hysteretic policy. In: Proceedings of the 6th International Congress on Ultra Modern Telecommunications and Control Systems (ICUMT 2014), Russia, pp. 634–640 (2014)

23. Samouylov, K., Gaidamaka, Y., Talanova, M., Pavlotsky, O.: Simulation of SIP-server with hysteretic input and loss-based overload control scheme. In: Proceedings of the 6th International Congress on Ultra Modern Telecommunications and Control Systems (ICUMT 2014), St. Petersburg, Russia, pp. 589–594 (2014)

24. Samouylov, K., Gaidamaka, Y., Abaev, P., Talanova, M., Pavlotsky, O.: Analytical modeling of rate-based overload control with token bucket traffic shaping on client side. In: Proceedings of the 29th European Conference on Modelling and Simulation (ECMS 2015). Digitaldruck Pirrot GmbH, Albena, pp. 669–674 (2015)

# The Synthesis of Service Discipline in Systems with Limits

Taufik Aliev[(✉)]

ITMO University, Kronverkski pr., 49, 197101 St. Petersburg, Russia
aliev@cs.ifmo.ru

**Abstract.** The problem of service discipline synthesis in queuing systems with constraints on the mean residence time of different classes of requests, or chances of exceeding the limits. Problem is solved for the class of disciplines with mixed priorities between requests of different classes - relative, absolute, or no priorities. An algorithm for prioritization, ensuring the implementation of the limitations specified for the minimal system performance is proposed. The problem is solved using analytical dependences obtained for the first two moments of the residence time distribution, and the approximation of the principles of distributions on two moments.

**Keywords:** Service discipline · Mixed priorities · Delay time · Limits · Device performance

## 1 Introduction

One of the main characteristics of the functioning of computer systems of different classes and computer networks is the delay time of data processing and transmission, which is named as residence time of the system in queuing systems used as models. The requirements for the quality of the functioning of such systems are formulated in a variety of limits on the residence times in the different classes of applications. For example, in data-processing systems, restrictions apply to the mean residence time of the queries in the system (the average limits), the excess of these limits does not lead to critical consequences. At the same time, in information and control systems, existing in a circuit of automatic control of technological equipment or mobile units, restrictions are imposed on the probability of exceeding the set value of residence time (probabilistic limit) which may lead to a sharp deterioration in the functioning of the system or even to the exit its failure. In general, the quality requirements of the system can be formulated as a combination of limits, where some classes have an average constraints, while others - probabilistic constraints. Applying limits ensuring correct application of the priority strategies of process control and data processing. Queuing system with one serving device and unlimited storage capacity, which receives the inhomogeneous flow of requests to be processed with a given capacity are widely used as models of such systems [1]. In this case, the problem

© Springer International Publishing Switzerland 2016
V. Vishnevsky and D. Kozyrev (Eds.): DCCN 2015, CCIS 601, pp. 151–156, 2016.
DOI: 10.1007/978-3-319-30843-2_16

of synthesis service disciplines comes to the distribution of priority to provide the desired limits on the residence time in the different classes of applications with minimal system performance.

## 2    Statement of a Problems

Load parameters for solving the problem of synthesis to a system with a non-uniform flow of applications used: the number of classes of applications - $H$, coming in from the intensities $\lambda_1, \ldots, \lambda_H$ and forms a simple flows; resource-processing applications each class, asked for at least three factors: average values $\theta_1, \ldots, \theta_H$ (commands or instructions executed by the processing of the application of the corresponding class); coefficients of variation $\nu_1, \ldots, \nu_H$ and third initial moments $\theta_1^{(3)}, \ldots, \theta_H^{(3)}$.

Suppose that the classes of applications are grouped so that the restrictions on the residence times $\tau_{u_1}, \ldots, \tau_{u_H}$ in the system for the first $H_1$ classes are defined in a probability, and for other grades - as mean limited to:

$$\Pr\left(\tau_{u_h} > u_h^*\right) \leq \delta_h^* \qquad (h = \overline{1, H_1}), \tag{1}$$

$$u_h \leq u_h^*, \qquad (h = \overline{H_1 + 1, H}) \tag{2}$$

where $u_h^*$ – allowable value of residence time in the application class $h$; $\delta_h^*$ – allowable probability of exceeding a specified limit $u_h^*$; $\Pr\left(\tau_{u_h} > u_h^*\right)$ – probability that the residence time in the application class $h$ ($h$-request) $\tau_{u_h}$ exceeds the allowable value $u_h^*$; $u_h = M[\tau_{u_h}]$ – average value (expectation) of residence time of $h$-request in system.

Resident times $\tau_{u_1}, \ldots, \tau_{u_H}$ depend on the system performance $V$ and service discipline. Obviously, for any restrictions to (1) and (2) can be performed at the expense of the system performance. At the same time the best solution will be the discipline which can perform with minimal restrictions performance.

The synthesis problem is solved service discipline in the classroom disciplines with mixed priorities, which are described by a matrix of priorities $Q = [q_{ij}(i, j = 1, \ldots, H)]$, where $q_{ij}$ set the priority of $i$-requests against $j$-requests: $0$ – no priority, 1 –relative priority (RP) 2 – absolute priority (AP) [1].

Thus, the problem of synthesis to systems with constraints is formulated as follows: to find the discipline of service of requests with mixed priority given that limits (1) and (2) are fulfilled with minimal performance V.

## 3    Calculated Identities

Resident time of $h$-requests ($h = \overline{1, H}$) in system $\tau_{u_h}$ in common consists of the waiting time of service start $\tau_{x_h}$ and processing time $\tau_{z_h}$, including the waiting time in the interrupted state: $\tau_{u_h} = \tau_{x_h} + \tau_{z_h}$. Then the expectation $u_h$ and the second moment $u_h^{(2)}$ of resident time of $h$-requests:

$$u_h = x_h + z_h; u_h^{(2)} = x_h^{(2)} + 2x_h z_h + z_h^{(2)}. \tag{3}$$

Values $x_h$, $z_h$ and $x_h^{(2)}$, $z_h^{(2)}$ are counted [1]:

$$
\left.
\begin{aligned}
x_h &= \frac{\sum_{i=1}^{H} r_4(i,k)\,\lambda_i\, b_i^{(2)}}{2(1-R_h^{(2)})\,(1-R_h^{(3)})}; \qquad
z_h = \frac{b_h}{1-R_h^{(1)}}; \\
x_h^{(2)} &= \frac{\sum_{i=1}^{H} r_4(i,k)\,\lambda_i\, b_i^{(3)}}{3(1-R_h^{(2)})^2\,(1-R_h^{(3)})} + \frac{\sum_{i=1}^{H} r_3(i,k)\,\lambda_i\, b_i^{(2)}\,\sum_{i=1}^{H} r_4(i,k)\,\lambda_i\, b_i^{(2)}}{2(1-R_h^{(2)})^2\,(1-R_h^{(3)})^2} \\
&\quad + \frac{\sum_{i=1}^{H} r_2(i,k)\,\lambda_i\, b_i^{(2)}\,\sum_{i=1}^{H} r_4(i,k)\,\lambda_i\, b_i^{(2)}}{2(1-R_h^{(2)})^3\,(1-R_h^{(3)})}; \\
z_h^{(2)} &= \frac{b_h^{(2)}}{(1-R_h^{(1)})^2} + \frac{\sum_{i=1}^{H} r_1(i,k)\,\lambda_i\, b_i^{(3)}}{(1-R_h^{(1)})^3},
\end{aligned}
\right\} \qquad (4)
$$

where $b_i^{(n)} = \theta_i^{(n)}/V^n$ – $n$ starting moment of service time of $i$-request ($i = \overline{1,H};\ n = 1,2,3$); $V$ – system performance; $R_h^{(g)} = \sum_{i=1}^{H} r_g(i,h)\,\lambda_i b_i$ – partial total loads; $r_g(i,h)$ – coefficients with values 0 or 1, depending on the values of the elements $q_{ih}$ and $q_{hi}$ of priority matrix and allow applications to allocate classes $i$ and $h$, with same priority class: $r_1(i,h) = 0,5\,q_{ih}(q_{ih}-1)$ – takes the value 1, if $i$-requests have absolute priority against $h$-requests; $r_2(i,h) = 0,5\,q_{ih}(3-q_{ih})$ – takes the value 1, if $i$-requests have relative priority or absolute priority against $h$-requests; $r_3(i,h) = 1-0,5\,q_{hi}(3-2q_{ih}+q_{hi})$ – takes the value 1, if $i$-requests have no priority or relative priority or absolute priority against $h$-requests; $r_4(i,h) = 1+0,5\,q_{hi}(1-q_{hi}+q_{ih})$ – takes the value 0 only if $h$-requests have absolute priority against $i$-requests.

## 4   The Algorithm Synthesis Service Discipline

The problem of synthesis of service discipline is reduced to the determination of values $q_{ij}$ of the priority matrix, at which the following restrictions (1) for all classes of applications. The analytical solution of the system of inequalities (1) is not possible because the number of different joint venture to even a small number of classes of applications is significant. In addition, the priority matrix should be correct [1] so the development of the appropriate discipline of service algorithm would not lead to ambiguity and uncertainty.

An example of incorrect priority matrix ($H = 3$):

$$
Q = \begin{array}{c|ccc}
 & 1 & 2 & 3 \\
\hline
1 & 0 & 1 & 1 \\
2 & 0 & 0 & 2 \\
3 & 0 & 0 & 0
\end{array}.
$$

Class 2 request enters the system, when device is operating with the class 3 request, and the queue contains one or several class 1 requests, which cannot interrupt the service of class 3 requests, as they have relative priority to class 3 ($q_{13} = 1$). This raises the following uncertainty. On the one hand, receiving the class 2 request shall abort the service of class 3 request, because it has absolute priority ($q_{23} = 2$) over class 3 requests. On the other hand, received class 2 request may not get service before class 1 requests are in a queue and have a relative priority to class 2 requests ($q_{12} = 1$). This uncertainty leads to ambiguity in constructing the algorithm the implementation of the queuing discipline. Any decision (to abort and not to abort the service of class 3 request) will result in a discipline not appropriate to the given priority matrix.

The rules of the correct formation of priority matrices were developed for canonical matrices, where $q_{ij} = 0$ for all $i \geq j$ $(i, j = 1, \ldots, H)$. Canonical priority matrix describes a queuing discipline in which request classes with a lower number have priority no lower than request classes with a higher number. These rules are:

1. if $q_{ij} = 1$ or $2$ then $q_{ik} \neq 0$ for all $k > j$ $(i, j, k = 1, \ldots, H)$;
2. $q_{i+1\,j} \leq q_{ij}$ $(i = 1, \ldots, H - 1; \, j = 1, \ldots, H)$;
3. if $q_{i\,i+1} = 0$ then $q_{i+1\,j} = q_{ij}$ for all $j = 1, \ldots, H$ $(i = 1, \ldots, H - 1)$.

The number of options for filling canonical priority matrix $\xi = 3^{H(H-1)/2}$, where $H$ – the number of request classes that determines the dimension of the matrix. However, this number must exceed incorrect priority matrices. So, if $H = 5$ then $\xi = 54049$ and the number of correct matrices is equal to 541, if $H = 6$ then the number of correct matrices is 4683, and if $H = 10$ then the number of correct matrices is more than 100 million.

A sequential scan of all possible matrices leads to time-consuming. In this case it is necessary to develop an effective and well-formalized synthesis algorithm for priority service disciplines based on a heuristic approach based on targeted iterating multiple disciplines of service with mixed priorities in general form, which allows to automate the process of setting priorities between different classes of requests.

The problem of synthesis of service disciplines is decided in two stages.

In the first stage, the problem reduces to the problem of the distribution of priorities for systems with medium restrictions by converting probabilistic constraints (1) in the middle limits.

To do this, considering the limitation (1) at the border, we write it as follows: $1 - U_h(u_h^*) = \delta_h^*$, where $U_h(\tau)$ – residence time distribution function $\tau_{u_h}$ in system of $h$-requests. Solving the resulting equation for the expectation, we find the estimate of restrictions on the mean residence time equivalent probabilistic limitation, as a function of the allowable of residence time $u_h^*$ and the probability of exceeding it $\delta_h^*$: $\tilde{u}_h = \varphi(u_h^*, \delta_h^*)$. Now probabilistic constraints (1) may be replaced by averages limitations:

$$u_h \leq \tilde{u}_h, \qquad\qquad (h = \overline{1, H_1}). \qquad\qquad (5)$$

In implementing the described approach, despite its simplicity, a problem arises in that the distribution function $U_h(\tau)$ depends on a service discipline which during synthesis are constantly changing. For evaluation of $\tilde{u}_h$ we assume that the distribution function $U_h(\tau)$ is exponential. Then $1 - U_h(u_h^*) = e^{-u_h^*/\tilde{u}_h} = \delta_h^*$, where we get: $\tilde{u}_h = -u_h^*/\ln \delta_h^*$. Suppose that $\delta_h^*$ is given as $\delta_h^* = 10^{-n}$, then $\tilde{u}_h = 0,435\, u_h^*/n$ $(h = \overline{1, H_1})$. As a result of this conversion problem of synthesis service discipline combined with restrictions reduced to the problem with average limitations include the following items.

1. No-priority service discipline Q was assigned as the initial service discipline.
2. For service discipline Q value of performance $V^Q_{\max}$ is determined, which is necessary for performing the specified constraints (5) and (2):

$$V^Q_{\max} = \max\{V^Q_1, \ldots, V^Q_H\}$$

where $V^Q_h$ – minimum value of performance, where $u_h = \tilde{u}_h$ for $h = \overline{1, H_1}$ and $u_h = u^*_h$ for $h = \overline{H_1 + 1, H}$ with service discipline with mixed priority Q.
3. For service discipline Q we rearrange class numbers $h_1, \ldots, h_H$ by counted performance values in descending order: $V^Q_{h_1} \geq \cdots \geq V^Q_{h_H}$, and selected two extreme classes: $k_1 := h_1$ $k_2 := h_H$.
4. For $k_1$ and $k_2$ classes we try to change the priority by increasing the priority of $k_1$ class against $k_2$ taking into account the requirements of priority matrix correctness.
5. If the change succeeded, then for new service discipline $Q'$ the value of the performance is determined as $V^{Q'}_{\max} = \max\{V^{Q'}_1, \ldots, V^{Q'}_H\}$ similarly to paragraph 2. If $V^{Q'}_{\max} < V^Q_{\max}$, then $Q := Q'$, $V^Q_{\max} := V^{Q'}_{\max}$ and we continue with item 3.
6. If we can't change the priority between $k_1$ and $k_2$ or for new service discipline $Q'$ performance is $V^{Q'}_{\max} \geq V^Q_{\max}$, then we need repeat items 3–4 consistently for $k_2 = h_{H-1}, h_{H-2}, \ldots, h_2$.
7. If the search for all combinations of pairs of classes for this scheme did not lead to a change in the matrix of priorities, the process will stop the search service discipline and as a result received last priority matrix Q.

On the second stage its necessary to check the implementation of probabilistic constraints (1), because the distribution function $U_h(\tau)$ may differs from exponential. These restrictions cannot be fulfilled, if the coefficient of variation of residence time $\alpha_h = \sqrt{u^{(2)}_h - u^2_h}/u_h$ is more than 1. In this case, to clarify the evaluation $\tilde{u}_h$ we perform the approximation $U_h(\tau)$ by two moments $u_h$ and $u^{(2)}_h$, calculated using the formula (3) and (4). As we use the two-phase distribution approximating representation hyperexponential distribution [2], for that we count $q_h \leq 2/(1 + \alpha^2_h)$ and expectations exponential phases $t'_h = [1 + \sqrt{(1 - q_h)(\alpha^2_h - 1)/(2q_h)}]u_h$ and $t''_h = \{1 - \sqrt{q_h(\alpha^2_h - 1)/[2(1 - q_h)]}\}\, u_h$.

We assume that the parameter distribution hyperexponential is $q_h = 2/(1 + \alpha^2_h)$. Then, by analogy with the exponential distribution we obtain a new estimate of the average limit:

$$\tilde{u}_h = -\frac{2\, u^*}{(1 + \alpha^2_h) \ln\{\delta^*_h(1 + \alpha^2_h)/2\}}.$$

After assuming changes in (5) for $h$-class requests of $\tilde{u}_h$, obtained under the assumption of exponential distribution, a new estimate obtained for hyperexponential distribution, you must rerun the items 2–6 of the first stage of the service discipline synthesis problem.

## 5    Conclusion

The proposed approach to the problem of the synthesis of service discipline with mixed priorities for systems with complex limits on the residence times in the different classes of applications allows for a specified quality of the system with a minimum performance.

## References

1. Aliev, T.I., Maharevs, E.: Queueing disciplines based on priority matrix. Sci. Tech. J. Inf. Technol. Mech. Opt. **6**(94), 91–97 (2014)
2. Aliev, T.I.: Approximation of probability distributions in queuing systems. Sci. Tech. J. Inf. Technol. Mech. Opt. **2**(84), 88–93 (2013)

# Relative Navigation for Node of Wireless Decentralized Network

D.A. Aminev[1][✉], R.F. Azizov[2], and D.Yu. Kudryavtsev[1,2]

[1] V. A. Trapeznikov Institute of Control Sciences of Russian Academy of Sciences,
65 Profsoyuznaya street, 117997 Moscow, Russia
`aminev.d.a@yandex.ru, dmt.kudr@gmail.com`
[2] National Research University Higher School of Economics, 20 Myasnitskaya street,
101000 Moscow, Russia
`radomir.azizov@gmail.com`

**Abstract.** The problem of determining the geographical coordinates of the wireless unit in the presence of a decentralized network of airborne navigation receivers on some other sites is researched. Similar solutions for locating a subscriber to GSM is considered, its shortcomings is revealed. A mathematical apparatus for the relative navigation based on multilateration is offered. An experiment to determine the geographic origin and the comparison of its results with the calculated data is carried. Considerations to restrict the mobility of nodes in different variants of the construction of the network is formulated.

**Keywords:** Navigation · Geographical coordinates · The wireless network

## 1 Introduction

The development of wireless transmission is a rapid pace [1], along with the navigation problem put their subscribers. In case of failure or absence of GPS signals, or in areas without GSM coverage, or to improve positioning accuracy relative navigation becomes necessary. This is especially true for decentralized networks [2,3], when the nodes are mobile and move in space, such as sensor networks or operative networks of tablet computers in places of emergencies [4].

There are two main methods for relative navigation subscriber in the network GSM [5] method of obtaining the time (Time of Arrival – TOA); time difference method (Enhanced Observed Time Difference, EOTD). The method of obtaining the time (TOA) [6] similar to the GPS satellite navigation technology and is based on the measurement of the delay in the shift of the frame with the signal from the base station to the phone where the distance to the base station. To determine the coordinates should be at least three simultaneous bearing. Calculations made by the operator of the triangulation algorithms. The time difference method (EOTD) [7] measured delay time difference signals from the two closest base stations, which is then converted to distance. Feature EOTD method is

© Springer International Publishing Switzerland 2016
V. Vishnevsky and D. Kozyrev (Eds.): DCCN 2015, CCIS 601, pp. 157–166, 2016.
DOI: 10.1007/978-3-319-30843-2_17

the need to integrate the mobile terminal computing module, which also passed the exact coordinates of the base stations, so blocks LMU base stations requires several times less than the TOA. EOTD has spread in networks with CDMA technology and is supported by some models of terminals to networks GSM. According to some estimates, the accuracy of the method EOTD exceed TOA.

The disadvantage of these methods is the high accuracy of determining the coordinates in a big city it is usually up to 400 m, in the regional center to a kilometer in rural areas and along the routes of 15–20 km. In addition, both methods require the installation of base stations LMU special module for calculating position places. The methods are applicable only in the area of GSM coverage.

Also known relative navigation method based on measurement of received signal power [8,9]. In the method of approximation conducted a distant relation of the type $\frac{1}{r^n}$ characteristics, where $n > 2$. This takes into account attenuation when passing the road and heterogeneity of the antenna pattern. This method is not tied to coverage GSM, but, despite the simplicity of calculations obtained by averaging feature provides low accuracy location.

A method of relative navigation unit wireless decentralized network based on measurements of the received signal power from a distant approximation characteristics dependent species $\frac{1}{r^2}$ and taking into account the position error.

## 2     Theoretical Basis of Relative Navigation

Block diagram of the geographic coordinate data from several nodes at the same time still represented in Fig. 1, one movable at different points in time (b) and distant feature (c).

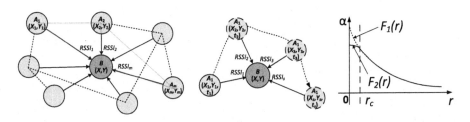

**Fig. 1.** Scheme geographical coordinate data from multiple fixed units (a), one movable (b) and distant characteristic (c)

In the case of (a), components $A_1, \ldots, A_m$ (beacons) with navigation data $(x_1, y_1), \ldots, (x_m, y_m)$, send your coordinates broadcast packets node $B$ (locator) with coordinates $(x, y)$ that has no navigation data, and calculates the levels of signals $RSSI_1, \ldots, RSSI_m$ [8]. In the case of (b) node $A_1$ (beacon) is moved on the plane and at times $t_1, \ldots, t_s$ sends its coordinates $(x_1, y_1), \ldots, (x_s, y_s)$ fixed node $B$ (locator), which calculates the coordinates $(x, y)$ of the signal levels

$RSSI_1, \ldots, RSSI_s$. It is also possible combined variant relative navigation event (a) with mobile nodes in the case (b), which increases the quality of the calculated data.

Thus, initial data for relative navigation coordinates are other network nodes (at least 2) and the ratio $RSSI$ of the received signal power level. Location determined by multilateration and methods of passive radar. It is necessary to take into account the dependence of the signal power of the distance (a distant characteristic), the presence or absence of obstacles in the signal path and the error location.

On the strength of the received signal is influenced by:

(a) the projective ("radiant") weakening;
(b) reflection and refraction (obstacles — terrain, buildings, structures, etc.);
(c) interference, diffraction, etc. (challenging obstacle and so forth.);
(d) the relative orientation of the transmitting and receiving antennas;
(e) absorption (rain, fog, smog, etc.).

Factor (a) the most dominant. As you know, in the free space capacity of the received electromagnetic wave from a point source is inversely proportional to the square of the distance from the source (Friis formula):

$$P^R = P(r) = P^T G^R G^T \left(\frac{\lambda}{4\pi r}\right)^2, \quad r > 0, \tag{1}$$

where $r$ — the distance between the beacon and locator, $P^T$ — power signal emitted by the antenna of the transmitter (beacon), $P^R$ — the signal strength received by the antenna receiver (locator), $G^T$ and $G^R$ — gains of transmitting and receiving antennas, respectively, $\lambda$ — the wavelength. We denote further $\alpha := P^R/P^T = u(P^R, P^T)$, $\beta := 10 \lg \alpha = RSSI = \nu(P^R, P^T)$ — the relative power of the received signal according to the linear and logarithmic scale, respectively. Then we get:

$$\alpha = F_1(r) = K_1/r^2, \quad r > 0, \tag{2}$$

$$\beta = 10 \lg(K_1/r^2) = K_2 - 20 \lg r, \quad r > 0, \tag{3}$$

where $F_1$ — a distant point source characteristics. Parameters link

$$K_1 = G^R G^T \left(\frac{\lambda}{4\pi}\right)^2 \text{ and } K_2 = 10 \lg K_1 \tag{4}$$

It can be considered permanent.

If the radiator — not a point but a small body (antenna), the pole at the $r = 0$ point is smoothed (near field), but far enough away from the body (far-field) can be regarded as a point and the "inverse square law" would be applied. It can be approximated by a distant features as follows:

$$\alpha = F_2(r) = K_1 \left(\exp(-ar^2) + \frac{1}{r^2} \exp(-b/r^2)\right), \quad r > 0, \tag{5}$$

where $F_2$ — distant antenna characteristics.

Since $F_2(r) \approx F_1(r)$ for $r \geq r_c$ the conventional boundary near and far zones antenna (Fig. 1c), and near-field problems locating antennas are not used, it is possible to restrict the formula (1) for $r \geq r_c$. Furthermore, if the transmitter and receiver are on the surface (the ground), then the reflection from it can be neglected, and also to use the formula (1).

Action factors (b) and (c) generally is unknown, but it can be evaluated on the basis of additional information about the terrain.

Factor (d) due to the heterogeneity of the diagrams of the transmitting and receiving antennas. If you can change the orientation of the antenna in space, to eliminate the influence of this factor requires information about their orientation. If such information is not available, you must either request a measurement at a specific antenna orientation (vertical, horizontal, etc.), or to carry out effective filtering incoming data from the beacon. It is desirable that the beacons are fairly evenly distributed over the neighborhood of the locator and the number is sufficiently large.

Action factor (e) manifested usually at sufficient remoteness of the transmitter and receiver from each other (several kilometers or more) and can be accounted for using the weather data.

## 2.1 Deterministic Models Locations on the Earth's Surface in the Absence of Obstacles

If the beacon is in the line of sight of the receiver, the distance to it is computed directly from the distant characteristic patterns, taking into account:

$$r = F^{-1}(\alpha) = \sqrt{\frac{K_1}{\alpha}} = \sqrt{K_1 \frac{PT}{PR}}, \tag{6}$$

$$r = 10^{\frac{K_2 - \beta}{20}}. \tag{7}$$

In this case, the locus is a circle of radius $r$, the center of which coincides with the beacon. With two beacons locus is a pair of points (the intersection of two circles), while the three — only one point (the intersection of the three circles).

If you know the coordinates $\{(x_k, y_k)\}_{k=1}^m$ of the locator beacons are the coordinates $(x, y)$ of the system (the problem is solved multilateration):

$$\begin{cases} (x - x_1)^2 + (y - y_1)^2 = r_1^2 \\ \quad\cdots\cdots\cdots\cdots \\ (x - x_m)^2 + (y - y_m)^2 = r_m^2 \end{cases} \tag{8}$$

If the coordinates of the beacons set geographic coordinates (latitude $\theta$ and longitude $\phi$), it is necessary to consider this curvilinear reference system. However, in a small neighborhood of any point other than the poles, it can be considered a Cartesian (linearized). The distance between the points in this neighborhood is using the metric

$$dp = \sqrt{(dx)^2 + (dy)^2} = R\sqrt{(\cos\theta d\phi)^2 + (d\theta)^2},$$

where $R$ — the average radius of the Earth (6367 km).

In view of this problem can be formulated as multilateration

$$\begin{cases} (\phi - \phi_1)^2 \cos^2\theta_1 + (\theta - \theta_1)^2 = (r_1/R)^2 \\ \quad\cdots\cdots\cdots\cdots\cdots \\ (\phi - \phi_m)^2 \cos^2\theta_m + (\theta - \theta_m)^2 = (r_m/R)^2 \end{cases} \tag{9}$$

## 2.2 Stochastic Model of Locations on the Earth's Surface in the Absence of Obstacles and Error Estimate

If the power emitted and received signal considered random variables (the process), the distance from the radar to the beacon will also be a random variable (process). At the same loci become "fuzzy" or blurred: the locus of beacons $L = \bigcap_k L_k$ — "fuzzy spot" where $L_k$ — "fuzzy ring" or locus of a single beacon (Fig. 2a).

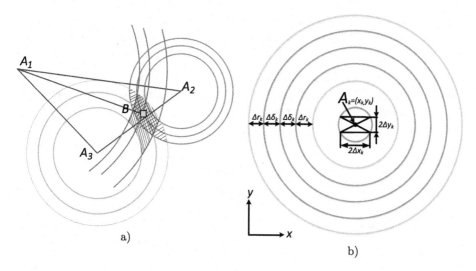

**Fig. 2.** The structure of the locus of the three beacons (a) and blur the "ideal" single locus beacon (b)

Calculation of the position locator is based on $RSSI$ values and coordinates of the beacons. Each of these parameters should be considered as a random process. Approximately stochastic component causing blur loci can be considered as additive noise. Options dispersion and determine the amount of noise amplitude blur. From the geometric point of view, blur locus locator for two reasons (Fig. 2b) uncertainty of the center of the circle (the coordinates of the beacon) and its radius (distance from the beacon).

Effective evaluation and analysis of errors output parameters (coordinates locator) can be obtained using total differentials of smooth functions in a deterministic model of the scene.

From (6) we get:

$$dr_k^{(1)} = (F^{-1})'(\alpha_k)d\alpha_k = -\frac{1}{2}\sqrt{\frac{K_1}{\alpha_k^3}}d\alpha_k \tag{10}$$

Since $\alpha_k = 10^{\beta_k/10}$, then $d\alpha_k = \frac{\ln 10}{10}\alpha_k d\beta_k$, then

$$dr_k^{(2)} = -\frac{\ln 10}{20}\sqrt{K_1}10^{-\beta_k/20}d\beta_k, \tag{11}$$

$$d\alpha_k = du(P_k^R, P_k^T) = \frac{1}{P_k^T}dP_k^R - \frac{P_k^R}{(P_k^T)^2}dP_k^T. \tag{12}$$

Here of

$$dr_k^{(3)} = -\frac{\sqrt{K_1}}{2}\left(\frac{P_k^T}{P_k^R}\right)^{3/2}\left(\frac{1}{P_k^T}dP_k^R - \frac{P_k^R}{(P_k^T)^2}dP_k^T\right), \tag{13}$$

or

$$dr_k^{(3)} = \frac{\sqrt{K_1}}{2}\left(\frac{1}{\sqrt{P_k^R P_k^T}}dP_k^T - \sqrt{\frac{P_k^T}{(P_k^R)^3}}dP_k^R\right). \tag{14}$$

Let $\Delta r$ — error (half precision interval estimation) to determine the distance of distant characteristic, $\Delta P^R$, $\Delta P^T$ — the power measurement error signal is sent and received, respectively, while

$$\Delta r = dr^{(1)} + o(\Delta\alpha), \tag{15}$$

$$\Delta r = dr^{(2)} + o(\Delta\beta), \tag{16}$$

$$\Delta r = dr^{(3)} + o\left(\sqrt{(\Delta P^R)^2 + (\Delta P^T)^2}\right). \tag{17}$$

Coordinates $k$-th beacon $(x_k, y_k)$ defined errors $\Delta x_k$, $\Delta y_k$, respectively (half precision interval estimates), so the error location is the beacon

$$\delta_k = \sqrt{(\Delta x_k)^2 + (\Delta y_k)^2}. \tag{18}$$

The total error in determining the distance to the beacon (half the thickness of conventional "fuzzy ring") equal to the sum of errors:

$$\Delta d_k = \Delta r_k + \delta_k. \tag{19}$$

A rough estimate of the error location by this method can be obtained by averaging the thickness of "fuzzy rings":

$$\Delta x := \frac{1}{\sqrt{2}m}\sum_{k=1}^{m}\Delta d_k, \qquad \Delta y := \frac{1}{\sqrt{2}m}\sum_{k=1}^{m}\Delta d_k. \tag{20}$$

Errors in determining the geographical coordinates of the locator on the longitude and latitude are equal

$$\Delta\phi := \frac{\Delta x}{R cos\theta}, \qquad \Delta\theta := \frac{\Delta y}{R}. \tag{21}$$

## 2.3   Location on the Plane in the Presence of Obstacles

If the beacon is hidden obstacles, the location and the reflection parameters are unknown, the error locations may be prohibitive. To mitigate the impact of obstacles, an effective space-time filtering of data from beacons. Under the temporal filtering is defined here as data processing of each beacon alone received over a certain period of time and under spatial — processing data received from the family of beacons, at the current time.

The essence of the spatial filtering that beacon transmits its position $(x_k, y_k)$ and power $P_k^T$ level, the weight $s_k$ assigned to each beacon. The smaller the distance to the beacon, the effect of interference and noise, the more $s_k$. Initially, weight unknown, as no known signal conditions (distance, obstacles, interference, noise, etc.). The initial assessment is carried out on the distant weight characteristic, but in the course of the weight is constantly recalculated. The problem multilateration beacon for the whole family and is calculated error in determining the coordinates and the distance to the beacons. Beacons are assigned weights according to the magnitude of the error distance. At the same time it is far from the hidden obstacles or locator beacons, the signal from which the noise is strongly distorted and noise, getting minimum weight. If beacons are in unfavorable conditions for ranging (in buildings, tunnels, forest, etc.), the network must constantly calculate their "internal" weight in relation to other beacons. For example, each beacon can calculate its location to other beacons and to compare the results with satellite navigation. The smaller the difference the higher "internal" weight.

# 3   Experiment

For the development of the technology has been developed relative navigation layout (Fig. 3) and an experiment on the ground in the field (3b) for the embodiment of a decentralized network in accordance with 1b.

Based layouts are receiving and transmitting device based on TI CC1101 transceiver and microcontroller control ST STM32F0, a set of specially designed programs for reception and transmission of navigation data. In addition, the program shows the level $RSSI$, dBm. The means used are two laptops with the operating system installed GNU/Linux, the two connected wireless transceiver module. The modules, developed by LLC "Open development" [10], are small-sized wireless devices with a USB interface and a spiral antenna [11]. The program runs on a laptop rf_sender beacon and sends the coordinates of the locator, rf_reciever - runs on a laptop locator receives the data shows the level of $RSSI$, calculates its own coordinates.

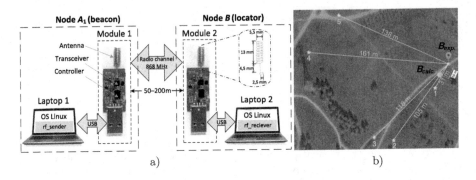

**Fig. 3.** The circuit layout (a) the location and the result on the ground (b)

Geographical coordinates $(\theta, \phi)$ of the moving area on the beacon $A$ determined through a GPS system and a fixed locator $B$ taken at time points $t_1, \ldots, t_6$ (Fig. 3b). The transmitter power is 16 mW, distance to $B$ the assembly varies in the range of 50–200 m. The conditional boundary between the near and far zones of the helical antenna is located, much less than 1 m. $RSSI$ is averaged over 10 measurements.

**Table 1.** Experimental data of reference points

| Number of beacon, $k$ | Latitude $\theta_k$, degrees | Longitude $\phi_k$, degrees | Error $\delta_k$, m | $RSSI$, dBm |
|---|---|---|---|---|
| 1 | 55.83386801 | 37.36731450 | 5 | −70,85 |
| 2 | 55.83335085 | 37.36654756 | 10 | −71,35 |
| 3 | 55.83338211 | 37.36618881 | 5 | −74,75 |
| 4 | 55.83424147 | 37.36499766 | 10 | −78,50 |
| 5 | 55.83461404 | 37.36554843 | 5 | −66,20 |

The resulting experimental radar coordinates of $B$ for GPS: latitude 55.83410991, 37.36756219, longitude with an error of 5 m.

## 3.1 Comparing the Experimental Results with the Calculated Values

When calculated in the first approximation consider the directivity pattern uniform, antenna gains believe unit, the accuracy of $RSSI$ believe zero. Below is a sequence of calculation.

(1) Calculate the parameters of the link $K_1 = 7{,}56 \cdot 10^{-4}$ m, and $K_2 = -31{,}2$ dBm the formula (4).

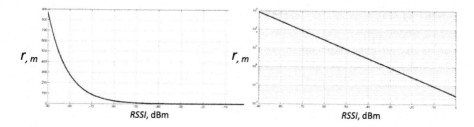

**Fig. 4.** Distant linear characteristics (a) and logarithmic (b) scale

(2) Find the distance $r$ from the beacon by the formula (7) and distant characteristic (Fig. 4).

(3) Solve the system of equations (6) for the unknowns $\phi$ and $\theta$ to obtain the coordinates of the point estimate locator and obtain $\theta \approx 55.83399999°$, $\phi \approx 37.36739999°$.

(4) To evaluate the error in determining the distance from the formulas (11) and (16), and accept: $\Delta\beta_k = 0$ and, therefore, $\Delta r_k = 0$.

(5) Estimate the total error in determining the distance to the beacon by the formula (19), where the error $\delta_k$ of GPS-location beacons given in Table 1.

(6) According to the formula (21) estimate error in determining the coordinates of the locator an average thickness of "fuzzy ring" (interval estimation accuracy coordinate locator) $\Delta x = \Delta d/\sqrt{2} \approx 5\,\text{m}$, $\Delta y = \Delta d/\sqrt{2} \approx 5\,\text{m}$.

The resulting calculated position locator for GPS are as follows: latitude $\theta \approx 55.83399999°$, longitude $\phi \approx 37.36739999°$, with an error of 5 m. The actual discrepancy between calculated and experimental data was $0.0002°$ $(0.72'')$ in latitude and longitude, which corresponds to 16 m. This discrepancy is due to the chaotic traffic nodes network space and a random arrangement of their antennae, which affects the radiation pattern and $RSSI$.

## 4    Conclusions

The proposed approach for the relative navigation allows to determine the geographical coordinates of the decentralized network nodes by the method of multilateration and evaluate the error location.

In the case of location on the plane in the presence of obstacles is recommended to use the space-time filtering and temporal filtering eliminates the effect of noise and space, ranging beacons in their accuracy and reliability, eliminates the effect of remoteness and interference caused by the weight of the adaptive system.

An indication of the importance of traffic nodes should be considered as a change of the distance between them at the time of the signal and its processing. This change in distance caused by the movement of nodes to each other and the finite velocity of propagation and signal processing, causing a regular error location. This error can be neutralized by extrapolation of the trajectories of network nodes.

Held in the field experiment on a specially-designed layouts allow to test these theoretical calculations and experimental results are compared with the calculated values confirmed the efficiency of the proposed method relative navigation.

## References

1. Tanenbaum, Je., Ujezeroll, D.: Komp'juternye seti.: 5-e izd. - SPb.: Piter, p. 960 (2012)
2. Azizov, R.F., Aminev, D.A., Ivanov, I.A., Uvajsov, S.U.: Organizacija svjazi na osnove prioritetov dlja uluchshenija JeMS pri informacionnom obmene v decentralizovannoj seti. Tehnologii jelektromagnitnoj sovmestimosti. 4(47), 5–8 (2013)
3. Azizov, R.F., Aminev, D.A., Uvajsov, S.U.: Programmnaja integracija jelementov uzla decentralizovannoj seti. In: Innovative Information Technologies: Materials of the International Scientific - Practical Conference. Part 3. - Praga, pp. 23–26 (2013)
4. Information Technology - Telecommunications and Information Exchange Between Systems - Local and Metropolitan Area Networks - Specific Requirements - Part 15.4: Wireless Medium Access Control (MAC) and Physical Layer (PHY) Specifications for Low-Rate Wireless Personal Area Networks (WPANs) - Amendment: Alternative Physical Layer Extension to Support the Japanese 950 MHz Band/Note: supplement to ANSI/IEEE 802.15.4-2006 *Approved 28 July 2009
5. De Groote, A.: GSM Positioning Control, p. 13. University of Fribourg, Switzerland (2005)
6. Yamasaki, R., Ogino, A., Tamaki, T., Uta, T., Matsuzawa, N., Kato, T.: TDOA location system for IEEE 802.11b WLAN. In: Proceedings of Wireless Communications and Networking Conference, vol. 4, pp. 2338–2343 (2005)
7. Jonsson, D.K., Olavesen, J.: Estimated accuracy of location in mobile networks using E-OTD. Master thesis in Information and Communication Technology, Grimstad, pp. 68 (2002)
8. Huang, X., Barralet, M., Sharma, D.: Accuracy of location identification with antenna polarization on RSSI. In: Proceedings of the International MultiConference of Engineers and Computer Scientists, IMECS 2009, vol. I. Hong Kong, 18–20 March 2009
9. Huang, T., Chen, Z., Xia, F., Jin, C., Liang, L.: A practical localization algorithm based on wireless sensor networks. In: 2010 IEEE/ACM Int'l Conference on Green Computing and Communications (GreenCom) & Cyber, Physical and Social Computing (CPSCom), pp. 50–54 (2010)
10. http://open-dev.ru — site of the Corp. "Open development"
11. Aminev, D., Azizov, R., Uvaysov, S.: Recommendations for the choice of antenna transceivers of decentralized self-organizing networks. V kn.: Innovacii na osnove informacionnyh i kommunikacionnyh tehnologij: materialy mezhdunarodnoj nauchno-tehnicheskoj konferencii/Otv. red.: I. A. Ivanov; pod obshh. red.: S. U. Uvajsova. M.: MIJeM NIU VShJe, pp. 480–481 (2013)

# Algorithm of State Stationary Probability Computing for Continuous-Time Finite Markov Chain Modulated by Semi-Markov Process

Alexander M. Andronov[1] and Vladimir M. Vishnevsky[2(✉)]

[1] Transport and Telecommunication Institute, Lomonosov, 1, Riga 1019, Latvia
`lora@mailbox.riga.lv`
[2] V.A. Trapeznikov Institute of Control Sciences of Russian Academy of Sciences,
65 Profsoyuznaya Street, Moscow 117997, Russia
`vishn@inbox.ru`

**Abstract.** This paper presents the description of the method for computing of state stationary probabilities for Markovian systems operating in the random environment. These systems are described as semi-Markov processes. The method generalizes some other results in this area.

**Keywords:** Semi-Markov processes · Stationary probability computing algorithm · Random environment

## 1 Introduction

Lately a great attention has been granted to probabilistic models, taking a random environment into consideration [4–8,10]. Generally it can be described by the so-called Markov-modulated processes [1,2,5,11]. Here the random environment is represented as a continuous-time finite Markov chain $X(t)$. For each state of this chain, there exists its own "inner" continuous-time Markov chain $Y(t)$. The state set of all inner Markov chains is the same, but the transition intensities are different.

We consider the following modification: random environment $X(t)$ will be presented by a semi-Markov process [6] instead of the Markov chain.

We present an efficient algorithm of stationary probabilities computing for the states of process $(X(t), Y(t))$. This algorithm allows solving various partial tasks, where a random environment has place [13].

The paper is organized as follows. The considered semi-Markov process $X(t)$ is described in the second section. There is considered an embedded Markov chain that is built on the moments of states changing. The expectation $S^{(i)}$ of the sojourn time is determined for each state $i$, and transition probabilities between states. The third section contains description of inner continuous-time Markov

---

This research was financially supported by the Ministry of Education and Science of the Russian Federation in the framework of the applied research project №14.613.21.0020 of 22.10.2014 (RFMEFI61314X0020).

V. Vishnevsky and D. Kozyrev (Eds.): DCCN 2015, CCIS 601, pp. 167–176, 2016.
DOI: 10.1007/978-3-319-30843-2_18

chain $Y(t)$. The expectations of the sojourn time in the states of $Y(t)$ during time $S^{(i)}$ is induced. Further the embedded discrete-time Markov chain $(X(n), Y(n))$ is considered, which is built on the time moments when semi-Markov process $X(t)$ changes own state: $n = 1, 2, ...$ There a stationary distribution of the sate probabilities is gotten. All these results allow calculate a stationary distribution of the sate probabilities for continuous-time chain $(X(n), Y(n))$ — Sect. 4. A partial case, when continuous-time Markov chain takes place instead of the semi-Markov process, is considered in the Sect. 5. The numerical example illustrates the suggested approach.

## 2  Semi-Markov Process as a Random Environment

Random environment is described by a semi-Markov process $X(t), t > 0$, with $k < \infty$ states $1, 2, ..., k$, and the matrix of transition intensities (hazard function) $\lambda(t) = (\lambda_{i,j}(t))_{k \times k}$ where $\lambda_{i,j}(t)$ is the transition intensity from state $i$ to state $j$ at time moment $t$, if the sojourn time in the state $i$ equals $t$ [6]. It is supposed that $\lambda_{i,j}(t) = 0$ for all $i$ and $t \geq 0$. Let $\Lambda_i(t)$ denote total transition intensity for the state $i$ at time moment $t$: $\Lambda_i(t) = \sum_{j=1}^{k} \lambda_{i,j}(t)$.

Let $S^{(i)}$ be a sojourn time of semi-Markov process in the state $i$. The expectation of this time

$$E(S^{(i)}) = \int_0^\infty \exp\left(-\int_0^t \Lambda_i(\tau)d\tau\right). \tag{1}$$

Let $E(S) = (E(S^{(1)}) \dots E(S^{(k)}))^T$.

Additionally we must known stationary probabilities of embedded Markov chain determined on the moments of states changing. Let $P_{i,j}$ be the one-step transition probability. Then

$$P_{i,j} = \int_0^\infty \lambda_{i,j}(t) \exp\left(-\int_0^t \Lambda_i(\tau)d\tau\right) dt. \tag{2}$$

Let $P = (P_{i,j})$ be the corresponding $k \times k$-matrix. If the stationary probability that new state is $j$ is $r_j$ and $r = (r_1, ..., r_k)$ is corresponding row-vector. Then we have a system of the linear equations

$$r = rP. \tag{3}$$

Further we suppose that distribution $r$ is known.

## 3  Inner Markov Chain

Now we consider continuous-time finite irreducible Markov chain $(Y(t), t > 0)$, having $m < \infty$ states $v = 1, 2, ..., m$. Let us introduce $m \times m$-matrix of transition intensities for $Y(t)$ if the random environment has the state $i$: $M^{(i)} = (M_{v,v^*}^{(i)})_{m \times m}$.

We denote $\mu_{\nu,i}$ an intensity of the output from state $\nu$ without changing the state $i$ of Markov chain: it is a sum of elements of the $\nu$-th row of matrix $M^{(i)}$:
$\mu_{\nu,i} = \sum\limits_{\nu^*=1}^{m} M_{\nu,\nu^*}^{(i)}$. Let $\mu_i = \left(\mu_{1,i}, ..., \mu_{m,i}\right)^T$.

Let $Q_{\nu,\nu^*}^{(i)}(t) = P\{Y(t) = \nu^* | Y(0) = \nu, X(\tau) = i, 0 \leq \tau \leq t\}$, $\nu, \nu^* = 1, ..., m$.

The expression of matrix $Q^{(i)}(t) = \left(Q_{\nu,\nu^*}^{(i)}(t)\right)_{m \times m}$ is very well known [3,9,11].
Let $\chi^{(i)} = \left(\chi_1^{(i)}, ..., \chi_m^{(i)}\right)^T$ and $\eta_1^{(i)}, ..., \eta_m^{(i)}; \tilde{\eta}_1^{(i)}, ..., \tilde{\eta}_m^{(i)}$ are the rows of matrix $N^{-1}$.
Then

$$Q^{(i)}(t) = \sum_{\zeta=1}^{m} \eta_\zeta^{(i)} xp(t\chi_\zeta^{(i)})\tilde{\eta}_\zeta^{(i)}, \ t \geq 0. \tag{4}$$

Let $U_{\nu,\nu^*}^{(i)}$ be a sojourn time of $Y(t)$ in state $\nu^*$ during interval $(0, S^{(i)})$ , if $Y(0) = \nu$ and $X(\tau) = i, \tau \in (0, S^{(i)})$. Let $U^{(i)} = \left(U_{\nu,\nu^*}^{(i)}\right)_{m \times m}$ be the corresponding matrix. Then:

$$E(U^{(i)}) = \int_0^{\infty} \exp\left(-\int_0^t \Lambda_i(\tau)d\tau\right) Q^{(i)}(t)dt$$

$$= \sum_{\zeta=1}^{k} \eta_\zeta^{(i)} \tilde{\eta}_\zeta^{(i)} \int_0^{\infty} \exp\left(-\int_0^t \Lambda_i(\tau)d\tau + t\chi_\zeta^{(i)}\right) dt$$

$$= \eta_1^{(i)} \tilde{\eta}_1^{(i)} \int_0^{\infty} \exp\left(-\int_0^t \Lambda_i(\tau)d\tau\right) dt$$

$$+ \sum_{\zeta=2}^{k} \eta_\zeta^{(i)} \tilde{\eta}_\zeta^{(i)} \int_0^{\infty} \exp\left(\int_0^t -\Lambda_i(\tau)d\tau + t\chi_\zeta^{(i)}\right) dt. \tag{5}$$

Note that expectation (1) can be got from (5). Actually, the first addend of the last row $\eta_1^{(i)} \tilde{\eta}_1^{(i)}$ is the matrix with the same rows. Each row is a vector of stationary distribution of the Markov chain. The sum of row elements equals one. Other addends contain row-vectors $\{\tilde{\eta}_\zeta^{(i)}\}$. The sum of every row elements equals zero. Therefore multiplying matrix $E(U^{(i)})$ from right by unit-vector we get the vector-column which all elements equal the right of (1).

## 4    Embedded Markov-Chain

Now we consider an embedded discrete-time Markov chain $(X(n), Y(n))$, $n = 0, 1, ...$, which is built on the time moments when semi-Markov process $X(t)$ changes own state. Here $X(n)$ is a new state of $X(t)$ immediately after the $n$-th transition. Note that state of $Y(t)$ does not change at time moment $t = S^{(i)}$.

Let $R_{\nu,\nu^*}^{(i,j)}$ be the probability that the new state of $X$ equals $j$: $R_{\nu,\nu^*}^{(i,j)} = P\{Y(t) = \nu^*, X(S^{(i)}) = j | Y(0) = \nu, X(\tau) = i, \tau \in (0, S^{(i)})\}$. Let $R^{(i,j)} = \left(R_{\nu,\nu^*}^{(i,j)}\right)_{m \times m}$ be the corresponding matrix. We have:

$$
\begin{aligned}
R^{(i,j)} &= \int\limits_{0}^{\infty} \lambda_{i,j}(t) \exp\left(-\int\limits_{0}^{t} \Lambda_i(\tau)d\tau\right) dt \\
&= \int\limits_{0}^{\infty} \lambda_{i,j}(t) \exp\left(-\int\limits_{0}^{t} \Lambda_i(\tau)d\tau\right) \sum_{\zeta=1}^{k} \eta_\zeta^{(i)} \exp\left(t\chi_\zeta^{(i)}\right) \tilde{\eta}_\zeta^{(i)} dt \\
&= \sum_{\zeta=1}^{k} \eta_\zeta^{(i)} \tilde{\eta}_\zeta^{(i)} \int\limits_{0}^{\infty} \lambda_{i,j}(t) \exp\left(-\int\limits_{0}^{t} \Lambda_i(\tau)d\tau + t\chi_\zeta^{(i)}\right) dt.
\end{aligned}
\tag{6}
$$

An unique eigenvalue equals zero, let it is $\chi_1$. Another are negative ones. The eigenvector $\nu_1$ corresponding to $\chi_1$, has identical elements, say $c$: $\nu_\bullet = (c, ..., c)^T$. It allows to rewrite 6 as follows:

$$
\begin{aligned}
R^{(i,j)} &= \eta_1^{(i)} \tilde{\eta}_1^{(i)} \int\limits_{0}^{\infty} \lambda_{i,j}(t) \exp\left(-\int\limits_{0}^{t} \Lambda_i(\tau)d\tau\right) dt \\
&+ \sum_{\zeta=2}^{k} \eta_\zeta^{(i)} \tilde{\eta}_\zeta^{(i)} \int\limits_{0}^{\infty} \lambda_{i,j}(t) \exp\left(-\int\limits_{0}^{t} \Lambda_i(\tau)d\tau + t\chi_\zeta^{(i)}\right) dt.
\end{aligned}
\tag{7}
$$

Therefore the one-step transition probabilities $R^{(i,j)}$, $i,j = 1, ..., k$, for embedded chain $(X(n), Y(n))$ are known. Let $h_\nu^{(i)} = P\{X = i, Y = \nu\}$ be the stationary probability of $(X, Y)$, and let $h^{(j)} = (h_1^{(i)}, ..., h_m^{(i)})$. Then

$$
h^{(i)} = \sum_{j=1} h^{(j)} R^{(j,i)}, \quad i = 1, ..., k.
\tag{8}
$$

Further we note that

$$
r_i = h^{(i)} ed(m), \quad i = 1, ..., k,
\tag{9}
$$

where $ed(m)$ is the $m$-column from units.

We solve systems (8) and (9) of linear equations by the following way. Let $W^{(i,j)}$ be the matrix $R(i,j)$ without the last row and the last column, $w(i,j)$ is the last row of the matrix $R(i,j)$ without the last element, $h_{m,i} = z_i$, $\tilde{h}^{(i)} = (h_{1,i}, ..., h_{m-1,i})$, $h^{(i)} = (\tilde{h}^{(i)} z_i)$. Let $I(m)$ be the identical matrix of order $m$. Then we have from (8) and (9):

$$
r_j = \tilde{h}^{(j)} ed(m-1) + z_j,
$$
$$
z_j = r_j - \tilde{h}^{(j)} ed(m-1),
$$
$$
\tilde{h}^{(j)} = \sum_{j \neq i} \left(\tilde{h}^{(j)} W^{(j,i)} + z_j w^{(j,i)}\right) = \sum_{j \neq} \left(\tilde{h}^{(j)} W^{(j,i)} + \left(r_j - \tilde{h}^{(j)} ed(m-1)\right) w^{(j,i)}\right),
$$
$$
\tilde{h}^{(i)} = \sum_{j \neq i} r_j w^{(j,i)} + \sum_{j \neq i} \left(\tilde{h}^{(j)} \left(W^{(j,i)} - ed(m-1) w^{(j,i)}\right)\right).
$$

This system can be solved by iterative procedure of recalculation of vectors $\{\tilde{h}^{(i)}\}$:

$$\tilde{h}^{(i)} = \sum_{j \neq i} r_j w^{(j,i)} + \sum_{j \neq i} \left( \tilde{h}^{(j)} \left( W^{(j,i)} - ed(m-1)w^{(j,i)} \right) \right), \quad i = 1, ..., k. \quad (10)$$

Let us remind, that $h^{(i)} = \left( \tilde{h}^{(i)} z_i \right) = \left( \tilde{h}^{(i)} r_i - \tilde{h}^{(i)} ed(m-1) \right)$.

## 5    Stationary Probabilities of the States

Let us remind that the embedded chain from Sect. 2 has the stationary distribution of state probabilities $r = (r_1, ..., r_k)$. Further $U_{\nu,\nu^*}^{(i)}$ is a sojourn time of $Y(t)$ in state $\nu^*$ during interval $(0, U^{(i)})$, if $Y(0) = \nu$ and $X(\tau) = i$, $\tau \in (0, U^{(i)})$. Expectations of $\{U_{\nu,\nu^*}^{(i)}\}$ are calculated by formula (5).

The stationary distribution of probabilities for the couple $(X, Y)$ is calculated as [6]

$$P\{X = i, Y = \nu^*\} = \frac{1}{rE(S)} \sum_{\nu=1}^{m} h_{\nu}^{(i)} E\left(U_{\nu,\nu^*}^{(i)}\right). \quad (11)$$

Let us compare the described approach with the classical one. The last consider $km \times km$-matrix of transition intensities of Markov chain $(X(t), Y(t))$. The stationary probabilities of the states are calculated with the aid of the corresponding inverse matrix. It is known that the inverse of matrices of big order isn't a simple computing procedure. Our approach works with $m \times m$-matrices only.

## 6    Partial Case: Markov Chain Instead of the Semi-Markov Process

In this case transition intensity $\{\lambda_{i,i}(t)\}$ do not depend on the time $t$. Then process $X(t), t > 0$, is continue-time Markov chain with $k < \infty$ states $1, 2, \cdots, k$, and the matrix of transition intensities $\lambda(t) = (\lambda_{i,j}(t))_{k \times k}$. As earlier we suppose that $\lambda_{i,i} = 0$ for all i. Let $\Lambda_i$ denote total transition intensity for the state $i$ at time moment $t$: $\Lambda_i = \sum_{j=1}^{k} \lambda_{i,j}$.

The expectation of the sojourn time $S^{(i)}$ of Markov chain $X(t)$ in the state $i$ equal $1/\lambda_i$. Let $E(S) = \left( E(S^{(1)}) \ldots E(S^{(k)}) \right)^T$

Further we must known stationary probabilities of embedded Markov chain determined on the moments of states changing. The one-step transition probability $u_{i,j}$ from state $i$ to state $j$ is of the form

$$u_{i,j} = \frac{\lambda_{i,j}}{\Lambda_i}, i, j = 1, \ldots, k. \quad (12)$$

Let $U = (u_{i,j})$ be the corresponding $k \times k$-matrix, the stationary probability of state $j$ be $r_j$ and $r = (r_1, \ldots, r_k)$ be the corresponding row-vector. Then we have a system of the linear equations

$$r = rU. \tag{13}$$

Further we suppose that distribution $r$ is known, for example by this way. Let us consider eigenvalues and eigenvectors of matrix $U - I(k)$; let $\Psi = (\psi_1, \ldots, \psi_k)$ be a matrix from the eigenvectors, $\Psi^{-1}$ be corresponding inverse matrix with rows $\tilde{\psi}_1, \ldots, \tilde{\psi}_k$. If the zero-eigenvalue has number $i^*$, then all rows of matrix $\psi_{i^*} \tilde{\psi}_{i^*}$ are vector $r$.

We have the following modification of formula (5):

$$E(U^{(i)}) = \sum_{\zeta=1}^{k} \eta_\zeta^{(i)} \tilde{\eta}_\zeta^{(i)} \int_0^\infty \exp(-\Lambda_i t + t\chi_\zeta^{(i)}) dt = \sum_{\zeta=1}^{k} \eta_\zeta^{(i)} \tilde{\eta}_\zeta^{(i)} \frac{1}{\Lambda_i - \chi_\zeta^{(i)}}. \tag{14}$$

Formula (6) has the following view:

$$R^{(i,j)} = \sum_{\zeta=1}^{k} \eta_\zeta^{(i)} \tilde{\eta}_\zeta^{(i)} \int_0^\infty \lambda_{i,j} \exp(-\Lambda_i t + t\chi_\zeta^{(i)}) dt = \sum_{\zeta=1}^{k} \eta_\zeta^{(i)} \tilde{\eta}_\zeta^{(i)} \frac{\lambda_{i,j}}{\Lambda_i - \chi_\zeta^{(i)}}. \tag{15}$$

Formulas (8) – (10) remain the same ones. Here $h_\nu^{(i)} = P\{X = i, Y = \nu\}, i = 1, \ldots, k, \nu = 1, \ldots, m$, are stationary probabilities of the states immediately after a jump of Markov chain $X(t)$. The vectors $h^{(i)} = \left(h_1^{(i)}, \ldots, h_m^{(i)}\right), i = 1, \ldots, k$, are calculated iteratively as it was described earlier.

Formula (11) for the stationary distribution of probabilities for the couple $(X, Y)$ becomes simpler:

$$P\{X = i, Y = \nu^*\} = \left(\sum_{i=1}^{k} r_i \frac{1}{\lambda_i}\right)^{-1} \sum_{\nu=1}^{m} h_\nu^{(i)} E(U_{\nu,\nu^*}^{(i)}). \tag{16}$$

## 7   Numerical Example

Considered semi-Markov process $X(t)$ has $k = 3$ states with the following transition intensities for the states.

A sojourn time of the $1^{st}$ state has Weibull distribution with parameters $a$ and $c$. The corresponding rate is expressed as ([12], p. 400)

$$\lambda V(x; a, c) = \frac{c}{a} \left(\frac{x}{a}\right)^{c-1}, x \geq 0.$$

The intensities of transition to states 2 and 3 are the following: $\lambda_{1,2}(x) = \lambda V(x; 2, 2), \lambda_{1,3}(x) = \lambda V(x; 3, 3), x \geq 0$.

A sojourn time of the $2^{nd}$ state has Erlang distribution with parameter $\lambda$ and $m$. The corresponding rate is expressed as

$$\lambda E(x; \lambda, m) = \frac{1}{(m-1)!} \lambda (\lambda x)^{m-1} \exp(-\lambda x), x \geq 0.$$

The intensities of transition to states 1 and 3 are the following: $\lambda_{2,1}(x) = \lambda E(x; 3, 2), \lambda_{2,3}(x) = \lambda E(x; 2, 2), x \geq 0$.

A sojourn time of the $3^{th}$ state has Simpson distribution with parameter $b$. The corresponding rate is expressed as

$$\lambda S(x; \lambda, m) = \begin{cases} \frac{4x}{b^2 - 2x^2}, & 0 \leq x < b/2, \\ \frac{2}{b-x}, & b/2 < x < b. \end{cases}$$

The intensities of transition to states 1 and 2 are the following: $\lambda_{3,1}(x) = \lambda S(x; 4), \lambda_{3,2}(x) = \lambda S(x; 6), 0 \leq x < b$.

The expectation of sojourn time of semi-Markov process in the state $i$ is calculated by formula (1). Vector of these values for different states is presented below:

$$E(S) = \left( E(S^{(1)}) \cdots E(S^{(k)}) \right)^T = \left( 1.565\ 0.496\ 1.778 \right)^T. \tag{17}$$

Matrix $P$ of one-step transition probabilities is calculated according to formula (2):

$$P = \begin{pmatrix} 0 & 0.753 & 0.247 \\ 0.648 & 0 & 0.352 \\ 0.743 & 0.257 & 0 \end{pmatrix}.$$

The eigenvalues $\{\nu_i\}$ and eigenvectors of matrix $P - I(3)$ are the following:

$$\nu = \begin{pmatrix} 0 \\ -1.609 \\ -1.391 \end{pmatrix}, \Psi = \begin{pmatrix} 0.686 & 0.563 & -0.213 \\ 0.616 & -0.793 & 0.789 \\ 0.386 & 0.230 & -0.576 \end{pmatrix}, \Psi^{-1} = \begin{pmatrix} 0.592 & 0.592 & 0.592 \\ 1.419 & -0.673 & -1.447 \\ 0.964 & 0.128 & -1.917 \end{pmatrix}.$$

We see that the zero eigenvalue has the first number $(i^* = 1)$, therefore

$$\Psi_{i^*} = \Psi_1 = \left( 0.686\ 0.616\ 0.386 \right)^T, \tilde{\Psi}_{i^*} = \tilde{\Psi}_1 = \left( 0.592\ 0.592\ 0.592 \right)^T.$$

The column of matrix $\Psi_{i^*} \tilde{\Psi}_{i^*}$ contain the same vector $r = \left( 0.406\ 0.365\ 0.229 \right)$ that gives stationary probabilities of states:

$$r = \left( 0.406\ 0.365\ 0.229 \right). \tag{18}$$

Further we set $m = 3$ for the state number of the inner Markov chain $Y(t)$. The $m \times m$-matrices of transition intensities for $Y(t)$ and for different state $i$ of the random environment are the following:

$$M^{(1)} = \left( M^{(1)}_{\nu,\nu^*} \right) = \begin{pmatrix} 0 & 0 & 0.3 \\ 0.1 & 0 & 0.1 \\ 0.3 & 0.1 & 0 \end{pmatrix},$$

$$M^{(2)} = \left( M^{(2)}_{\nu,\nu^*} \right) = \begin{pmatrix} 0 & 0.2 & 0.3 \\ 0.2 & 0 & 0.1 \\ 0.2 & 0.1 & 0 \end{pmatrix},$$

$$M^{(3)} = \left( M^{(3)}_{\nu,\nu^*} \right) = \begin{pmatrix} 0 & 0.4 & 0.1 \\ 0.3 & 0 & 0.2 \\ 0 & 0.2 & 0 \end{pmatrix}.$$

Formula (4) gives the nonstationary probabilities of $Y(t)$'s states $Q^{(i)}_{\nu,\nu^*}(t) = P\{Y(t) = \nu^* \mid Y(0) = \nu, X(\tau) = i, 0 \le \tau \le t\}, \nu, \nu^* = 1, \ldots, m$. For example, we have the following eigenvalues and eigenvectors for the second state:

$$\nu = \begin{pmatrix} 0 \\ -0.741 \\ -0.459 \end{pmatrix}, N = \begin{pmatrix} -0.577 & -0.722 & 0.160 \\ -0.577 & -0.143 & -0.811 \\ -0.577 & 0.677 & 0.563 \end{pmatrix}.$$

The expectation $E\left( U^{(i)}_{\nu,\nu^*} \right)$ of the sojourn time of $Y(t)$ in state $\nu^*$ till a departure of the state $i$, if $Y(0) = \nu$ and $X(0) = i$ is calculated by formula (5). For example, for the second state of $X(t)$ we have the following matrix of $\{E\left( U^{(i)}_{\nu,\nu^*} \right)\}$:

$$E\left( U^{(i)} \right) = E\left( U^{(i)}_{\nu,\nu^*} \right) = \begin{pmatrix} 0.423 & 0.030 & 0.043 \\ 0.029 & 0.449 & 0.017 \\ 0.029 & 0.017 & 0.450 \end{pmatrix}. \tag{19}$$

Note that the element sum of each row equals 0.497, it is the expectation of sojourn time at state $i = 2$ of $X(t)$ during single visit (see formula (1)).

Now we consider two-dimensional Markov chain $(X(t), Y(t)), t > 0$. For example, we present the matrix (6) for the states $i = 2, j = 1$ and $j = 3$ of embedded Markov chain:

$$R^{(2,1)} = \left( R^{(2,1)}_{\nu,\nu^*} \right) = \begin{pmatrix} 0.522 & 0.051 & 0.043 \\ 0.050 & 0.567 & 0.017 \\ 0.050 & 0.029 & 0.568 \end{pmatrix}, R^{(2,3)} = \left( R^{(2,3)}_{\nu,\nu^*} \right) = \begin{pmatrix} 0.281 & 0.029 & 0.042 \\ 0.028 & 0.306 & 0.017 \\ 0.028 & 0.017 & 0.307 \end{pmatrix}.$$

Let us note that element sum of each row for the first matrix equals 0.648 and the same for the second matrix equals 0.352.

Further we illustrate the above described iterative procedure (8) using matrices. Matrices $W^{(2,1)}, W^{(2,3)}$ and vectors $w^{(2,1)}, w^{(2,3)}$ are the following:

$$W^{(2,1)} = \begin{pmatrix} 0.522 & 0.051 \\ 0.050 & 0.567 \end{pmatrix}, W^{(2,3)} = \begin{pmatrix} 0.281 & 0.029 \\ 0.028 & 0.306 \end{pmatrix},$$

$$w^{(2,1)} = \begin{pmatrix} 0.050 & 0.029 \end{pmatrix}, w^{(2,3)} = \begin{pmatrix} 0.028 & 0.017 \end{pmatrix}.$$

Initial vectors $\tilde{h}^{(i)} = (h_{1,i}, h_{2,i})$, $i = 1, 2, 3$, are presented by matrix

$$\tilde{H} = \begin{pmatrix} \tilde{h}^{(1)} \\ \tilde{h}^{(2)} \\ \tilde{h}^{(3)} \end{pmatrix} = \begin{pmatrix} 0.196 & 0.100 \\ 0.100 & 0.200 \\ 0.100 & 0.029 \end{pmatrix}$$

After 5 iterations, values of $\{\tilde{h}^{(i)}\}$ become constants, namely

$$\tilde{H} = \begin{pmatrix} \tilde{h}^{(1)} \\ \tilde{h}^{(2)} \\ \tilde{h}^{(3)} \end{pmatrix} = \begin{pmatrix} 0.111 & 0.130 \\ 0.105 & 0.097 \\ 0.092 & 0.134 \end{pmatrix}$$

The stationary probabilities $h_\nu^{(i)} = P\{X = i, Y = \nu\}$ for the embedded Markov chain are calculated as $h^{(i)} = \left( \tilde{h}^{(i)} r_i - \tilde{h}^{(i)} ed(m-1) \right)$. They are presented by matrix

$$\tilde{H} = \begin{pmatrix} h^{(1)} \\ h^{(2)} \\ h^{(3)} \end{pmatrix} = \begin{pmatrix} 0.111 & 0.130 & 0.165 \\ 0.105 & 0.097 & 0.163 \\ 0.092 & 0.134 & 0.003 \end{pmatrix} \tag{20}$$

The stationary probability distribution of the states for the continuous-time Markov chain $(X(t), Y(t))$, $t > 0$, are calculated by formula (11). Terms of this formula are determined by 18–20. The result is presented by the matrix

$$P = (P_{i,\nu}) = (P\{X = i, Y = \nu\}) = \begin{pmatrix} 0.168 & 0.155 & 0.197 \\ 0.043 & 0.040 & 0.065 \\ 0.122 & 0.165 & 0.046 \end{pmatrix}.$$

Addition of the rows and the columns give us the marginal probability distribution for the states of semi-Markov process $X(t)$ and Markov chain $Y(t)$:

$$(P\{X = i\}) = (0.333 \ 0.360 \ 0.308), (P\{Y = \nu\}) = \begin{vmatrix} 0.520 \\ 0.148 \\ 0.333 \end{vmatrix}.$$

## 8  Conclusion

The described method gives a uniform approach to computing of the state stationary probabilities for Markovian systems operating in random environment. The last are described as semi-Markov processes. The method generalizes some another results in this area.

**Acknowledgment.** This research was financially supported by the Ministry of Education and Science of the Russian Federation in the framework of the applied research project №14.613.21.0020 of 22.10.2014 (RFMEFI61314X0020).

# References

1. Andronov, A.M.: Markov-modulated birth-death processes. Autom. Control Comput. Sci. **45**(3), 123–132 (2011)
2. Andronov, A., Gertsbakh, I.B.: Signatures in Markov-modulated processes. Stoch. Models **30**, 1–15 (2014)
3. Bellman, R.: Introduction to Matrix Analysis. Mcgraw-Hill Book Company Inc., New York (1969)
4. Kim, C.S., Dudin, A., Klimenok, V., Khramova, V.: Erlang loss queueing system with butch arrivals operating in a random environment. Comput. Oper. Res. **36**(3), 674–697 (2009)
5. Kim, C.S., Klimenok, V., Mushko, V., Dudin, A.: The BMAP/PH/N retrial queueing system operating in Markovian random environment. Comput. Oper. Res. **37**(7), 1228–1237 (2010)
6. Grabski, F.: Semi-Markov Processes: Application in System Reliability and Maintenance. Elsevier, Amsterdam (2014)
7. Kopoci'nska, I., Kopoci'nski, B.: On system reliability under random load of elements. Aplicationes Math. **XVI**(1), 5–14 (1980)
8. Kopoci'nska, I.: The reliability of an element with alternating failure rate. Zastosowania Matematki (Aplicationes Math.) **XVIII**(2), 187–194 (1984)
9. Kijima, M.: Markov Processes for Stochastic Modelling. Chapman & Hall, London (1997)
10. Leri, M.: Forest fire model for configuration graphs in a random environment. In: Rasch, D., Melas, V., Pilz, J., Moder, K., Spangl, B. (eds.) Eighth International Workshop on Simulation. Book of Abstracts. Institute of Applied Statistics and Computing, University of Natural Resources and Life Sciences, Vienna, pp. 125–126 (2015). ISBN 978-3-900932-28-2
11. Pacheco, A., Tang, L.C., Prabhu, N.U.: Markov-Modulated Processes & Semiregenerative Phenomena. World Scientific, New Jersey (2009)
12. Sleeper, A.: Six Sigma Distribution Modeling. McGraw Hill, New York (2007)
13. Vishnevsky, V.M., Semenova, O.V., Sharov, S.Y.: Modeling and analysis of a hybrid communication channel. Autom. Remote Control **72**, 345–352 (2013)

# Approximate Description of Dynamics of a Closed Queueing Network Including Multi-servers

Svetlana Anulova[(⊠)]

V.A. Trapeznikov Institute of Control Sciences of Russian Academy of Sciences,
65 Profsoyuznaya Street, Moscow, Russia
anulovas@ipu.rssi.ru
http://www.ipu.ru/

**Abstract.** We investigate a closed network consisting of two multi-servers with $n$ customers. Service requirements of customers at a server have a common cumulative distribution function. The state of the network is described by the following state parameters: for each multi-server and for the queue empirical measures of the age of customers being serviced/waiting in the queue multiplied by $n^{-1}$. The approximation of a single multi-server dynamics is currently studied by famous scientists H. Kaspi, K. Ramanan, W. Whitt et al. We find approximation for a network, but only in discrete time.

A motivation for studying such systems is that they arise as models of computer data systems and call centers.

**Keywords:** Multi-server queues · GI/G/n queue · Fluid limits · Mean-field limits · Strong law of large numbers · Measure-valued processes · Call centers

## 1 Introduction

### 1.1 Review of Investigated Contact Centers Models

In the last ten years an extensive research in mathematical models for telephone call centers has been carried out, cf. [2–6,8–17]. The object has been expanded to more general customer contact centers (with contact also made by other means, such as fax and e-mail). One of important relating questions is the dynamics of multi-server queues with a large number of servers. In order to describe the object efficiently the state of the model must include: (1) for every customer in the queue the time that he has spent in it and (2) for every customer in the multi-server the time that he has spent after entering the service area, that is being received by one of the available servers.

This work was supported by RFBR grant No. 14-01-00319 "Asymptotic analysis of queueing systems and nets".

© Springer International Publishing Switzerland 2016
V. Vishnevsky and D. Kozyrev (Eds.): DCCN 2015, CCIS 601, pp. 177–187, 2016.
DOI: 10.1007/978-3-319-30843-2_19

The focus of research was on multi-server queues with a large number of servers, because it is typical of contact centers. For such queues were found fluid limits with the number of servers tending to infinity. Notice that such a limit is a deterministic fumction of time with values in a certain measure space, or in a space containing such a component. These developed deterministic fluid models provided simple first-order performance descriptions for multi-server queues under heavy loads.

## 1.2    A New Model for Contact Centers

We suggest here a more suitable model for contact centers. The number of customers is fixed. Customers may be situated in two states: normal and failure. There is a multi-server which repairs customers in the failure state. The repair time/the time duration of a normal state is a random variable, independent and identically distributed for all customers. Now "the arrival process" in the multi-server does not correspond to that of the previous $G/GI/s + GI$ model. For a large number of customers and a suitable number of servers calculate the number of current failures, so much as an approximation. This is a continuation of our work [1], where a single multi-server in a network was functioning.

We confine ourselves to a discrete time model. W. Whitt has written a very interesting seminal article [15], in a simple discrete case. About 150 authors have cited it and made generalizations to the continuous time. But their results do not enclose Whitt's discrete time ones. Walsh Zuñiga [12] with his results most close to discrete time admits only discrete time service but his arrival process is continuous.

In Whitt's article [15] the idea of the convergence proof is true and very lucid, and the proof is clearly presented. But Whitt has not covered all cases in his proofs, and at that Whitt does not accurately point to it. Only the main case is examined: the number of serviced customers is not zero and does not exceed the number of customers in the queue. Also when he proves convergence of the number of customers being served for a given time $b(t, k)$ in (6.33) he omits the case $b_s(n - 1, k - 1)$ tends to zero. We have transferred his proof technique to our new network model and filled all these misses.

## 1.3    Problem Origin

Consider a closed network consisting of $n$ customers. They may be situated in two states: normal and failure. A multi-server repairs customers in the failure state. The repair time (resp., the time duration of a normal state) is a random variable, independent and identically distributed for all customers. For a large number of customers and a suitable number of servers we shall calculate the number of current failures, so much as an approximation.

Now we give a rigorous description of this model.

Consider a closed network consisting of $n$ customers and two multi-servers. Multi-server 1 (further denoted MS1) consists of $n$ servers (for the customers in the normal state), the time they service a customer has distribution $G^1$. Multi-server 2 (further denoted MS2) consists of $s_n n$ servers with a number $s_n \in (0, 1)$ (for the customers in the failure state), the time they service a customer has distribution $G^2$. The distributions $G^1, G^2$ are discrete: they are concentrated on $\{1, 2, \ldots\}$. Service times are independent for both servers and all customers. We will investigate the behavior of the net as $n \to \infty$, namely, we shall establish a stochastic-process fluid limit. It will be done only in a special case: discrete time $t = 0, 1, 2, \ldots$

We begin with a simple example of functioning of this network.

*Example 1.* Let at time $t = 0$ all $n$ customers be in a normal state. Each customer switches over to the failure state according to the distribution function $G^1$ and tries to enter multi-server 2. The early failure customers can do it, but with time growing multi-server 2 may become fully occupied. Then the failure customers create a queue, waiting for the first available server in multi-server 2. Recall that a server becomes afresh available with time distribution $G^2$.

In this example and everywhere further we demand:

**Assumption 1.** *Customers are served in order of their arrival to MS2 or to its queue (FCFS) by the first available server.*

Denote the number of customers at a moment $t = 0, 1, \ldots$ in MS1 (resp., MS2) by $B_n^1(t)$ (resp., $B_n^2(t)$) and the number of customers in the queue $Q_n(t) = n - B_n^1(t) - B_n^2(t)$. These quantities must be defined more exactly. Namely,

$$B_n^i(t) = \sum_{k=0}^{\infty} b_n^i(t, k), \ i = 1, 2, \ \text{and} \ Q_n(t) = \sum_0^{\infty} q_n(t, k)$$

with $b_n^i(t, k)$ being the number of customers in the multi-server $i$ at the moment $t$ who have spent there time $k$, $i = 1, 2$, and $q_n(t, k)$ being the number of customers in the queue at the moment $t$ who have been there precisely for time $k$. $b_n^i(t, k)$ may also be interpreted as the number of busy servers at time $t$ in the multi-server i that are serving customers that have been in service precisely for time $k$, $i = 1, 2$.

At the same time moment $t \in \{1, 2, \ldots\}$ multiple events can take place, so we have to specify their order.

We must create a fictitious queue for the MS1—in fact this multi-server is so large ($n$ servers), that any customer of the whole quantity $n$ trying to enter the MS1 at once finds a free server in it.

At the time moment $t$ the parameters $b^1, b^2, q$ are taken from the previous time $t - 1$ and processed to the current situation.

For both multi-servers:

– first, customers in service are served;
– second, the served customers move to another multi-server queue, to the end of it;

– third, waiting customers in queue move into service of the multi-server according to Assumption 1.

Customers enter service in MS2 whenever a server is available, so that the system is work-conserving; i.e. we assume that $Q_n(t) = 0$ whenever $B_n^2(t) < s_n n$, and that $B_n^2(t) = s_n n$ whenever $Q_n(t) > 0$, $t = 0, 1, 2, \ldots$. This condition can be summarized by the equation

$$(s_n - B_n^2(t)/n)Q_n(t) = 0 \text{ for all } t \text{ and } n.$$

**Notations.** Let $\sigma_n^i(t)$ be the number of service completions in MSi at time moment $t = 1, 2, \ldots$, $i = 1, 2$. Denote for $k = 1, 2, \ldots$ $G^{i;c}(k) = 1 - G^i(k)$ and $g^i(k) = G^i(k) - G^i(k-1)$, $i = 1, 2$. Symbol $\Rightarrow$ means convergence of the network state characteristics to a constant in probability as the index $n$ denoting the number of cusomers tends to infinity.

**Theorem 1 (The Discrete-Time Fluid Limit).** *Suppose that for each $n$, the system is initialized with workload characterized by nonnegative-integer-valued stochastic processes*

$$b_n^i(0, k), \ i = 1, 2, \ and \ q_n(0, k), \ k = 0, 1, 2, \ldots,$$

*satisfying*

$$B_n^1(0) + B_n^2(0) + Q_n(0) = n, \tag{1}$$

$$B_n^2(0) \le s_n n, \ and \ (s_n n - B_n^2(0))Q_n(0) = 0 \tag{2}$$

*for each $n$ w.p.1. Suppose that $s_n \to s \in (0, 1)$ and*

$$\frac{b_n^i(0, k)}{n} \Rightarrow b^i(0, k), \ i = 1, 2, \ and \ \frac{q_n(0, k)}{n} \Rightarrow q(0, k) \ for \ k = 0, 1, 2, \ldots \tag{3}$$

*as $n \to \infty$, where $s$ is a constant and $b^i(0, k), i = 1, 2$, and $q(0, k)$ are deterministic functions. Moreover, suppose that for each $\epsilon > 0$ and $\eta > 0$, there exists an integer $k_0$ such that for $n = 1, 2, \ldots$*

$$\mathbf{P}\left(\sum_{k=k_0}^{\infty} \frac{b_n^i(0, k)}{n} > \epsilon\right) < \eta, \ i = 1, 2, \ and \ \mathbf{P}\left(\sum_{k=k_0}^{\infty} \frac{q_n(0, k)}{n} > \epsilon\right) < \eta. \tag{4}$$

*Then, as $n \to \infty$,*

$$\frac{b_n^i(t, k)}{n} \Rightarrow b^i(t, k), \ i = 1, 2, \tag{5}$$

$$\frac{q_n(t, k)}{n} \Rightarrow q(t, k), \tag{6}$$

$$\frac{\sigma_n^i(t)}{n} \Rightarrow \sigma^i(t), \ i = 1, 2, \tag{7}$$

*for each $t \ge 1$ and $k \ge 0$, where $(b^1, b^2, q, \sigma^1, \sigma^2)$ is a vector of deterministic functions (all with finite values).*

*Further, for each* $t = 0, 1, \ldots$

$$\frac{B_n^i(t)}{n} \equiv \frac{\sum_{k=0}^{\infty} b_n^i(t, k)}{n} \Rightarrow B^i(t) \equiv \sum_{k=0}^{\infty} b^i(t, k), \ i = 1, 2, \tag{8}$$

$$\frac{Q_n(t)}{n} \equiv \frac{\sum_{k=0}^{\infty} q_n(t, k)}{n} \Rightarrow Q(t) \equiv \sum_{k=0}^{\infty} q(t, k), \tag{9}$$

*with*

$$B^1(t), B^2(t), Q(t) \geq 0, \ B^1(t) + B^2(t) + Q(t) = 1, \tag{10}$$

$$B^2(t) \leq s, \ and \ (s - B^2(t))Q(t) = 0. \tag{11}$$

*The evolution of the vector* $(b^1, b^2, q, \sigma^1, \sigma^2)(t)$, $t = 0, 1, 2 \ldots$, *proceeds with steps of* $t$ *in the following way. As we go from time* $t - 1$ *to* $t$, *there are two cases, depending on whether* $B^2(t - 1) = s$ *or* $B^2(t - 1) < s$.

*Case 1.* $B^2(t - 1) = s$. *In this first case, after moment* $t - 1$ *asymptotically all servers are busy and in general there may be a positive queue. In this case,*

$$\sigma^i(t) = \sum_{k=1}^{\infty} b^i(t - 1, k - 1) \frac{g^i(k)}{G^{i;c}(k - 1)}, \tag{12}$$

$$b^i(t, k) = b^i(t - 1, k - 1) \frac{G^{i;c}(k)}{G^{i;c}(k-1)}, \ k = 1, 2, \ldots, \ i = 1, 2, \tag{13}$$

$$b^1(t, 0) = \sigma^2(t), \tag{14}$$

$$b^2(t, 0) = \min\{\sigma^2(t), Q(t - 1) + \sigma^1(t)\}, \tag{15}$$

*and finally* $q$ *is determined with the help of an intermediate queue* $q'$,

$$q'(t, 0) = \sigma^1(t), \ q'(t, k) = q(t - 1, k - 1), \ k = 1, 2, \ldots : \tag{16}$$

$$if \ \sigma^2(t) = 0 \ then \ q(t, k) = q'(t, k), \ k = 0, 1, \ldots, \tag{17}$$

$$if \ \sigma^2(t) \geq \sum_{k=0}^{\infty} q'(t, k) \ then \ q(t, k) = 0, \ k = 0, 1, \ldots, \tag{18}$$

$$if \ 0 < \sigma^2(t) < \sum_{k=0}^{\infty} q'(t, k) \ then \ with \tag{19}$$

$$c(t) = \min\{i \in \{0, 1, \ldots\} : \sum_{k=i}^{\infty} q'(t, k) \leq \sigma^2(t)\}, \tag{20}$$

$$q(t, k) = \begin{cases} 0 \ for \ k \geq c(t), \\ \sum\limits_{i=c(t)-1}^{\infty} q'(t, i) - \sigma^2(t) \ for \ k = c(t) - 1, \\ q'(t, k) \ for \ k < c(t) - 1. \end{cases}$$

*Case 2.* $B^2(t - 1) < s$. *In this second case, after the time moment* $t - 1$ *asymptotically all servers are not busy so that there is no queue. As in the first case, Eqs. (12), (13), and (14) hold. Instead of (15),*

$$b^2(t, 0) = \min\{s - B^2(t - 1) + \sigma^2(t), \sigma^1(t)\}. \tag{21}$$

*Then,*

$$q(t, k) = 0 \ for all \ k > 0 \ and \ q(t, 0) = \sigma^1(t) - b^2(t, 0). \tag{22}$$

**Proof.** For $t = 0$ conditions (3) and (4) imply the convergence presented in (8) and (9), and conditions (1), (2) provide properties (10), (11). Thus the proof of the Theorem is reduced to the following Lemma:

**Lemma 1.** *Suppose for a given* $t \in \{1, 2, \ldots\}$ *Eqs.* (5), (6) *hold for* $t - 1$. *Then all the statements of the Theorem hold for* $t$.

We divide the proof of this Lemma to several Lemmas below.

**Lemma 2.** *Fix* $n \in \{1, 2, \ldots\}$. *If for given* $\epsilon > 0$ *and* $\eta > 0$ *and integer* $k_0$ *holds*

$$\mathbf{P}(\sum_{k=k_0}^{\infty} \frac{b_n^i(0, k)}{n} > \epsilon) < \eta, \ i = 1, 2, \ and \ \mathbf{P}(\sum_{k=k_0}^{\infty} \frac{q_n(0, k)}{n} > \epsilon) < \eta, \tag{23}$$

*then for every* $t = 1, 2, \ldots$

$$\mathbf{P}(\sum_{k=k_0+t}^{\infty} \frac{b_n^i(t, k)}{n} > \epsilon) < \eta, \ i = 1, 2, \ and \ \mathbf{P}(\sum_{k=k_0+t}^{\infty} \frac{q_n(t, k)}{n} > \epsilon) < \eta. \tag{24}$$

*Proof.* Evidently, for any $t = 1, 2, \ldots, \ k = 0, 1, \ldots$ w.p.1

$$b_n^i(t, k + t) \le b_n^i(0, k) \ and \ q_n(t, k + t) \le q_n(0, k).$$

**Corollary 1.** *Condition* (4) *holds not only for time* $t = 0$, *but for any time* $t = 1, 2, \ldots$, *i.e., for each* $t, \epsilon > 0$ *and* $\eta > 0$, *there exists an integer* $k_0$ *such that for* $n = 1, 2, \ldots$

$$\mathbf{P}(\sum_{k=k_0}^{\infty} \frac{b_n^i(t, k)}{n} > \epsilon) < \eta, \ i = 1, 2, \ and \ \mathbf{P}(\sum_{k=k_0}^{\infty} \frac{q_n(t, k)}{n} > \epsilon) < \eta. \tag{25}$$

In a network consisting of $n \in \{1, 2, \ldots\}$ customers denote by $\sigma_n^i(t, k)$ the number of customers served in MSi, $i = 1, 2$, at time moment $t$ who had been in service for time $k$ at this time moment $t$, for $t, k \in \{1, 2, \ldots\}$.

**Lemma 3.** *If for given* $t, k \in \{1, 2, \ldots\}$

$$\frac{b_n^i(t - 1, k - 1)}{n} \Rightarrow b^i(t - 1, k - 1),$$

*then*

$$\frac{\sigma_n^i(t, k)}{n} \Rightarrow b^i(t - 1, k - 1)\frac{g^i(k)}{G^{i;c}(k - 1)}, \ k = 1, 2, \ldots, \ i = 1, 2,$$

$$\frac{b_n^i(t, k)}{n} \Rightarrow b^i(t, k) = b^i(t - 1, k - 1)\frac{G^{i;c}(k)}{G^{i;c}(k - 1)}, \ k = 1, 2, \ldots, \ i = 1, 2.$$

*Proof.* Set $i = 1, 2$. For each $n, t, k \geq 1$, we can represent $\sigma_n^i(t, k)$ and $b_n^i(t, k)$ as random sums of IID Bernoulli random variables; in particular,

$$\sigma_n^i(t, k - 1) = \sum_{i=1}^{b_n^i(t-1, k-1)} X_i \tag{26}$$

and

$$b_n^i(t, k) = \sum_{i=1}^{b_n^i(t-1, k-1)} (1 - X_i), \tag{27}$$

where $X_i$ assumes the value 1 if the $i$th customer among those in service at time moment $t - 1$ that have been in the system for time $k - 1$ is served at time moment $t$, and assumes the value 0 otherwise. Thus, $X_i, i \geq 1$, is a sequence of IID random variables with

$$\mathbf{P}(X_1 = 0) = \frac{g^i(k)}{G^{i;c}(k - 1)}, \ k = 1, 2, \ldots, \ i = 1, 2.$$

Apply now Appendix Lemma 5.

**Corollary 2.** *If for given* $t \in \{1, 2, \ldots\}, i \in \{1, 2\}$

$$\frac{b_n^i(t - 1, k - 1)}{n} \Rightarrow b^i(t - 1, k - 1), \ k = 1, 2, \ldots,$$

*then limit (7) holds:*

$$\frac{\sigma_n^i(t)}{n} \Rightarrow \sigma^i(t). \tag{28}$$

*Proof.* This is guaranteed by Lemma 2.

Now we shall calculate the queue after it has filled the free servers. For a given $n \in \{1, 2, \ldots\}$ we move $\sigma_n^2(t)$ customers from the end of queue, that is, the customers who have spent the longest time in the queue (if there are fewer customers in the queue, we move they all).

Define an intermediate queue $q_n'$:

$$q_n'(t, 0) = \sigma_n^1(t), \ q_n'(t, k) = q_n(t - 1, k - 1), \ k = 1, 2, \ldots \tag{29}$$

Obviously,

$$\text{if } \sigma_n^2(t) = 0 \text{ then } q_n(t, k) = q_n'(t, k), \ k = 0, 1, \ldots, \tag{30}$$

$$\text{if } \sigma_n^2(t) \geq \sum_{k=0}^{\infty} q_n'(t, k) \text{ then } q_n(t, k) = 0, \ k = 0, 1, \ldots, \tag{31}$$

$$\text{if } 0 < \sigma_n^2(t) < \sum_{k=0}^{\infty} q_n'(t, k) \text{ then with} \tag{32}$$

$$c_n(t) = \min\{i \in \{0, 1, \ldots\} : \sum_{k=i}^{\infty} q_n'(t, k) \leq \sigma_n^2(t)\}, \tag{33}$$

$$q_n(t,k) = \begin{cases} 0 \text{ for } k \geq c_n(t), \\ \displaystyle\sum_{i=c_n(t)-1}^{\infty} q_n'(t,i) - \sigma_n^2(t) \text{ for } k = c_n(t) - 1, \\ q_n'(t,k) \text{ for } k < c_n(t) - 1. \end{cases}$$

*Remark 1.* As $q_n'(t,k) = q_n(t-1, k-1)$,

$$\frac{q_n'(t,k)}{n} \Rightarrow q'(t,k), \ k = 1, 2, \ldots$$

It is easy to understand other presentations of $q_n$ and $q$:

**Lemma 4.** *For $i = 0, 1, 2, \ldots$*

$$\sum_{k=i}^{\infty} q_n(t,k) = (\sum_{k=i}^{\infty} q_n'(t,k) - \sigma_n(t)) \vee 0, \tag{34}$$

$$\sum_{k=i}^{\infty} q(t,k) = (\sum_{k=i}^{\infty} q'(t,k) - \sigma(t)) \vee 0. \tag{35}$$

It follows straightforward:

$$\sum_{k=i}^{\infty} \frac{q_n(t,k)}{n} = (\sum_{k=i}^{\infty} \frac{q_n'(t,k)}{n} - \frac{\sigma_n(t)}{n}) \vee 0. \tag{36}$$

Setting in Appendix Lemma 7

$$a_n(k) = \frac{q_n'(t,k)}{n} \text{ and } \theta_n = \frac{\sigma_n(t)}{n},$$

$$a_n^-(k) = \frac{q_n(t,k)}{n}, \ a^-(k) = q(t,k) \text{ and } \theta = \sigma(t),$$

and using Remark 1 and Corollary 2, we obtain convergence in Eqs. (6), (7).

## 2 Appendix

**Lemma 5.** $\xi_n, n = 1, 2, \ldots$, *is a random variable with values in* $\{0, 1, 2, \ldots\}$, *and* $\frac{\xi_n}{n} \Rightarrow \xi$ *(a constant). Let* $X_i, i = 1, 2, \ldots$, *be IID random variables with values in* $\{0, 1\}$. *Then*

$$\lim_{n \to \infty} \frac{1}{n} \sum_{i=1}^{\xi_n} X_i = \xi \mathbf{P}(X_1 = 1). \tag{37}$$

*Proof.* As $\sum_{i=1}^{\xi_n} X_i \le \xi_n$, $n = 1, 2, \ldots$, the proposition is evident for $\xi = 0$. If $\xi > 0$, then $\lim_{n \to \infty} \xi_n = \infty$ and (weak LLN)

$$\frac{1}{n}\sum_{i=1}^{\xi_n} X_i = \frac{\xi_n}{n}\frac{\sum_{i=1}^{\xi_n} X_i}{\xi_n} \xrightarrow[n \to \infty]{} \xi \mathbf{E}X_1 = \xi \mathbf{P}(X_1 = 1).$$

**Assumption 2.** *For $n = 1, 2, \ldots$ there exists a sequence $a_n$ of nonnegative random variables $a_n(k)$, $k = 0, 1, 2, \ldots$ with finite deterministic limits $a_n(k) \Rightarrow a(k)$, $k = 0, 1, 2, \ldots$*

*For each $\epsilon > 0$ and $\eta > 0$, there exists an integer $k_0$ such that for $n = 1, 2, \ldots$*

$$\mathbf{P}(\sum_{k=k_0}^{\infty} a_n(k) > \epsilon) < \eta. \tag{38}$$

**Lemma 6.** *If Assumption 2 holds then*

$$\mathbf{P}(\sum_{k=0}^{\infty} a_n(k) < \infty) = 1, \ n = 1, 2, \ldots, \ and \ \sum_{k=0}^{\infty} a(k) < \infty. \tag{39}$$

*Furthermore, for $i=0,1,2,\ldots$*

$$\sum_{k=i}^{\infty} a_n(k) \Rightarrow \sum_{k=i}^{\infty} a(k). \tag{40}$$

*Proof.* Whitt has introduced this Assumption and derives from it partially the statement of the Lemma for $i = 0$ without proof (see the first paragraph in the proof of Theorem 6.1 [15]).

**Lemma 7.** *Suppose that Assumption 2 holds and nonnegative random variables $\theta_n$, $n = 1, 2, \ldots$, with deterministic limit $\theta$, $\theta_n \Rightarrow \theta$, are given.*

*Define new sequences[1] $a_n^-$, $n \in \{1, \ldots\}$, $a^-$ by the following equations sequences: for $i = 0, 1, 2, \ldots$*

$$\sum_{k=i}^{\infty} a_n^-(k) = (\sum_{k=i}^{\infty} a_n(k) - \theta_n) \vee 0, \tag{41}$$

$$\sum_{k=i}^{\infty} a^-(k) = (\sum_{k=i}^{\infty} a(k) - \theta) \vee 0. \tag{42}$$

*These $a_n^-$, $n \in \{1, \ldots\}$, $a^-$ sequences satisfy:[2]*

$$a_n^-(k) \Rightarrow a^-(k), \ k = 0, 1, 2, \ldots, \ and \ \sum_{k=0}^{\infty} a_n^-(k) \Rightarrow \sum_{k=0}^{\infty} a^-(k). \tag{43}$$

---

[1] They describe queues after customers leave them and go to the emptied servers.

[2] Even starting from an arbitrary $k$.

*Proof.* According to Lemma 6,

$$\text{for } i = 0, 1, 2, \dots \quad \sum_{k=i}^{\infty} a_n(k) \Rightarrow \sum_{k=i}^{\infty} a(k). \tag{44}$$

As function $f(x, y) = (x - y) \vee 0$ is continuous, this and convergence of $\theta_n$, $n = 1, 2, \dots$, guarantees convergence

$$\text{for } i = 0, 1, 2, \dots \quad \left( \sum_{k=i}^{\infty} a_n(k) - \theta_n \right) \vee 0 \Rightarrow \left( \sum_{k=i}^{\infty} a(k) - \theta \right) \vee 0. \tag{45}$$

what is identical to

$$\sum_{k=i}^{\infty} a_n^-(k) \Rightarrow \sum_{k=i}^{\infty} a^-(k). \tag{46}$$

Finally, for $i = 0, 1, 2, \dots$

$$a_n^-(i) = \sum_{k=i}^{\infty} a_n^-(k) - \sum_{k=i+1}^{\infty} a_n^-(i) \Rightarrow \sum_{k=i}^{\infty} a^-(k) - \sum_{k=i+1}^{\infty} a^-(k) = a^-(k). \tag{47}$$

## 3   Conclusion

We have investigated a model without abandonment in the queue, although the necessary details of the customers age in the queue are provided. Since such a behavior in the queue is universally recognized, we intend to consider it next. As customers in a closed network cannot abandon it, probably we shall choose instead a similar version of "nonpersistent customers", see [7].

## References

1. Anulova, S.V.: Age-distribution description and "fluid" approximation for a network with an infinite server. In: International Conference "Probability Theory and its Applications", Moscow, pp. 219–220. 26–30 June 2012. (M.: LENAND)
2. Brown, L., Gans, N., Mandelbaum, A., Sakov, A., Shen, H., Zeltyn, S., Zhao, L.: Statistical analysis of a telephone call center: a queueing-science perspective. J. Am. Stat. Assoc. **100**(469), 36–50 (2005)
3. Dai, J., He, S.: Many-server queues with customer abandonment: a survey of diffusion and fluid approximations. J. Syst. Sci. Syst. Eng. **21**(1), 1–36 (2012). http://dx.doi.org/10.1007/s11518-012-5189-y, and http://link.springer.com/article/10.1007/s11518-012-5189-y
4. Gamarnik, D., Goldberg, D.A.: On the rate of convergence to stationarity of the M/M/n queue in the Halfin-Whitt regime. Ann. Appl. Probab. **23**(5), 1879–1912 (2013)
5. Gamarnik, D., Stolyar, A.L.: Multiclass multiserver queueing system in the Halfin-Whitt heavy traffic regime: asymptotics of the stationary distribution. Queueing Syst. **71**(1–2), 25–51 (2012)

6. Kang, W., Pang, G.: Equivalence of fluid models for Gt/GI/N+GI queues. ArXiv e-prints, February 2015. http://arxiv.org/abs/1502.00346
7. Kang, W.: Fluid limits of many-server retrial queues with nonpersistent customers. Queueing Systems (2014). http://gen.lib.rus.ec/scimag/index.php?s=10.1007/s11134-014-9415-9
8. Kaspi, H., Ramanan, K.: Law of large numbers limits for many-server queues. Ann. Appl. Probab. **21**(1), 33–114 (2011)
9. Koçağa, Y.L., Ward, A.R.: Admission control for a multi-server queue with abandonment. Queueing Syst. **65**(3), 275–323 (2010)
10. Pang, G., Talreja, R., Whitt, W.: Martingale proofs of many-server heavy-traffic limits for Markovian queues. Probab. Surv. **4**, 193–267 (2007). http://www.emis.ams.org/journals/PS/viewarticle9f7e.html?id=91&layout=abstract
11. Reed, J.: The $G/GI/N$ queue in the Halfin-Whitt regime. Ann. Appl. Probab. **19**(6), 2211–2269 (2009)
12. Zuñiga, A.W.: Fluid limits of many-server queues with abandonments, general service and continuous patience time distributions. Stochast. Process. Appl. **124**(3), 1436–1468 (2014)
13. Ward, A.R.: Asymptotic analysis of queueing systems with reneging: A survey of results for FIFO, single class models. Surv. Oper. Res. Manage. Sci. **17**(1), 1–14 (2012). http://www.sciencedirect.com/science/article/pii/S1876735411000237
14. Whitt, W.: Engineering solution of a basic call-center model. Manage. Sci. **51**(2), 221–235 (2005)
15. Whitt, W.: Fluid models for multiserver queues with abandonments. Oper. Res. **54**(1), 37–54 (2006). http://pubsonline.informs.org//abs/10.1287/opre.1050.0227
16. Xiong, W., Altiok, T.: An approximation for multi-server queues with deterministic reneging times. Ann. Oper. Res. **172**, 143–151 (2009). http://link.springer.com/article/10.1007/s10479-009-0534-3
17. Zhang, J.: Fluid models of many-server queues with abandonment. Queueing Syst. **73**(2), 147–193 (2013). http://link.springer.com/article/10.1007/s11134-012-9307-9

# Probability Characteristic Algorithm of Upstream Traffic in Passive Optical Network

Gely Basharin and Nadezhda Rusina[✉]

Peoples Friendship University of Russia, Moscow, Russia
gbasharin@sci.pfu.edu.ru, rusina_nadezda@inbox.ru

**Abstract.** Nowadays, access network evaluation is being conducted in both directions: high bit rate access development for providing high quality of service and decrease length of cooper wiring in local line networks. It's paid special attention to the networks, which are based on optical and optoelectronic components. Passive optical network (PON) is an all optical network, which is based on passive optical components only and excludes the conversion of electrical signal into optical form and *vice versa*. This paper is concerned with the algorithm for calculation of call blocking probability in upstream traffic multiservice model for PON considering the functioning process of optical network units and the principle of wavelength dynamic distribution. These results are used in the blocking probability analysis of the model.

**Keywords:** Passive Optical Network (PON) · Optical Line Terminal (OLT) · Optical Network Unit (ONU) · Upstream · Wavelength Division Multiplexing (WDM) · Time Division Multiple Access (TDMA) · Blocking probability

## 1 Introduction

PON is an optical access network, which supports transmitting various classes of the network traffic between optical line terminal (OLT) and optical network units (ONU) through a passive optical multiplexer/demultiplexer, which combines/splits spectral channels through a fiber [1–6]. A spectral channel is a data transmission channel related with transmitter and receiver units and making data transmission through assigned wavelength. OLT is located in a central office (CO) connecting PON with metropolitan area network (MAN) or wide area network (WAN). ONU is located on the subscriber side.

According to the time division multiple access (TDMA) technology [4,5] ONU may be in the ON-state and transmits data through its earlier assigned time domain, or ONU may be in the OFF-state, when no data transmission occurs.

According to the wavelength division multiplexing (WDM) technology [1–6], PON employs a set of wavelengths $W$ to transmit upstream traffic from ONU to OLT.

WDM-TDMA PON architecture is shown in the Fig. 1.

V. Vishnevsky and D. Kozyrev (Eds.): DCCN 2015, CCIS 601, pp. 188–198, 2016.
DOI: 10.1007/978-3-319-30843-2_20

Each ONU uses a reconfigurable laser and can transmit upstream traffic in the selected range of wavelengths. The dynamically allocating wavelengths mechanism allows increasing the system capacity, providing highly scalable. As well as WDM-TDMA PON provides new users and required quality of service level through existing frequency plan.

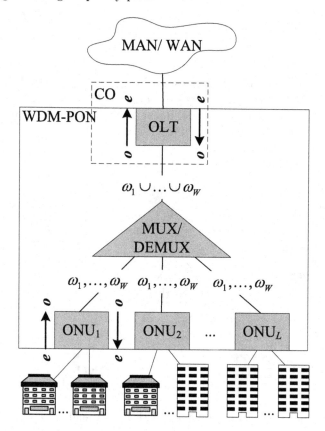

**Fig. 1.** WDM-TDMA PON architecture.

Let the WDM-TDMA PON with dynamic distribution of $W$ wavelengths, which we propose is called the Network I.

The Network I solves the problem of wavelengths limited number $W \leq L$ distribution between ONU finite number $L$. If there is no free wavelength in OLT at the ONU activation moment, the data transmitting is blocked in the ONU assigned time domain.

## 2 Optical Network Units Confunction in the Network I Fragment

Examine ONU confunction model in the Network I fragment with $L$ ONU. $l$-requests to give a wavelength arrive at according to a Poisson process with mean arrival rate $\kappa_l$.

**Table 1.** The model parameters.

| Parameters | Description |
| --- | --- |
| $L$ | Number of optical network units |
| $W, W \leq L$ | Number of upstream traffic wavelengths |
| $\kappa_l, l = \overline{1, L}$ | $l$-requests arrival intensity to assign a wavelength |
| $\nu_l, l = \overline{1, L}$ | ONU$_l$ transition rate from the ON- to OFF-state |

$\nu_l$ is a mean service time of $l$-requests, which is exponentially distributed. In other words $\nu_l$ is ONU transition intensity from the ON- to OFF-state. The parameters of the ONU confunction model in the Network I fragment are presented in the Table 1. The model is shown in the Fig. 2.

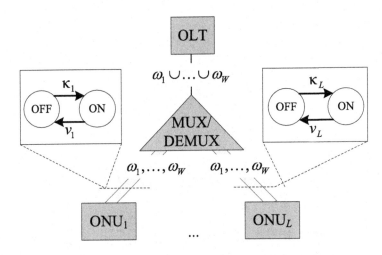

**Fig. 2.** The ONU confunction model with $L$ ONU and $W$ wavelengths.

The state vector of the system

$$\mathbf{n} := (n_l)_{l=\overline{1,L}}, n_l \in \{0, 1\}. \tag{1}$$

The state space of the system

$$\Omega := \{\mathbf{n} | n_\bullet \in \{0, 1, \ldots, W\}, W \leq L\}, n_\bullet := \mathbf{1}^T \mathbf{n} = \sum_{i=1}^{L} n_i. \tag{2}$$

Then,

$$|\Omega| = \sum_{w=0}^{W} \binom{L}{w}. \tag{3}$$

The state subspace when $ONU_l$ is in the ON-state or can pass into the ON-state at the activation moment

$$\Omega_{l,ON} := \{\mathbf{n} \in \Omega | n_l = 1\} \cup \{\mathbf{n} \in \Omega | n_l = 0, n_\bullet < W\}, l = \overline{1,L}. \tag{4}$$

Then,

$$|\Omega_{l,ON}| = 2 \sum_{w=0}^{W-1} \binom{L-1}{w}. \tag{5}$$

The state subspace when there is no available wavelength on OTL at the $ONU_l$ activation moment

$$\Omega_{l,OFF} := \{\mathbf{n} \in \Omega | n_l = 0, n_\bullet = W\}, l = \overline{1,L}. \tag{6}$$

Then,

$$|\Omega_{l,OFF}| = \binom{L-1}{W}. \tag{7}$$

The queue system functioning is described by a stationary Markov process $\mathbf{X}(t) := (X_l(t))_{l=\overline{1,L}}, t > 0$. $X_l(t) = 1$ means, that $ONU_l$ is in the ON-state, $X_l(t) = 0$ means, that $ONU_l$ is in the OFF-state.

**Theorem 1.** *The stationary Markov process* $\mathbf{X}(t)$, $t > 0$, *has got the stationary probability distribution, which does not depend on the starting probability distribution and is multiplicative*

$$p(\mathbf{n}) = G^{-1} \prod_{l=1}^{L} a_l^{n_l}, \mathbf{n} \in \Omega, G = \frac{1}{p(\mathbf{0})} = \sum_{\mathbf{n} \in \Omega} \prod_{l=1}^{L} a_l^{n_l}, a_l := \frac{\kappa_l}{\nu_l}, l = \overline{1,L}. \tag{8}$$

The Theorem 1 proof is similar to that presented in [1].

The nonblocking probability of the data transmission for $ONU_l$, $l = \overline{1,L}$, is calculated by the formula

$$p(\Omega_{l,ON}) = \sum_{\mathbf{n} \in \Omega_{l,ON}} p(\mathbf{n}) := \alpha_l. \tag{9}$$

## 3    Computation Algorithm for the Nonblocking Probability

Based on the model described in the previous section, we need to solve an allocation problem of $W$ wavelengths between finite number of $L$ ONU. We apply a convolution Buzen type algorithm [11], which is widely used at the teletraffic theory nowadays [1,12].

Let us denote

$g(L, w)$ - average number of variants that the first $L$ ONU use w wavelengths;

$g_{-l}(w)$ - average number of variants that optical network units use w wavelengths, and $ONU_l$ is in the OFF-state.

$\alpha_l$ is a nonblocking probability of the data transmission for ONU$_l$, $l = \overline{1, L}$, and is calculated by the formulas [7,8]

$$\alpha_l = 1 - G^{-1}g_{-l}\left(W\right), G = \sum_{w=0}^{W} g\left(L, w\right), \tag{10}$$

$$g_{-l}(w) = \begin{cases} 1, & w = 0, \\ g\left(L, w\right) - a_l g_{-l}\left(w - 1\right), & w = \overline{1, W}, \end{cases} \tag{11}$$

$$g(l, w) = \begin{cases} 0, & l = 0, \quad w = \overline{1, W}, \\ 1, & l = \overline{0, L}, w = 0, \\ g\left(l - 1, w\right) + a_l g\left(l - 1, w - 1\right), & l = \overline{1, L}, w = \overline{1, W}, \end{cases} \tag{12}$$

where $a_l := \frac{\kappa_l}{\nu_l}, l = \overline{1, L}$.

$\alpha_l, l = \overline{1, L}$, is one of the key performance factors of the Network I.

## 4    Upstream Traffic Transmission Model of the Network I

Let us consider the upstream traffic transmitting process in the Network I fragment with $L$ ONU [9]. Every ONU$_l$ has a finite memory buffer of size $R_l$ units, $0 < R_l < \infty$, $l = \overline{1, L}$. The analyzed queue system supports $K$ types of service classes. $k$-calls arrive at ONU$_l$ according to a Poisson process with a mean arrival rate $\lambda_{l,k}$, $0 < \lambda_{l,k} < \infty$, $l = \overline{1, L}$, $k = \overline{1, K}$. Arrival streams are independent in total for each ONU$_l$. A $k$-call is realized by allocating a specific number of units, $b_k$, $0 < b_k \leq \min R_l$, $l = \overline{1, L}$, $k = \overline{1, K}$. The $k$-call holds $b_k$ units in the ONU$_l$ buffer until it is serviced, then it unbuffers immediately after the completion of service, thus idling the wavelength. A service policy for each ONU$_l$ is "first come first serve" (FCFS).

A new $k$-call, $k = \overline{1, K}$, will be rejected by ONU$_l$, $l = \overline{1, L}$, if there are no more than $R_l - b_k$ free units in the buffer. The $k$-call leaves the system and has no effect on the stream arrival rate.

$\mu_k$, $0 < \mu_k < \infty$, $k = \overline{1, K}$, is a mean service time of a $k$-call in ONU$_l$, which is exponentially distributed. This model of the upstream traffic transmitting will encode $\underset{\mathbf{\Lambda, b}}{\mathbf{M}} \left| \underset{\mu}{\mathbf{M}} \right| \mathbf{1} \left| \mathbf{R} \right.$ [1,10].

However, the model described above does not account for the fact that there may not be free wavelengths on OLT at the ONU$_l$ activation moment. This fact leads to blocking the data transmitting in the ONU$_l$ time domain. Then the $k$-call service rate, taking into consideration the fact and the results of the previous section, will be

$$\alpha_l \mu_k, l = \overline{1, L}, k = \overline{1, K}. \tag{13}$$

The queue system scheme of the upstream traffic transmitting from ONU to OLT is shown in the Fig. 3.

The queue system in the Fig. 3 is described using the following parameters.

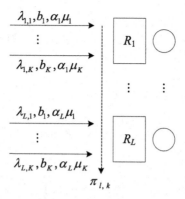

**Fig. 3.** The queue system scheme.

The state matrix of the system

$$\mathbf{M} := (m_{l,k})_{l=\overline{1,L},k=\overline{1,K}}, m_{l,k} \in \left\{0, 1, \ldots, \left\lfloor \frac{R_l}{b_k} \right\rfloor \right\}. \tag{14}$$

The state space of the system

$$S := \{\mathbf{M}|0 \le \sum_{k=1}^{K} b_k m_{l,k} \le R_l, l = \overline{1,L}\}. \tag{15}$$

The state subspace of $k$-calls received by $ONU_l$

$$S_{l,k} := \{\mathbf{M} \in S| \sum_{k=1}^{K} b_k m_{l,k} \le R_l - b_k\}, l = \overline{1,L}, k = \overline{1,K}. \tag{16}$$

The state subspace of $k$-calls blocked by $ONU_l$

$$\bar{S}_{l,k} = S \backslash S_{l,k} = \{\mathbf{M} \in S| \sum_{k=1}^{K} b_k m_{l,k} > R_l - b_k\}, l = \overline{1,L}, k = \overline{1,K}. \tag{17}$$

## 5    Formula Derivation for the Blocking Probability Calculation

The queue system functioning is described by a two-dimensional stationary Markov process $\mathbf{Y}(t) := (Y_{l,k}(t))_{l=\overline{1,L},k=\overline{1,K}}$, $t > 0$, where $Y_{l,k}(t)$ is the number of $k$-calls in $ONU_l$.

$E_{l,k}$ is a matrix of $L \times K$, where at the intersection of the $l^{th}$ row and $k^{th}$ column is "1" and the rest matrix elements are "0". The transition rate diagram for any $ONU_l$ is shown in the Fig. 4.

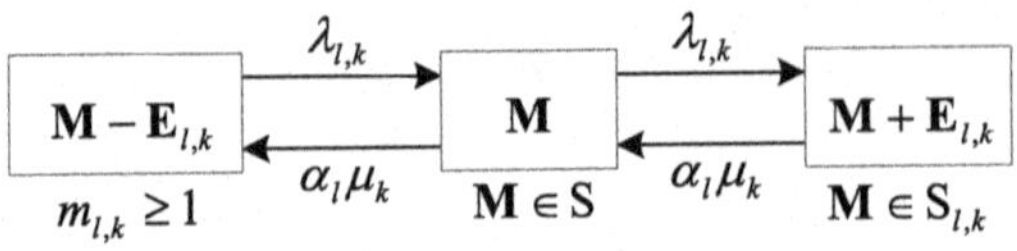

Fig. 4. The state transition scheme for $ONU_l$, $l = \overline{1,L}$, and $k$-call, $k = \overline{1,K}$.

**Theorem 2.** *The stationary Markov process* $\mathbf{Y}(t)$, $t > 0$, *has got the stationary probability distribution,which does not depend on the starting probability distribution and is multiplicative*

$$p(\mathbf{M}) = G^{-1} \prod_{l=1}^{L} \frac{1}{\alpha_l^{m_{l,\bullet}}} \prod_{k=1}^{K} \rho_{l,k}^{m_{l,k}},$$

$$G = \frac{1}{p(\mathbf{0})} = \sum_{\mathbf{M} \in S} \prod_{l=1}^{L} \frac{1}{\alpha_l^{m_{l,\bullet}}} \prod_{k=1}^{K} \rho_{l,k}^{m_{l,k}}, \qquad (18)$$

$$\mathbf{M} \in S, \rho_{l,k} := \frac{\lambda_{l,k}}{\mu_k}, k = \overline{1,K}, m_{l,\bullet} := \sum_{k=1}^{K} m_{l,k}, l = \overline{1,L}.$$

The Theorem 2 proof is similar to that presented in [1].
The blocking probability of $k$-calls for $ONU_l$

$$\pi_{l,k} = \sum_{\mathbf{M} \in \bar{S}_{l,k}} p(\mathbf{M}), l = \overline{1,L}, k = \overline{1,K}. \qquad (19)$$

# 6    Computation Algorithm for the Blocking Probability

Based on the model described in the previous section, we need to solve an allocation problem of buffer limited capacity $R_l$ between finite number of $K$ call types for each $ONU_l$, $l = \overline{1,L}$. We apply a convolution Buzen type algorithm [11], which is widely used at the teletraffic theory nowadays [1,12].

It is necessary to calculate the norming quantity to calculate the blocking probability of $k$-calls for $ONU_l$.

Let us denote $g_l(k,r)$ - average number of variants that the first $(1,\ldots,k)$ call types use all of the $r$ units of the $ONU_l$ buffer.

**Theorem 3.** *Norming quantity* $G$ *(18) calculated by the formulas*

$$G = \prod_{l=1}^{L} \sum_{r=0}^{R_l} g_l(K,r), \qquad (20)$$

$$g_l(k,r) = \begin{cases} 0, & k = \overline{1,K}, r < 0, \\ 0, & k = 0, \quad r = \overline{1,R_l}, \\ 1, & k = \overline{1,K}, r = 0, \\ g_l(k-1,r) + \frac{\rho_{l,k}}{\alpha_l} g_l(k,r-b_k), & k = \overline{1,K}, r = \overline{1,R_l}, \end{cases} \qquad (21)$$

*where* $l = \overline{1,L}$, $\rho_{l,k} := \frac{\lambda_{l,k}}{\mu_k}$, $k = \overline{1,K}$.

The blocking probability of $k$-calls for $ONU_l$ in the Network I

$$\pi_k = \frac{1}{G} \sum_{r=R-b_k+1}^{R} g(K,r), k = \overline{1,K}. \tag{22}$$

# 7   Numerical Analysis Example

Consider the dependence of the nonblocking probability of the data transmission for $ONU_1$ and $ONU_2$ and the number of ONU in the ONU confunction models of the Network I using one and two wavelengths. The model Parameters values are represented in the Table 2.

Table 2. The model parameters.

| Parameters | Value | Parameters | Value | Parameters | Value |
|------------|-------|------------|-------|------------|-------|
| $L$ | 16 | $a_6$ | 0.4 | $a_{12}$ | 0.22 |
| $a_1$ | 0.1 | $a_7$ | 1 | $a_{13}$ | 0.36 |
| $a_2$ | 0.2 | $a_8$ | 0.6 | $a_{14}$ | 0.54 |
| $a_3$ | 0.15 | $a_9$ | 0.25 | $a_{15}$ | 0.78 |
| $a_4$ | 0.3 | $a_{10}$ | 0.35 | $a_{16}$ | 0.8 |
| $a_5$ | 0.45 | $a_{11}$ | 0.65 | | |

The chart of the nonblocking probability of the data transmission $\alpha_1$ and $\alpha_2$ from the number of ONU $L$ is shown in the Fig. 5.

The nonblocking probability of the data transmission $\alpha_l$, $l = \overline{1,L}$, decreases with increasing the number of ONU $L$, because each ONU needs a wavelength to transmit the data.

When the number of wavelengths $W$ and the number of ONU $L$ are equal to two then the nonblocking probability of the data transmission for $ONU_1$ and $ONU_2$ is equal to one, because each ONU uses its own wavelength.

Consider the dependence of the nonblocking probability of the data transmission for $ONU_1$, $ONU_7$ and $ONU_{14}$ and the number of wavelengths $W$ in the model of the Network I with $L = 16$ ONU. The model Parameters values are represented in the Table 2.

The chart of the nonblocking probability of the data transmission $\alpha_1$, $\alpha_7$ and $\alpha_{14}$ from the number of wavelengths $W$ is shown in the Fig. 6.

The nonblocking probability of the data transmission $\alpha_l$, $l = \overline{1,L}$, increases with increasing the number of wavelengths $W$. If there is $L$ wavelengths in the Network I, each ONU uses its own wavelength to transmit the data and the nonblocking probability of the data transmission $\alpha_l$ is equal to one.

The nonblocking probability of the data transmission $\alpha_1$ is less then $\alpha_{14}$, because the parametr value of $a_1$ is less then $a_{14}$.

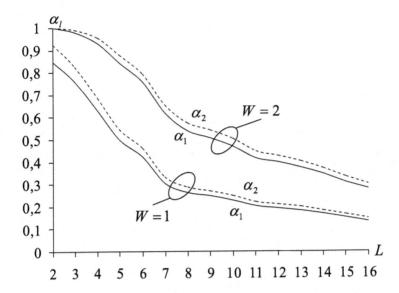

**Fig. 5.** The chart of the nonblocking probability $\alpha_1$ and $\alpha_2$ from $L$.

**Table 3.** The model parameters.

| Parameters | Value |
|---|---|
| $K$ | 2 |
| $L$ | 16 |
| $W$ | $\overline{0,16}$ |
| $R_l, l = \overline{1,L}$ | 18 |
| $b^T$ | (1;2) |
| $\rho_{11}, \rho_{12}$ | 0.5, 0.25 |
| $\rho_{lk}, l = \overline{2,L}, k = \overline{1,K}$ | 1 |

Consider the dependence of the $k$-calls blocking probability for ONU$_l$, $l = \overline{1,L}$, and the number of wavelengths $W = \overline{0,16}$ allocated for the upstream traffic transmitting process from $L$ ONU to OLT in the Network I. The buffer capacity is fixed, $R_l = 18$, $l = \overline{1,L}$. The model Parameters values are represented in the Table 3.

The chart of $\pi_{11}$ and $\pi_{12}$ from the number of wavelengths $W$ is shown in the Fig. 7.

Parameter values $\alpha^T$ are calculated according to formulas (10, 11 and 12) based on the ONU co-operation model in the Network I. This model Parameters values are represented in the Table 2.

The $k$-calls blocking probability for ONU$_l$, $l = \overline{1,L}$, decreases with increasing the number of wavelengths allocated in the upstream traffic transmitting model, because $\alpha_1$ the nonblocking probability of the data transmission for ONU$_1$ increases.

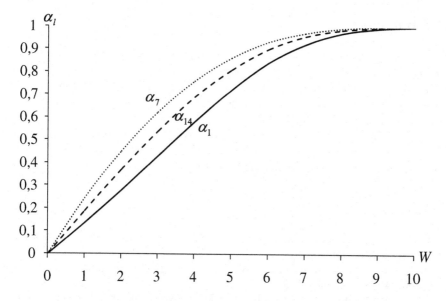

**Fig. 6.** The chart of the nonblocking probability $\alpha_1$, $\alpha_7$ and $\alpha_{14}$ from $W$.

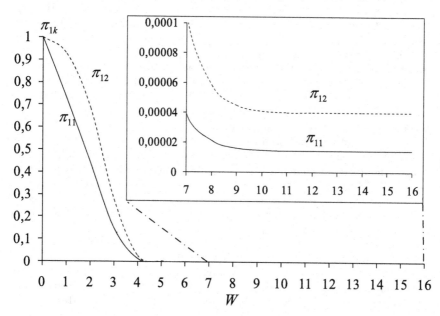

**Fig. 7.** The chart of the blocking probabilities $\pi_{11}$ and $\pi_{12}$ from $W$.

The blocking probability of 1-calls is less then the blocking probability of 2-calls for $ONU_1$, because 1-calls need one buffer unit and 2-calls need two buffer units in the system.

## 8   Conclusion

In this paper, the mathematical model of the upstream multiservice traffic transmitting from $L$ ONU to OLT in the Network I is presented. The algorithm for calculating of the blocking probability is proposed. In addition, there is the numerical analysis example in which the role of $W$ is shown to select the optimal functioning behavior of the ONU.

The authors recommend the use of the concepts described in this paper for the algorithm development to calculate of the blocking probability for the upstream priority traffic transmitting model in the Network I.

## References

1. Basharin, G.P.: Lectures on mathematical teletraffic theory. The 3rd publication. PFUR, Moscow (2009)
2. Listvin V.N., Treschikov V.N.: DWDM-systems. The 2nd publication. TECH-NOSPHERA, Moscow (2015)
3. Efimushkin, V.A., Savangukov, I.M.: Resource Allocation in Optical Transport Networks. CRIC, Moscow (2010)
4. Ramaswami, R., Sivarajan, K.N., Sasaki, G.H.: Optical Networks A practical Perspective, 3rd edn. Morgan Kaufmann Publishers Inc., San Francisco (2010)
5. Mukherjee, B.: Optical WDM Networks. Springer, New York (2006)
6. Siva Ram Murthy, C., Gurusamy, M.: WDM Optical Networks: Concepts Design and Algorithms. Prentice Hall PTR, Upper Saddle River (2002)
7. Basharin, G.P., Gaidamaka, Y.V., Rusina, N.V.: Probability characteristic computation algorithm of ONUs functioning in PON. Vestn. PFUR **2**, 28–32 (2015)
8. Naymov, V., Samyilov, Y., Yarkina, N.V.: Teletraffic theory of multiservice networks. Monography. PFUR, Moscow (2008)
9. Basharin, G., Rusina, N.: Multirate loss model for optical network unit in passive optical networks. In: Vishnevsky, V., Kozyrev, D., Larionov, A. (eds.) Distributed Computer and Communication Networks. Communications in Computer and Information Networks, vol. 279, pp. 219–228. Springer, Cham (2014)
10. Basharin, G.P.: Maintenance of two streams with relative priority for a fully accessible system with a limited waiting space. In: Proceedings of the AS of the USSR, Technical Cybernetics, vol. 2, pp. 72–86 (1967)
11. Busen, J.P.: Computational algorithms for closed queueing networks with exponential servers. Commun. ACM **16**, 527–531 (1973)
12. Basharin, G.P., Gaidamaka, Y.V., Samouylov, K.E.: Mathematical theory of teletraffic and its application to the analysis of multiservice communication of next generation networks. Autom. Control Comput. Sci. **47**(2), 62–69 (2013)

# Redundant Distribution of Requests Through the Network by Transferring Them Over Multiple Paths

V.A. Bogatyrev[(⊠)] and S.A. Parshutina

ITMO University, Kronverksky Pr. 49, 197101 St. Petersburg, Russia
vladimir.bogatyrev@gmail.com, svetlana.parshutina@gmail.com
http://en.ifmo.ru

**Abstract.** The research focuses on opportunities of multipath routing for providing reliability and fault-tolerance of distributed computing systems, given that requests are distributed dynamically through the network. There are proposed models for non-redundant and redundant search for available servers, which are ready to serve incoming requests. Server unavailability occurs due to faults, temporary shutdown, or overload with requests arriving from the network. An unavailable server rejects incoming requests, which thereafter underlie the flow of repeated requests, or those sent to other servers time and again, until they are finally accepted. Transferring requests repeatedly and, in case of redundant search, polling several servers simultaneously cause the increase in the network load. It is found that searching for an available server over multiple paths at once results in the decreased requests' time in the system, provided that server availability and the intensity of request flow are not high.

**Keywords:** Reliability · Cluster · Request · Redundancy · Routing · Queueing systems

## 1 Introduction

Reliability and fault-tolerance of distributed computing systems are conventionally achieved by applying duplicate computing and communication resources. Owing to the use of redundant network components, requests are distributed adaptively among network links and computing nodes that are in order at the moment. Such approach relies on the comprehensive system design and assumes partial loss of requests due to the network failures. In the critical systems, however, any loss of requests may cause the whole logic of computations to fail, implying considerable consequences in terms of money, safety, and others.

The study focuses on distribution of requests through highly reliable networks by using the mechanisms of multipath routing. Routing is referred to as the choice of the route that meets specified requirements for the current dataflow. In case of single-path routing, requests are sent over a single path, *the best* one

© Springer International Publishing Switzerland 2016
V. Vishnevsky and D. Kozyrev (Eds.): DCCN 2015, CCIS 601, pp. 199–207, 2016.
DOI: 10.1007/978-3-319-30843-2_21

from a certain point of view. In case of multipath routing, there exists a set of appropriate paths from the sender (the source of requests) to their receiver and each request is transmitted over several paths simultaneously.

The use of multipath routing mechanisms serves to avoid overloads and utilize network resources in the optimal way. The considerable part of the models that involve those routing techniques addresses the problem of network load balancing, like [1]. Nevertheless, multipath routing is also discussed in relation to network survivability, recovery, security, and energy efficiency [2].

Employing multipath routing techniques suggests that data packets are distributed over multiple paths either evenly or on grounds of one or more criteria. In the former case, Equal-Cost Multi-Path (ECPM) routing strategies are implemented, which means that there are no less than two routes having equal minimum cost of data transmission. The decision about the next hop, or the path to forward a packet along, is taken on the basis of a number of algorithms. For example, a Hash-Threshold algorithm involves performing a hash over the packet header fields and utilizing the resulting threshold value, while the Round-Robin principle implies choosing the least recently used next-hop [3]. In the latter case, packets are distributed over paths according to a certain criterion or a combination of several criteria. It can be bandwidth, requests' time in the system, or some other metrics, which serves as a basis for calculating special values. Those values, or indices, are assigned to all possible routes and thus rank them by preference, depending on the metrics. One more criterion that deserves attention is fault statistics, which underlies the idea to prefer more reliable paths to less reliable ones.

Contemporary models and methods for providing reliability of distributed computing systems are expected to rest upon the notion of Quality of Service (QoS) [4]. QoS is to describe the overall performance of a given network, in particular its most important features, such as the data transfer rate (DTR), the reliability of the network and its components, the IP packet transfer delay (IPTD), the IP packet delay variation (IPDV), or jitter, and the IP packet loss ratio (IPLR). Path reliability and IPTD, also called *time in the system* in this paper, are of special interest for current research.

Although ways to provide reliability and fault-tolerance of computing networks have already been widely discussed, for example in [5–11], the opportunities offered in this regard by dynamic distribution of data packets-requests through the network, on the basis of multipath routing, remain understudied for today.

This research paper discusses how to provide reliability of distributed computing systems using redundant requests, spread through the network over multiple paths. It should be noted that current problem has nothing to do with choosing *the best* route out of all possible ones; instead, it deals with utilizing numerous paths, not necessarily of the same cost, contemporaneously. Thus, the study explores the methods of improving reliability by forwarding multiple requests, or better to say multiple copies of the same request, along multiple paths. It results, on the one hand, in the increased network load and hence in the increased packet transfer delay, but on the other hand, in the enhanced chance of a packet-request being delivered within a specified time interval.

There are proposed two ways of transmitting requests through a given network by polling servers, which are possible receivers of the requests, in a serial and in a simultaneous manner. These ways underlie the corresponding models, which take into account the increased network load occurring because of simultaneous transmission of requests over several routes and the existence of the flow of repeated (unserved) requests – the result of polling unavailable servers.

## 2  Problem Statement and System Design

The focus of current study is to discuss and compare two different ways of distributing requests through a given network by using multipath routing. In this paper, these are called serial and parallel server search (or polling). It is assumed that requests are being transferred over numerous disjoint routes, which is the simplest case for modeling. Each path is a single-channel non-preemptive M/M/1 queueing system with the infinite queue [12–15]. All calculations are carried out using Mathcad 14.

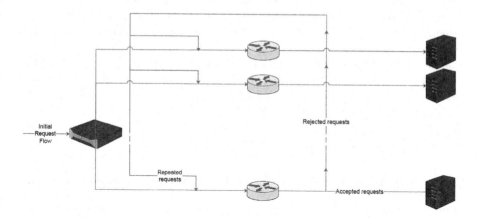

**Fig. 1.** The general scheme of searching for an available server. The router accepts the Initial Request Flow and forwards incoming requests along one or a few appropriate disjoint paths, with one switch (or router) on each way to the corresponding server. Requests underlie the flow of either Accepted requests, which are to leave the system after being processed by the server, or Rejected requests, turning into Repeated requests when being distributed through the network again.

The proposed models take into account the possibility of loss of requests and the probability $p$ of server availability. Since server availability is not guaranteed, there arises a problem of finding an available server. A server is considered to be available and, as a result, ready to serve incoming requests if it is in working order, powered up, and not overloaded with the requests having already come from the network. While the arrived requests are being processed or waiting on the server's queue before entering service, the senders of those requests are notified of success. If the server is busy or unavailable due to other reasons, a request is rejected and its sender receives a notice of refusal, whereupon the request is resent to another server over some different path. Servers are supposed to be accessed repeatedly in a random or a Round-Robin fashion until the success is achieved.

The general scheme of distribution of requests over $n$ paths, with taking into account the flow of rejected requests, which later resent to other servers, is shown in Fig. 1. Requests forming the initial flow are forwarded along one or several routes simultaneously. If a request is transmitted over one of all appropriate paths at a time, it is serial search for available servers. Transferring a request over two or more routes contemporaneously indicates parallel, or *redundant*, search. On the way to one of $n$ possible servers, each request passes through one switch (or router) and gets into the flow of either accepted requests or rejected ones. The accepted requests continue to the respective server and leave the system after having been processed. The rejected requests underlie the flow of repeated requests and subsequently merge with the flow of initial requests, newly arrived in the system. The period of time a request resides in the system, including the time spent on all unsuccessful efforts to be delivered and the successful distribution over one of the routes in the end, is called requests' *time in the system* in this paper. It is chosen as the criterion of efficiency of the systems under consideration.

The queueing model for distribution of requests through the network under consideration is represented in Fig. 2. Here, $\Lambda$ stands for the intensity of the initial flow of requests, which is divided into $\lambda_1, \lambda_2, ..., \lambda_n$ flows; $R_1, R_2, ..., R_n$ do for switches (or routers) from 1 to $n$; $P_1, P_2, ..., P_n$ are probabilities of servers' $S_1, S_2, ..., S_n$ being available for incoming requests. In this paper, the probabilities are considered to be the same and equal to $p$.

## 2.1   Non-redundant Distribution of Requests Through the Network

Non-redundant distribution of requests implies transmitting a request over one appropriate route at once. The intensity of request flow $\Lambda_0$ is determined by the intensity of the initial request flow $\Lambda$, comprised of requests first entering the system, and the intensity of the flow of repeated requests, rejected by unavailable servers and returned to the system. The intensity $\Lambda_0$ is calculated as

$$\Lambda_0(\Lambda) = \Lambda p \sum_{i=1}^{\infty} i(1-p)^{i-1} = \frac{\Lambda}{p} \, , \tag{1}$$

where $p$ is the probability of server availability, or the server's readiness to serve a request, having been distributed through the network; $i(1-p)^{i-1}$ is the expected value of the number of repeated efforts to do this. The resulting $\frac{\Lambda}{p}$ is the symbolic evaluation of the primary expression; here and further on in this paper, produced in the Mathcad 14 environment.

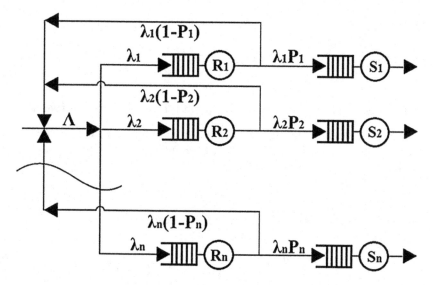

**Fig. 2.** The queueing model of the system, where $\lambda_1, \lambda_2, ..., \lambda_n$ are intensities of the flows of distributed requests, $\lambda_1 P_1, \lambda_2 P_2, ..., \lambda_n P_n$ are ones of accepted requests, and $\lambda_1(1 - P_1), \lambda_2(1 - P_2), ..., \lambda_n(1 - P_n)$ are ones of rejected requests.

Probability $p$ depends not only on the server's serviceability, defined through its availability, but also on the correctness of transferring requests through the network. Thus,

$$p = k_a PB, \tag{2}$$

where $k_a$ is the coefficient of server availability; $P$ is the probability that network facilities involved in transmitting requests do not fail. Furthermore,

$$P = e^{-\lambda T}, \tag{3}$$

where $-\lambda$ is the failure rate for those network facilities. Finally,

$$B = (1 - b)^L \tag{4}$$

is the failure rate in bits for transferring a packet-request of length L-bit through the network.

Time delay, which arises from each of those efforts, when using (1), is represented by

$$T(\Lambda) = \frac{v}{1 - \frac{(\Lambda_0(\Lambda) + \alpha\Lambda)v}{n}} = \frac{vnp}{p(n + \alpha v\Lambda) - v\Lambda}. \tag{5}$$

Here, $v$ is the average time of transmitting requests through the network, without taking into account waiting on the queue; $t$ is the time of response from a given server (whether or not it is available); $n$ is the number of possible routes from the source of requests to one of the servers. As for $\alpha$, it is the share of requests coming from other sources; consequently, $\alpha\Lambda$ is the intensity of request flow from the sources other than the one under consideration.

Time in the system, or average residence time, $T(\Lambda)$ – replaced with (5) – is

$$T_0(\Lambda) = p \sum_{i=1}^{\infty} i(1-p)^{i-1}(T(\Lambda) + t) = \frac{vt\Lambda + p[\alpha vt\Lambda - n(v+t)]}{p[v\Lambda + p(\alpha v\Lambda - n)]}. \qquad (6)$$

## 2.2  Redundant Distribution of Requests Through the Network

Requests are distributed through the network in a redundant manner when each of them is sent to $k$ servers over $k$ routes simultaneously. The probability $r$, saying that at least one server is available, is

$$r = 1 - (1-p)^k. \qquad (7)$$

The intensity of request flow $\Lambda_1$, when (7) substitutes for $r$, is

$$\Lambda_1(\Lambda) = \Lambda r \sum_{i=1}^{\infty} i(1-r)^{i-1} = \frac{\Lambda}{1-(1-p)^k}. \qquad (8)$$

Time delay arising from each effort to deliver a request, granting that $\Lambda_1$ can be replaced with (8) is

$$T(\Lambda) = \frac{v}{1 - \frac{(k\Lambda_1(\Lambda) + \alpha\Lambda)v}{n}} = \frac{vnr}{r(n + \alpha v\Lambda) - kv\Lambda}. \qquad (9)$$

Time in the system is calculated based on (9) as follows:

$$T_1(\Lambda) = r \sum_{i=1}^{\infty} i(1-r)^{i-1}(T(\Lambda) + t) = \frac{kvt\Lambda + r[\alpha vt\Lambda - n(v+t)]}{r[kv\Lambda + r(\alpha v\Lambda - n)]}. \qquad (10)$$

## 3  Calculations and Evaluation

The proposed ways of distributing requests through the network are to be evaluated by comparing them for efficiency, described in terms of time in the system. For that purpose, some values should be assigned to the parameters of the models under consideration, like the following ones: $v = 1$ s, $t = 0.1$ s, $n = 10$ routes as well as servers, $\alpha = 0$.

In simplest case, it is assumed that requests originate from a single source – the one in question, what is indicated with the value of $\alpha$.

The redundancy order $k$ varies from 2 to 10 routes, while server availability $p$ ranges between 0 and 1.

Figure 3 illustrates an example calculation of time in the system $T$, which depends on the intensity of request flow $\Lambda$ and is computed for three cases:

- non-redundant distribution of requests,
- redundant distribution when the redundancy order $k$ equals to 2, and
- redundant distribution when $k$ equals to 3.

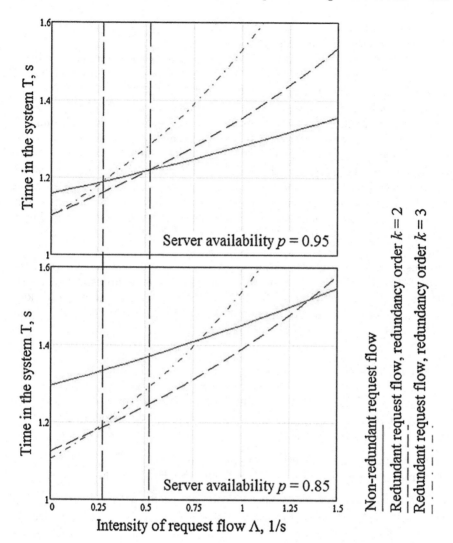

**Fig. 3.** Time in the system, or residence time of requests in the system, $T$, depending on the intensity of request flow $\Lambda$, with server availability $p = 0.95$ (*up*) and $p = 0.85$ (*down*). Three ways of distributing requests through the network are illustrated, when requests are transferred over one path at a time (*solid line*), over two paths (*dashed line*) or over three paths (*dash-dotted line*) simultaneously.

Here, server availability $p$ is chosen to be $p = 0.95$ and $p = 0.85$ for purposes of both plausibility and visualization.

It is clear that the decrease of server availability from 0.95 to 0.85 leads to the increase of time in the system $T$ in any case. Nevertheless, the change of $T$ is more significant in case of non-redundant distribution of requests. Therefore, the non-redundant system is more preferable than the other two redundant ones in case $p = 0.95$ and $\Lambda >= 0.5$ 1/s but is less attractive as compared with them when $p = 0.85$ and $\Lambda < 0.5$ 1/s.

It should also be noted that the redundant system with the redundancy order $k = 3$ produces worse results than with $k = 2$ in case $p = 0.95$ but proves to be better than the latter in case $p = 0.85$ and $\Lambda < 0.25$ 1/s.

To conclude, the simultaneous use of several routes is pointless in case of high server availability but, with its value decreasing, the efficiency of redundant search starts to increase. However, this will continue until a certain point (the threshold value), which is represented by the intersection of the functions of non-redundant and redundant search in Fig. 3. On exceeding the threshold value, the intensity of the flow of requests – of the initial ones as combined with the repeated ones – all sent over multiple paths, causes the dramatic rise of time in the system. Moreover, the lower server availability is, the greater the number of those routes is, which are needed to detect an available server in the shortest time.

Thus, an interesting problem arises, which is to optimize the choice of the number of redundant paths, needed to distribute a request through the network, with the use of multipath routing techniques. Solving this problem means making use of the discussed strategies of transferring requests over $k$ paths by applying them in turn to non-uniform flows of requests.

## 4   Conclusion

In this paper, there are proposed models for fault-tolerant distributed computing systems with adaptive distribution of requests through the network. The models rest upon multipath routing techniques and describe the ways of polling servers:

- one at a time in case of serial (non-redundant) search and
- several servers at once in case of parallel (redundant) search.

A server is supposed to be available, i.e. ready for serving requests, or unavailable due to its possible faults and temporary shutdown as well as overload with the requests, which have already been accepted by the server and are either being processed or waiting on its queue to be serviced.

Time in the system, or average residence time of requests in the system, takes into account the flow of repeated requests (previously rejected by unavailable servers) and is chosen as the criterion of efficiency of the systems in question.

It is found that non-redundant search for an available server, when each request is sent over only one appropriate path at a time, while the others are regarded to be reserve, turns out to be the best option if server availability is high. The less server availability is, the more preferable redundant search is, when each request is being transmitted over multiple paths simultaneously. However, it is true only until the intensity of request flow is less than a certain threshold value. Increasing the number of routes used to search for an available server (the redundancy order) gives little advantage in case server availability is high but proves significant if the availability is decreasing.

**Acknowledgments.** The study is a part of two research projects: "Methods and models for providing the integrated security and stable functioning of computer systems" (State Registration Number is 610481) and "Methods for designing the key systems of information infrastructure" (State Registration Number is 615869) of ITMO University, St. Petersburg, Russia.

# References

1. Gunnar, A., Johansson, M.: Robust load balancing under traffic uncertainty tractable models and efficient algorithms. Telecommun. Syst. **48**, 93–107 (2011)
2. Banner, R., Orda, A.: Multipath Routing Algorithms for Congestion Minimization. CCIT Report No. 429, Dept. Electr. Eng., Technion, Haifa, Israel (2004). http://www.ee.technion.ac.il/people/ron/Congestion.pdf
3. Hopps, C.: Analysis of an Equal-Cost Multipath Algorithm. IETF, RFC 2992 (2000). http://www.ietf.org/rfc/rfc2992
4. McDysan, D.: QoS and Traffic Management in IP and ATM Networks. McGraw-Hill, New York (2000)
5. Andreev, S., Saffer, Z., Turlikov, A.: Delay analysis of wireless broadband networks with non real-time traffic. In: Sacchi, C., Bellalta, B., Vinel, A., Schlegel, C., Granelli, F., Zhang, Y. (eds.) MACOM 2011. LNCS, vol. 6886, pp. 206–217. Springer, Heidelberg (2011)
6. Bogatyrev, V.A.: Exchange of duplicated computing complexes in fault tolerant systems. Autom. Control Comput. Sci. **46**(5), 268–276 (2011)
7. Bogatyrev, V.A., Bogatyrev, S.V., Golubev, I.Y.: Optimization and the process of task distribution between computer system clusters. Autom. Control Comput. Sci. **3**, 103–111 (2012)
8. Bogatyrev, V.A.: Fault tolerance of clusters configurations with direct connection of storage devices. Autom. Control Comput. Sci. **45**(6), 330–337 (2011)
9. Bogatyrev, V.A.: An interval signal method of dynamic interrupt handling with load balancing. Autom. Control Comput. Sci. **34**(6), 51–57 (2000)
10. Bogatyrev, V.A.: Protocols for dynamic distribution of requests through a bus with variable logic ring for reception authority transfer. Autom. Control Comput. Sci. **33**(1), 57–63 (1999)
11. Bogatyrev, V.A., Bogatyrev, A.V.: Functional Reliability of a Real-Time Redundant Computational Process in Cluster Architecture Systems. Autom. Control Comput. Sci. **49**(1), 46–56 (2015)
12. Kleinrock, L.: Queueing Systems, vol. 1: Theory. Wiley Interscience, New York (1975)
13. Kleinrock, L.: Queueing Systems, vol. 2: Computer Applications. Wiley Interscience, New York (1976)
14. Vishnevsky, V.M.: Teoreticheskie osnovy proektirovaniya komputernykh setey (Theoretical Fundamentals for Design of Computer Networks). Tekhnosfera, Moscow (2003) (in Russian)
15. Aliev, T.I.: Osnovy modelirovaniya diskretnykh system (Fundamentals of simulation of discrete systems). SPbSU ITMO Publ., St. Petersburg (2009) (in Russian)

# FPGA-Prototyping with Advanced Fault Injection Methodology for Tolerant Computing Systems Simulation

Oleg Brekhov$^{(\boxtimes)}$, Alexander Klimenko, Konstantin Kordover, and Maksim Ratnikov

Moscow Aviation Institute (National Research University), 125993 Moscow, Russia
{obrekhov,m.o.ratnikov}@mail.ru,
{a.v.klimenko,k.a.kordover}mai.ru

**Abstract.** The definition of architecture fault-tolerant high-performance VLSI-based distributed computing systems is an important task and requires reasonable modeling at the design stage. One of the methods of generating faults and failures is the method of fault injection. The paper proposes a advanced fault injection technique, involving the use of determine the most likely negative externalities and the response of the system to them. This gives the possibility to determine the threat list, i.e. a list of elements, all of which were violated. In this article the concept of universal hardware-software complex FPGA-based prototyping to simulate the operation of the system and monitor of the device under test work correctness "on the fly". This approach has more flexibility compared to the traditional and allows you to use more complex patterns in the modeling. It also allows simulation of operation fault-tolerant systems in terms of fault injection.

## 1 Introduction

When designing fault-tolerant computing systems imposes high requirements failure resiliency. In these systems, the failure of one element can lead to loss of efficiency of the system. This defines one of the most important challenges facing the developers of such systems providing failure resiliency of the constituent elements. This is especially important for embedded systems of unmanned space vehicles, operating in conditions of space radiation. Therefore, the design and testing steps of fault tolerant systems development have their own characteristics. So these steps are carried out in three stages:

(1) development and functional testing of the system without failures protection subsystems;
(2) development and implementation of failures protection subsystems and re-conduct functional testing of system;
(3) checking operation of the system when adding faults and failures.

One way to perform this verification is a method of fault injection with simulation of FPGA-based-prototyping of a computer system.

© Springer International Publishing Switzerland 2016
V. Vishnevsky and D. Kozyrev (Eds.): DCCN 2015, CCIS 601, pp. 208–223, 2016.
DOI: 10.1007/978-3-319-30843-2_22

The article proposed technique for modeling tolerant computing systems based on the FPGA prototype using the method of fault injection due to external and internal factors. The article also provides requirements for a universal software-hardware complex for evaluation of failure resiliency projects circuits computing systems.

**The Structure of the Article:**

- The methodology of modeling
- The concept of universal hardware-software complex
- Conclusions.

## 2    The Methodology of Modeling

### 2.1    Description of the Concept of Fault Injection and Its Extension

Traditionally, the fault injection during simulation or imitation is a process by which at a given point in model time at specified nodes of the chip, there is a forced distortion of stored values or behavior of the node (fixing a particular value, introducing delay, etc.). The choice of node (element) and time for fault injection is carried out using a pseudorandom generator or sequential switching of pre-selected combinations of nodes/time pairs. This approach allows to determine the response of the simulated system to such distortions, but may not correspond to actual impacts. Also it does not always give the answer to the question about the necessary changes in the system or recommendations to developers.

This article proposes a methodology for advancing the notion of "fault injection in process modeling/simulation system". Therefore, as the data source for the fault injection will use the threat model. Threat it is the event which led to the change in the behavior of a fragment or fragments of the chip. Each threat defines a set of fragments of a chip, whose behavior will be changed, and the time and duration of this change. All threats are determined by the parameters of the environment (the types, intensity and direction of radiation in the environment), parameters of the internal environment (production technology of the chip and the chip topology). Accordingly, the threat defines the basic parameters of distortion and performed with this distortion simulation shows the system's response to it (the appearance of failure/rejection).

Conducting such simulations with different source data with the accumulation of statistics will ensure the identification of the most vulnerable fragments of the chip and guidance for developers. Thus, the developer receives recommendations about definition of architecture tolerant system and method of fault tolerance providing, manufacturing techniques of chips and design of computing systems.

The Fig. 1 shows the block diagram methodology of modeling.

Here external influences model, threats model, fault injection module and firmware generator are software, and these elements we will discuss further.

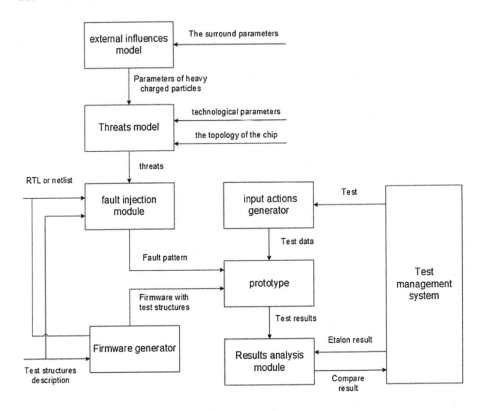

**Fig. 1.** The modeling system structure

Test management system, input actions generator, prototype and results analysis module are hardware.

Test management system — this is a computer system with appropriate software, which allows you to select the test, controls the operation of the prototype and its environment and receives data from the Results analysis module.

Input actions generator — this is a hardware device that provides transmission of input data in accordance with the selected test.

Results analysis module — this is hardware device "on the fly" compares the data coming from the prototype and benchmark results. In case of divergence between the real and the benchmark result the start time of the discrepancy and its duration, is transmitted to the test management system.

Prototype — is a functional model of the simulated target system implemented in FPGA basis.

Consider that Input actions generator, Prototype Results analysis module are implemented in one or more FPGAs that are interconnected and controlled by Test management system.

When passing the first two steps of development and testing stage, the mechanisms for fault injection are not involved. The system is tested on a set consisting only of Test management system, Input actions generator, Prototype and

Results analysis module. In this methodology, these steps do not have any special features and is held normal job with the FPGA-based ASIC prototype.

Then we consider the functioning of the system in the third stage, i.e., using the mechanism of fault injection.

## 2.2   The Purpose of the Simulation

The purpose of the simulation is the obtained data of system response to different external conditions (primarily on the changes in the intensity and direction of radiation exposure). Based on these data it is possible to compare functionally identical systems, but implemented on different technologies or using different ways to provide fault tolerance. Also, using the modeling results we can draw conclusions about the need for additional structural protection.

Statistics obtained from simulation results of different systems will enable to develop a list of recommendations for developers in the choice of construction technologies, architectural ways to ensure failure resiliency, etc.

## 2.3   The Input Data for Modeling

As mentioned earlier, the input data for the simulation are:

- the chip description
- the parameters of the environment
- technological parameters of IC
- the topology of the chip
- description of the test structures
- a set of test inputs and reference results.

In preparation for the modeling will be determined by the following data:

- the direction, energy and type of heavy charged particles
- threat parameters.

Consider the following source data and requirements in more detail.

The chip description - detailed functional description of the system given in the form of netlist or synthesized HDL.

The parameters of the environment are characteristics of the external environment: radiation conditions, temperature, vibration, etc. The exact list of parameters depends on the used model of external influences. So, for example, when the simulation of space-based systems as the parameters of the environment will require: the parameters of the target orbit, the parameters of the radiation field in the orbit, the parameters of sources of ionizing radiation, information about the space craft design and its orientation relative to the radiation source.

The direction, energy and type of heavy charged particles is calculated from data about the environment and are used to determine the impact of the environment on the chip.

Data on technological parameters is determined by the production technology of the chip. Such data can include production technology of the chip, the size of the transistor, etc.

Topology data — data about placement of system components on the chip. They are required to identify the elements which have been adversely affected. To simplify the modeling it is proposed to divide the chip into squares of equal area. Then, for each of the resulting squares to make a list of elements that are in it. The size of the square is set on the basis of computational capabilities and the required accuracy of the simulation.

The example of partitioning shown in Fig. 2. If the element is in the two squares at the same time, it will be included in the list of affected elements for both squares. So, when particles in square the K1 or in the K2 square the operation of the element will be distorted.

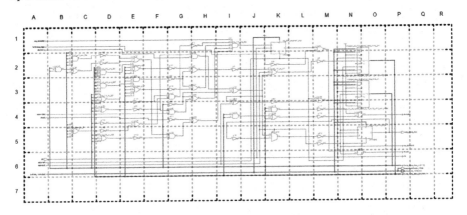

**Fig. 2.** Example of system split

Threats parameters are calculated on the basis of data about the effects of technological parameters and the topology of the chip. Each threat is the list of elements affected, time, duration of exposure and type of exposure (latching values, setting a certain value, introducing delay).

Description of the test structures is a synthesized description of the test structure that will allow on command from the control device to distort the chip.

As discussed earlier in the modeling involved only the memory elements (Fig. 3), so that all elements involved in the test should be replaced by a test structure, examples of which are shown in figures (Figs. 4, 5 and 6). In the simulation using test patterns of the following types:

- fixed output value (Fig. 4) — after activating the OUTP_DATA set to the constant (0, 1 or a value stored in the element at the time of activation), which persists until deactivating.
- a specified output value (Fig. 5) — after activation the OUTP_DATA is set to the value that was an input port TEST_DATA. In the ongoing simulation

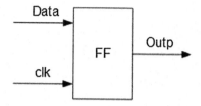

**Fig. 3.** Base Flip-Flop element

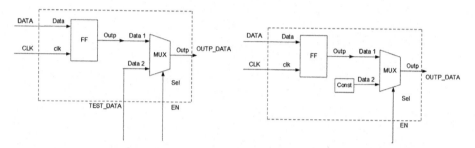

**Fig. 4.** Test structure with constant error value

**Fig. 5.** Test structure with variable error value

input value of TEST_DATA, and accordingly the output OUTP_DATA may vary.

– the introduction of the delay (Fig. 6) — after activation of the test structure passes a value from input DATA with a delay of 0.5–1 clock cycle input clock.

To manage test structures is used, the EN input.

All test patterns are controlled by shift registers, which are called test chain. Such test chains can be one or more. Examples of source systems and as test elements and the test chain is shown in the Figs. 7 and 8. For corruption of the system the prototype is suspended in a test chain are transferred signs of activation and, if necessary, test data. Then a prototype system is resumed. To simulate the recovery of the element chip is performed similar actions - suspension system prototype, the introduction of a new test sequence (which disables test patterns) and prototype resume.

A set of test inputs and etalon results are streams of input values that will be passed to the system-prototype and the expected responses to these values. All the differences obtained with etalon data recorded with further classification depending on the duration (failed/refused).

## 2.4 Used Models

The modeling process uses two main models describing the internal and external parameters of the chip, and providing information about possible threats to the system.

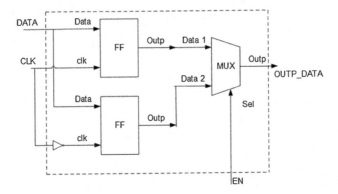

**Fig. 6.** Test structure with delay

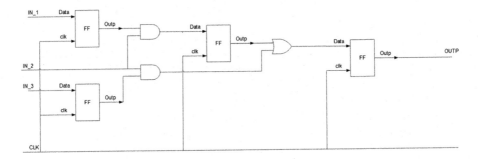

**Fig. 7.** Analyzing system before test structure install

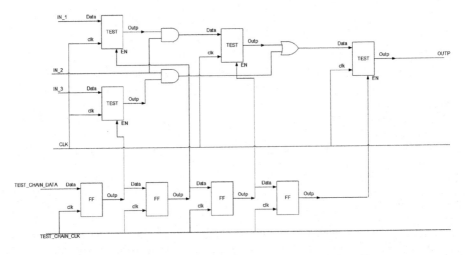

**Fig. 8.** Analyzing system after test structure install

**External Influences Model.** Is used to determine the most likely parameters of the negative effects acting on the chip. In this paper, we consider only radiative effects related to exposure of heavy charged particles in the chip included in the spacecraft. Thus, the result of the use of this model are the most probable parameters of heavy charged particles. The main parameters are: the number of particles per unit time, the particle energy, the angle of incidence on the chip, the type of particles. An example of such models are models of the distributions of fluxes of charged particles Skobeltsyn Institute of Nuclear Physics Lomonosov Moscow State University Model [1], CREME [2], AP8 [3].

**Threats Model.** It used to determine the parameters of the threats for the chip. Based on the most probable parameters of heavy charged particles, obtained from the model of external influences, manufacturing techniques of microcircuits, the size of chip elements, element type chip, the chip topology and spacecraft design at the conclusion that the number of elements affected.

Main stages of calculation of parameters of threats:

– evaluation of the impact of heavy charged particles on the spacecraft design with specific parameters;
– assessment of the impact of heavy charged particles on the chip and the total area of exposure.

Examples of such analysis see [4–6].

To simplify calculations the chip is divided into squares of equal area. Then for each of the squares is determined by a list of constituent elements. Depending on the technology, energy, heavy charged particles and angle of incidence resulting from the impact may affect one or more squares (Fig. 9).

All squares caught in the area of impact are considered victims, and the elements entering into their composition — have been adversely affected. The duration of negative impact is determined by the production technology of chips and let particles. Obviously, the smaller the area of each of the squares the more accurate and complex calculation. The resulting list of items which have been adversely affected is used to generate the parameters of the threat.

To obtain the most complete and accurate picture of system response to changes in the tests may be carried out until, all the squares have been checked. But this would require a significant amount of time, so it is recommended to use one of the methods to identify the most vulnerable elements.

**Methods to Identify the Most Vulnerable Elements.** A simulation model for all possible squares in the chip is not always necessary, because the squares can be empty (if the chip is not completely filled) or in a square can is items not having a significant impact on the results of work such as test nodes, blocks, gathering statistics, etc. To reduce the number of calculations is proposed using the following methods: static analysis of the netlist, dynamic analysis of the netlist, recommendations from the developer.

Static analysis netlist involves building a dependency and disseminating data tree from each of memory elements. Obviously, the element, the value of which

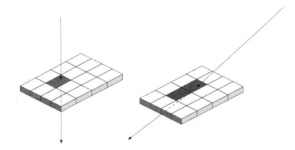

**Fig. 9.** Versions of the trajectories of heavy charged particles and damaged units

depends on the value of the largest amount of other elements is the most vulnerable, i.e. dangerous for the system.

Netlist dynamic analysis is carried out in the process which counts the number of items that changed their state by changing the state of the analyzed element.

Developer guide suggest to create the list of elements, a failure in which, in the opinion of the developer would lead to the most serious negative consequences.

### 2.5   The Stages of the Simulation

**(1) The Creation of FPGA Prototype.** In this work it is considered that the functional model of the system is part of the original data (developed earlier). In this case, the first step of the algorithm creates an FPGA-prototype. I.e. functional model of the ASIC is implemented in a FPGA included in the system simulation.

**(2) Determination of Parameters of the External Environment.** Based on the requirements presented to the system, are determined by the operating conditions of the circuits in terms of radiation exposure.

**(3) The Calculation of the Most Probable Parameters of Heavy Charged Particles.** Based on the available models are determined by the parameters of heavy charged particles. Namely the direction, intensity, let. Depending on the objectives of the simulation can be defined a few sets of parameters.

**(4) The Selection of Elements Directly Involved in Modeling.** Using the results of the analysis netlist and/or information transferred by the developer the functional model and topology data analyzed VLSI define the list of analyzed elements (or the set of squares that will participate in the simulation).

**(5) Identification of Major Threats.** Based on defined parameters of heavy charged particles, the list of simulated elements, the characteristics of the topology will form a set of threats, i.e. the list of distortions that will be implemented in the FPGA in the process. Each element of this list consists of the following fields: identification of item, time of the beginning of distortion, type of distortion, time of distortion ending.

**(6) Generation of Template Failure.** After the list of members subjected to distortion is necessary it is necessary to generate the pattern fault. Fault injection module matches the name of the item from the list of threatened and bit(s) in the test chain, corresponding to elements in the FPGA-prototype. The values of these bits is set in accordance with the type of impact. Also formed the chain that turns off the test mode after the end of exposure (if exposure was short-term). For one test can be defined several templates fault, if in accordance with the model of threats for this test, there were several distortions in the work.

**(7) Testing.** At this stage, simulates the operation of the system under specified conditions. In the process of testing the FPGA-prototype serves the input data and constantly compares the actual output data and the reference set by the test management system. For making threats at the specified moment of model time, the system is being suspended, setting the values of the test of chain, then work the FPGA-prototype continues until the next threat or the end of the current exposure of the threat (i.e., disables test mode for the corresponding elements).

**(8) Evaluation of Test Results.** According to the test results, it is concluded that deviations from normal operation as a result of each of the impacts. Accordingly, we identify the most exposure (most often resulting in unacceptable consequences). Based on these data can be given recommendations on the necessary changes in the computer system.

## 2.6   Evaluation of Simulation Results

The results of the simulation is the ranking impact of internal and external parameters of the system and system architecture.

As a result may change the way to ensure failure resiliency, technology chips, design a computer system (e.g. the introduction of additional protection solution). Also produced a set of statistics for the purpose of compiling a list of recommendations for developers of computing systems space-based. The resulting simulation data can also be used in the radiation tests of the chip to reduce the amount of time required to test and determine the most dangerous for the chip parameters negative impact.

# 3   Hardware-Software Simulation Complex (HSSC)

## 3.1   HSSC Structure

Block diagram of the HSSC is presented in Fig. 10.

The basis of the complex is a workstation carrying a software implemented core modules of the modeling system. N additional FPGA-based expansion cards, combined with a workstation in a single system with a high-speed PCIe bus, are generally designed for the direct emulation of the target chip prototypes. Special interconnects are organized between the expansion cards to support various modes of target chip projects simulation.

**Simulation Modes.** Now we define three main simulation modes:

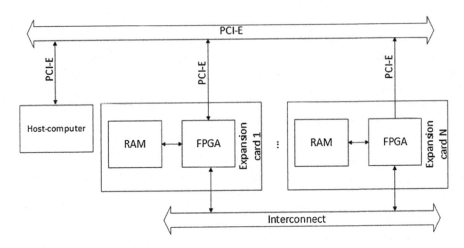

**Fig. 10.** Structural scheme of hardware-software complex

1. Parallel simulation of the target VLSI projects.
2. Master-slave pattern of the FPGA-based expansion cards interaction.
3. Combined.

**Parallel Simulation of the Target VLSI Projects.** In the first mode, all expansion cards are used for independent parallel simulation of target VLSI projects. The internal structures of the FPGAs for all cards will be the same (Fig. 11): In addition to the target VLSI project the following modules are hardware implemented:

– The Fault-Injection Module, that carries out direct fault injection to the VLSI project accordingly to the fault injection pattern.
– Stimulus provider, imitating external signals for the target VLSI project
– The Data Acquisition Module, which reads the values of the outputs of the target VLSI project and sends the data to the software-based simulation evaluation module for data processing
– The Control Unit, which coordinates all the modules implemented on FPGA, starts and ends simulation and data exchange with the workstation.

RAM blocks, of the expansion cards allow to carry out temporary storage of fault patterns, stimulus and output values of the chip. In general, this allows to offload computing means of the workstation, and gives an opportunity to perform long-term simulation without interruption.

In the parallel simulation mode it is possible to implement different scenarios:

– Parallel simulation of the different fault-tolerant architectures of the same target chip under the influence of the same external influences. In this case the problem of determining the most effective target VLSI architectures, providing the required level of fault tolerance and minimal hardware costs, can be solved.

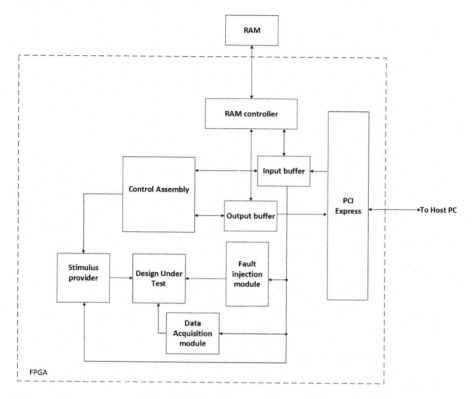

**Fig. 11.** The internal structure of the FPGA in parallel mode simulation tolerant architectures

- Parallel modeling various functional modules of a computer system under the influence of the same or different external influences. The problem of determining the most vulnerable element of the system for a given model of external influences can be solved by implementing this scenario. The simulation results in this case will allow to ensure additional measures for providing the required level of tolerance for the most vulnerable elements of the system. It is important that the implementation of this scenario allows to apply different models of external influences for each functional module separately, which can be useful when simulating distributed computing systems.
- Parallel modeling of the VLSI projects with structural redundancy implemented. It is well-known technique to implement for the individual functional modules or the whole chips that have a specific purpose, structural redundancy methods to provide the desired service life. The most commonly used method is a triple modular redundancy (TMR), which imply implementing three copies of the desired module/chip, which outputs are voted by the principle 2 valid of 3 participated. The described simulation mode allows to evaluate the level of tolerance of such redundant systems, as well as provide an opportunity to identify the root cause of faults and failures of such systems.

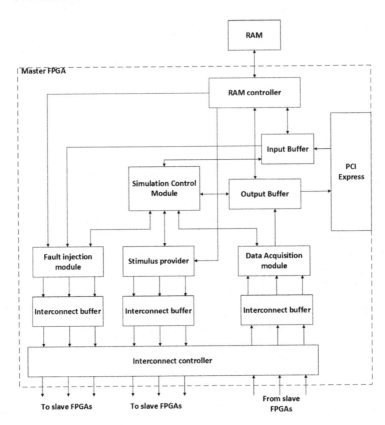

**Fig. 12.** The internal structure of the master-FPGA in simulation systems, with TMR by chip

**Master-Slave Pattern of the FPGA-Based Expansion Cards Interaction.** Mode 2 allows to pick up one or more FPGAs, called masters, just for the implementation of simulation providing modules. The rest of the FPGAs are used only for the target VLSI projects implementation. This approach can be used in modeling VLSI projects with a large number of components that provide a high filling percentage of the FPGA. Depending on a given percentage, as well as the performance and throughput of FPGA interconnect interface used, various realizations of this approach are possible. Figures 12 and 13 shows an embodiment of the internal structure of the master and slave FPGAs implementations in the case of realization all the simulation providing modules in a one FPGA. In this case, the control unit controls the simulation of VLSI projects implemented in N-1 slave FPGAs. Furthermore, the data acquisition, fault-injection and stimulus provider modules are implemented in master FPGA and interact with all the VLSI projects. Among the alternative embodiments of the lead FPGA can be, for example, the architecture which implies delegation of the fault injection stimulus generation functions to the two control (master) FPGA. This simulation mode allows to implement the same scenarios that suit mode 1, with the only difference in the number of concurrent

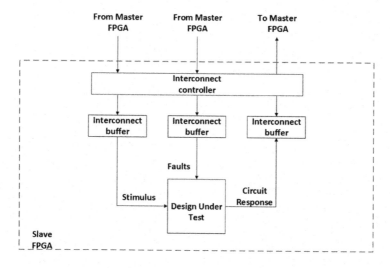

**Fig. 13.** The internal design of the slave FPGA

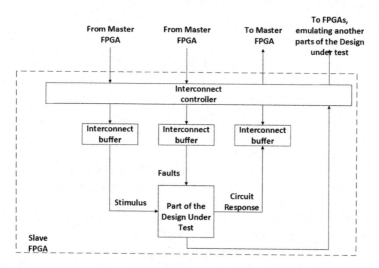

**Fig. 14.** The internal design of the slave FPGA for huge ASIC prototyping

simulated systems will depend on the number of master FPGAs for each slave, as well as the number of slaves.

**Combined Simulation Mode.** Combined simulation mode enables efficient parallel simulation of VLSI projects of various sizes. For example, for VLSI projects with a large number of elements the master - slave mode can be used, and untapped FPGAs can be used to model projects of smaller volumes and have an internal structure for the parallel simulation mode.

When simulating VLSI projects of large volumes the situation when a project could not be implemented via the single FPGA may arise. In this case, by using

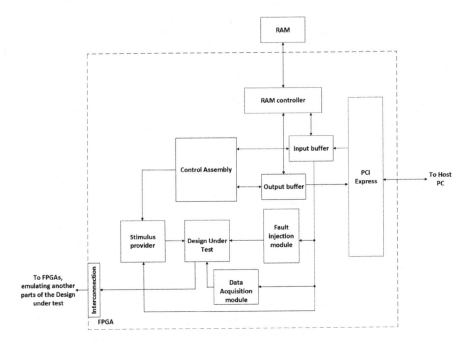

**Fig. 15.** The internal design of the slave FPGA for huge ASIC prototyping for parallel modeling

different partition algorithms the project composition is made to give the opportunity to implement the project via multiple FPGAs. In this case, depending on the target VLSI project volume master-slave or parallel simulation modes could be used. In the first case, the internal structure of the slave FPGA will have the form shown in Fig. 14. In the case of parallel simulation scheme is also somewhat more complicated (Fig. 15).

It is worth noting that the implementation of the calculations for models of external influences and the emergence of threats due to the large volume of data to be processed requires considerable computing power that could not be implemented via separate HSSC. Solving this problem requires the development of efficient algorithms of necessary calculations parallelization and the usage of high-performance computing servers or networked systems of distributed computing. The question of choosing the effective implementation of this approach is not covered in this paper.

## 4    Conclusions

(1) In this paper, the method of modeling fault-tolerant computing systems based on FPGA prototypes using the advanced fault-injection method is proposed.

(2) This method takes into account the parameters of the environment (types, intensity and direction of radiation in the environment) and the parameters of

the internal environment (chip production technology and the topology of the chip). Chip simulation, carried in the presence of specific injected threats shows the behavior of the system under specified conditions.

(3) The concept of the integrated HSSC with defined simulation modes, the constituent elements of the complex and a set of required software, including models of the environment (external factors) and the occurrence of threats (appearance of failures depending on internal factors) is proposed.

(4) The three-stage modeling concept, that combines functional testing and failures causes modeling is also proposed. It comprises:

1. functional test of the system without taking into account means of tolerating faults and failures
2. functional test of the system with implement means of providing fault tolerance
3. verification of system performance in the presence of faults and failures.

The proposed method gives an opportunity to determine the impact of each of the external and internal factors arising during operation on it's performance and provides recommendations for the further system's fault tolerance level improvement.

**Acknowledgments.** This research was conducted under Federal special-purpose program "Research and development on priority directions of scientific and technological complex of Russia in 2014–2020" under contract RFMEFI57715X0161.

# References

1. Panasyuk, M.I., Novikov, L.S.: Space model. In: The Impact of the Space Environment on Spacecraft Materials and Equipment, vol. 2, Moscow (2007)
2. Tylka, A.J., Adams Jr., J.H., Boberg, P.R., Brownstein, B., Dietrich, W.F., Flueckiger, E.O., Petersen, E.L., Shea, M.A., Smart, D.F., Smith, E.C.: CREME96: a revision of the cosmic ray effects on micro-electronics code. IEEE Trans. Nucl. Sci. **44**(6), 2150–2160 (1997)
3. Gaffey, J., Bilitza, D.: NASA/National space science data center trapped radiation models. J. Spacecraft Rockets **31**(2), 172 (1994)
4. Weller, R.A., Mendenhall, M.H., Reed, R.A., Schrimpf, R.D., Warren, K.M., Sierawski, B.D., Massengill, L.W.: Monte Carlo simulation of single event effects. IEEE Trans. Nucl. Sci. **57**(4), 1726–1746 (2010)
5. Mendenhall, M.H., Weller, R.A.: A probability-conserving cross-section biasing mechanism for variance reduction in Monte Carlo particle transport calculations. Nucl. Instrum. Meth. A **667**(1), 38–43 (2012). doi:10.1016/j.nima.2011.11.084
6. Barak, J.: Simple calculations of proton SEU cross sections from heavy ion cross sections. IEEE Trans. Nucl. Sci. **53**(6), 3336–3342 (2006)

# Space Simulation Wireless Broadband Network in Electromagnetic Interference, and Obstructions

Victor M. Churkov[✉]

Moscow State Institute of Electronics and Mathematics
Higher School of Economics (MIEM HSE), Moscow, Russia
wchur@mail.ru

**Abstract.** The paper presents a mathematical model, the technology, tools and simulation results broadband heterogeneous network under different broadband noise spectrum, the intensity of natural and fabricated obstacles. The features of the joint use of existing and emerging technologies of broadband networks in various ranges of the radio spectrum.

**Keywords:** Mathematical model · Wireless networks and modeling tools · Broadband networks · Electromagnetic interference · System interference · Natural and artificial obstacles · The radio frequency spectrum · The coverage area

## 1 The Spatial Mathematical Model of the Service Area of a Wireless Broadband Network

Famous figures CINR and RSSI measure the quality of the radio signal in the client equipment. CINR (Carrier to Interference + Noise Ratio) - the ratio of carrier signal level to noise level is used in telecommunications for assessing the quality of the signal and connection.

If the signal quality is above a certain standard for the norm. RSSI (Received Signal Strength Indication) - the power level of the received signal can be used to approximate the signal quality estimate.

Indicator CINR measured periodically in the channel and radio signal receiver is transmitted source for analysis and decision-making on how to connect and uniquely defines the client's network speed.

Thus, CINR is an integral indicator of the quality of the radio signal, taking into account:

natural and artificial obstacles; internal and external noise, electromagnetic interference; dynamics of change in all the factors of time and space.

It is obvious that the use of the index CINR fundamentally necessary as you would a broadband network and a qualitative design and development of

V. Vishnevsky and D. Kozyrev (Eds.): DCCN 2015, CCIS 601, pp. 224–229, 2016.
DOI: 10.1007/978-3-319-30843-2_23

its infrastructure, taking into account the relative position coordinate effectively and to minimize the number of elements.

External factors - the desired signal level, noise, and electromagnetic disturbance largely determine the bandwidth of the radio channel, which can significantly reduce the data transmission rate and, consequently, the quality of the traffic [1].

Spatial model of the wireless network comprises a matrix parameter CINR, consider the quality of the radio as a function of given above influencing factors defining network traffic at any point.

Formation of the matrix takes place based on the results of an experimental study of the parameters of the radio signal CINR in the network coverage area.

Mathematical model [2] of the spatial organization of the coverage of wireless broadband network comprises a plurality of functions, variables and constants which: $U$ - universal set of elements of the system wireless network (coverage):

S – A lot of system coverage subsets (regions);
C – The set of centers of areas;
R – Radius of the area;
$(x, y, z)$ – Spatial coordinates;
$u(x, y, z)$, $c(x, y, z)$, $s(x, y, z)$ – Elements of the set;
$m, n$ – Cardinality of the set;
$\rho(x, y, z)$ – An indicator of the effectiveness of the radio signal coverage.

Let $U, C, S, R, (x, y, z), \rho(x, y, z)$ — variable mathematical model:

$U, R, (x, y, z), \rho(x, y, z)$ – Asked variables;
$C, S$ – Expanded variables.

Let the sets $U, C, S$ associated functional dependency.

We introduce additional features $f, \phi, \psi, |\nu_u|$ specifying constraints and performance indicators for coverage, operating:

$f$ – Computation $C \subset U$ and subject to the limitations of the selected performance criteria (for example $|S_j| \to \max \to |U|$);
$\phi$ – Calculation of the elements subsets $U_i$;
$\psi$ – Calculation of the subset $U_j \setminus U_{j+1}$;
$|\overrightarrow{\nu_u}|$ – Calculation of magnitude of the vector in the coordinates $U$;

Express $f, \phi, \psi, |\nu_u|$ through the original variables:

$$\begin{cases} f\left[\phi(u, |\overrightarrow{\nu_u(x, y, z)}|, R, \rho(x, y, z), \psi(U))\right]; \\ \phi\left(u, |\overrightarrow{\nu_u(x, y, z)}|, R, \rho(x, y, z), \psi(U)\right); \\ |\overrightarrow{\nu_u(x, y, z)}|; \\ \psi(U); \end{cases}$$

Then the desired sets $C, S$ can be written as a function of variables $U, R, (x, y, z), \rho(x, y, z), f, \phi, \nu_u, \psi$; following system of equations.

$$\begin{cases} C = \left\{ u : f\left[\phi(u, |\overrightarrow{\nu_u(x,y,z)}|, R, \rho(x,y,z), \psi(U))\right] \right\}; \\ S = \left\{ u : \phi(c, |\overrightarrow{\nu_u(x,y,z)}|, R, \rho(x,y,z), \psi(U)) \right\}. \end{cases} \tag{1}$$

We introduce the function $\xi$ convolution common to the mathematical description of the sets $C, S$. Function $\xi = (|\overrightarrow{\nu_u(x,y,z)}|, R, \rho(x,y,z), \psi(U))$ calculates the occurrence of an element $u_i$ in the set $\psi(U)$ with parameters $R$ for $\rho(x,y,z)$ all elements $\psi(U)$. Then the system (1) can be minimized.

$$\begin{cases} C = \left\{ u : f\left[\phi(u, \xi)\right] \right\}; \\ S = \left\{ u : \phi(c, \xi) \right\}. \end{cases} \tag{2}$$

Thus, the system (2) is a generalized mathematical model of the spatial organization of wireless coverage. The model provides the calculation of coverage areas and centers for wireless network infrastructure based on defined performance criteria and constraints.

## 2  Simulation of Electromagnetic Parameters

Figures 1, 2, 3 - to select the network technology and suitable models of equipment;
Safety tasks;
anti-interference;
determining speed limits, covering radius cells;

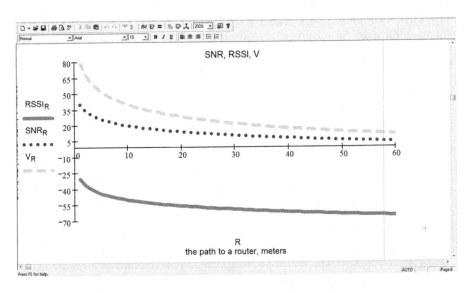

**Fig. 1.** Distribution of SNR and network speed under nominal and operating noise.

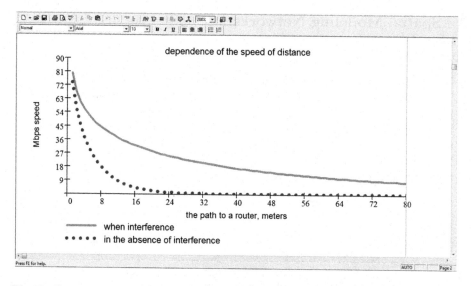

**Fig. 2.** Comparison of the speed of the network at different levels of interference at different distances from the access point.

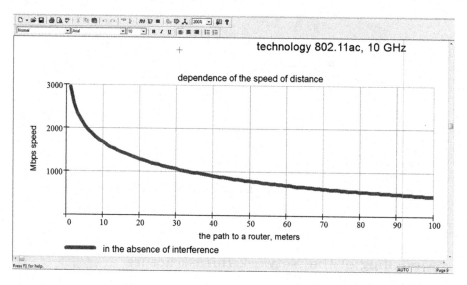

**Fig. 3.** The speed of the network when the standard 802.11n 5 GHz, standard 802.11ac 10 GHz.

frequency of training WDS mode;

Ensure the sustainability of the network traffic to fluctuations and other external and internal loads; ensure scalability;

Modeling parameters of the latest and advanced networking technologies to address issues of development, forecasting and analysis of fundamental solutions.

## 3   Spatial Modeling Networks

Spatial modeling networks

Figure 4 - for the formation of a mathematical model of the spatial coverage;
Visual analysis and processing machine to determine the coordinates of access
points for networks with complex spatial configuration;
Determining the structural network diagrams required frequency resources,
the number of cells;
Equipment selection - bridges, repeaters, repeaters, routers, routers, and other
tools for network infrastructure.

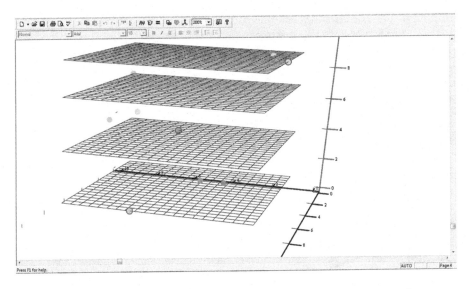

**Fig. 4.** Result of 3D modeling space coverage heterogeneous network.

## 4   Modelling of the Network Infrastructure

Modelling of the network infrastructure

Figure 5 - for the construction of multi-level network optimization;
localization, verification under-interference, obstacles and predictable sporadic
impacts;
the formation of design decisions on the system and application levels;
providing a carrier mode traffic with QoS.

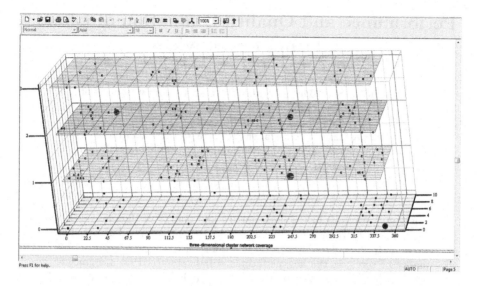

**Fig. 5.** Resulted space simulation coverage while minimizing infrastructure.

## 5   Conclusion

The presented models and tools ensure the full, high-quality design and testing of broadband wireless networks and systems for existing and future applications.

At the initial stage of the deployment of complex wireless networks, modernization and operation of these funds increase reliability; the stability of the technical parameters, functional stability.

Reduces the complexity, cost, recovery time during operation and maintenance.

## References

1. Vishnevsky, V.M.: Broadband wireless data transmission networks, p. 592. Technosphera, Moscow (2005)
2. Churkov, V.M., Kosilov, N.A., Zhdanov, V.M.: Optimization of spatial models wireless sensor networks. In: Proceedings of Seventh International Conference Information and Telecommunication Technologies Intelligent System, Schweiz, 03–09 July 2010, pp. 9–16 (2010)

# Performance and Quality Evaluation of Cloud Information Services

Vadim Efimov[1], Serg Mescheryakov[2(✉)], and Dmitry Shchemelinin[1]

[1] RingCentral Inc., San Mateo, CA, USA
{vadim.efimov,dmitry.shchemelinin}@ringcentral.com
[2] St. Petersburg Polytechnic University, St. Petersburg, Russia
serg-phd@mail.ru

**Abstract.** Analysis of the performance and quality of the information services, which are provided globally over the Internet cloud distributed infrastructure, is given. For integral evaluation of the performance and quality of services, the key performance indicators (KPIs) are considered, computed, and compared with the IT industry level. Examples of real world implementation of change management KPIs at RingCentral Company, which allowed improving the quality of service availability to a higher level of 99.998 %, are presented.

**Keywords:** Cloud information services · Service availability · Quality of service · Key performance indicators

## 1 Introduction

At the International Congress on Ultra-Modern Telecommunications and Control Systems (ICUMT-2014) [1], the authors presented new adaptive technique of using Zabbix open-source monitoring solution [2] for integration of change management and real-time operations in a big globally distributed cloud infrastructure at RingCentral Company (RC). RC provides over 150 cloud telecommunication services for 300K+ business companies in North America, West Europe and East Asia [3]. Given big cloud environment consisted of 10K+ Virtual Machines (VMs) and taking into account 40 % annual growth, it is important to evaluate the Level of Service Availability (SLA) and the Quality of Service (QoS) in real time using worldwide known Key Performance Indicators (KPIs).

This paper describes new KPI metrics, which are automatically computed based on the integrated data model proposed in [4].

## 2 Fault Detection and Remediation of RC Services

Most of modern telecommunication companies provide not only the Internet access, but also Infrastructure as a Service (IaaS) and Software as a Service (SaaS), which are cloud based and virtualized.

© Springer International Publishing Switzerland 2016
V. Vishnevsky and D. Kozyrev (Eds.): DCCN 2015, CCIS 601, pp. 230–237, 2016.
DOI: 10.1007/978-3-319-30843-2_24

At RC, the entire globally distributed infrastructure is monitored 24/7 by special Global Network Operating Center (GNOC) with the help of Zabbix monitoring system and custom dashboards. All the monitoring data and system events on remote hosts are tracked and stored via proxy into Zabbix centralized database (DB). Network Monitoring Console (NMC) and Service Oriented Dashboard (SOD) are specifically designed as web applications to visualize all the critical alerts in real time on a single integrated screen. The latest values are verified by Zabbix system for troubleshooting and triggering the alarm when a certain threshold is exceeded, while the historical data is effectively used to analyze the trends, figure out repeatable issues, predict and prevent the anomalies beforehand.

Table 1 shows various data sources and corresponding fault detection methods used in RC.

**Table 1.** Fault detection methods in RC services.

| Fault detection methods and data sources | RC average statistics |
| --- | --- |
| Auto-alerts from Zabbix, NMC, SOD | 90 % |
| Notifications from external providers and partners | 4 % |
| Hourly tests executed half-manually by GNOC | 3 % |
| Email or call from customer support | 2 % |
| Other sources | 1 % |

In case of any anomaly detection in IaaS/SaaS, GNOC is 24/7 responsible for the following actions:

1. Acknowledge the anomaly using NMC, SOD, or other detection methods and sources listed in Table 1.
2. Communicate to the other team members and notify of the problem. For that purpose, the RC Company uses the worldwide popular documenting system JIRA [5] as well as in-house real time tracking web application Incident Management Portal (IMP) with its own DB of incidents.
3. Restore the SaaS using predefined workaround in case the problem is well-known, for example reboot problematic VM or redirect the workload to a standby computing resource in the cloud.
4. Escalate the problem to corresponding technical support engineering if IaaS/SaaS cant be restored by means of GNOC.

All the changes to production IaaS/SaaS, either during remediation process or as a part of planned upgrade, are reflected in separate in-house web application Change Management Portal (CMP) with its own DB linked to the projects in Roadmap information system.

For particular chronic problems, which are related to external vendors and do not depend on RC internal resources, the auto-remediation procedures are implemented. In this case the action items from GNOC are not required, JIRA document is auto-generated for future statistical analysis, and the auto-notification is sent as well. Good solution for auto-remediation is described in [6] where capacity management of Java applications, which are running under heavy workload in server pool and often lead to a well-known problem of Java memory leak, is introduced to prevent the entire SaaS outage and evaluate the required redundancy of cloud computing resources.

## 3    Key Performance Indicators of Incident Management

For the purpose of performance and QoS evaluation, new KPI metrics are introduced. Main limitation of KPI computing is that RC management processes, including IMP, CMP, JIRA, Zabbix, Roadmap, etc., are tracked by and implemented as independent applications with separated DBs that need to be integrated. Since the monitoring system is linked with and configured for all remote hosts, including production IaaS/SaaS and IMP/CMP/JIRA management systems, Zabbix was found to be the best option for integration of different data sources into centralized Zabbix DB and further automation of KPI computations.

For that reason, custom SQL scripts are developed and scheduled to run daily from Zabbix proxy against corresponding remote DBs, returning KPI values for each of IMP/CMP/JIRA metrics back to Zabbix DB. For example, the following SQL query is implemented to find out and count the repeatable incidents for the last 30 days:

```
select
  ifnull(sum(tags_count)-count(*),0) as dupes_count
from
  (select i.short_desc,
    (select group_concat(t.tag_name)
     from inc_has_tags h
       left outer join service_tags t
       on h.service_tags_id_tag = t.id_tag
     where h.incident_inc_id = i.inc_id
     group by h.incident_inc_id
    ) as tags,
    count(*) as tags_count
   from incident i
   where create_start_date < curdate()
     and create_start_date >= adddate(curdate(),-30)
   group by i.short_desc, tags
   having count(*) > 1
  ) as dupes_list;
```

Table 2 shows sample report generated in Zabbix with KPI metrics for IMP and their daily values at RC.

Having KPI data collected in centralized DB separately for IMP/CMP/JIRA, Zabbix then allows creating new integral metrics of arbitrary complexity, which are computed locally on Zabbix side without additional remote querying.

**Table 2.** Sample Zabbix daily report for IMP KPI at RC Company.

| IMP KPI metric name | Latest daily value |
| --- | --- |
| % incidents auto-detected by Zabbix | 84% |
| % incidents escalated to customer support | 5% |
| % incidents as a result of GNOC hourly testing | 11% |
| % incidents from other sources | 0% |
| Total number of incidents | 19 |
| Number of repeatable incidents | 3 |
| Average customer impact, minutes | 6 |
| Maximum number of affected customers | 0 |

The following world-class KPI metrics are implemented in Zabbix to evaluate QoS and performance of incident management at RC Company (Fig. 1) [1]:

1. Time to Detect (TTD). Typically, TTD is a term of control engineering, which identifies the time between a fault has actually occurred and the moment it is recognized by the monitoring system [7]. Specifically at RC, TTD is a control reaction of GNOC to Zabbix alert triggering on NMC/SOD or incident auto-submission to IMP, which may not be fast enough due to proxy delay in data transfer. TTD has SLA of 2 min as maximum and is determined by the time to acknowledge alert in Zabbix.

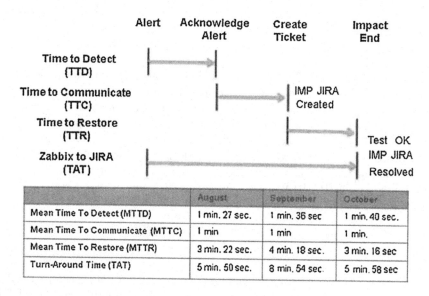

| | August | September | October |
| --- | --- | --- | --- |
| Mean Time To Detect (MTTD) | 1 min. 27 sec. | 1 min. 36 sec | 1 min. 40 sec. |
| Mean Time To Communicate (MTTC) | 1 min | 1 min | 1 min. |
| Mean Time To Restore (MTTR) | 3 min. 22 sec. | 4 min. 18 sec. | 3 min. 16 sec |
| Turn-Around Time (TAT) | 5 min. 50 sec. | 8 min. 54 sec. | 5 min. 58 sec |

**Fig. 1.** Incident management scorecard and monthly KPI statistics at RC Company.

2. Time to Communicate (TTC). TTC is a nonverbal communication time to notify the involved people [8]. At RC Company, TTC is defined as the time to create IMP/JIRA ticket, either automatically or manually by GNOC, and escalate to corresponding executor depending on the type and severity of the incident. SLA is 2 min for both tracking systems but actually is less than a minute due to Zabbix-to- and IMP-to-JIRA automation.

3. Time to Restore/Repair (TTR). TTR is a basic measure of the corrective maintenance downtime, which is required to repair/restore the failed IaaS/SaaS [9]. Some IaaS/SaaS providers interpret TTR as Time to Response/React that is not correct because restoring process takes a certain time to execute and should include post-restore testing.

   In practice, daily TTD, TTC and TTR are often computed to weekly and monthly average values that is usually a part of SLA [10], which is compared with the other IT companies and is declared as 99.99 % standard for 24/7 IaaS/SaaS availability. Sample monthly average KPI numbers at RC are shown in Fig. 1 and are evaluated using the following expressions:

$$\text{MTTD} = \sum (T_2 - T_1)/N, \tag{1}$$

$$\text{MTTC} = \sum (T_3 - T_2)/N, \tag{2}$$

$$\text{MTTR} = \sum (T_4 - T_3)/N, \tag{3}$$

   where MTTD — Mean Time to Detect; MTTC — Mean Time to Communicate; MTTR — Mean Time to Restore; T1 — actual start of downtime on remote host; T2 — the moment the alert is detected and acknowledged by GNOC; T3 — the time when IMP/JIRA incident created and properly escalated; T4 — actual start of uptime, meaning that IaaS/SaaS has been restored and tested and IMP/JIRA status is Resolved/Closed; N — total number of incidents.

4. Turnaround Time (TAT). TAT is a world-class category of performance engineering [11] to evaluate overall effectiveness of system operations. For performance management in general, TAT means the total time to fulfill a request [12]. Particularly in computing, TAT stands for the total time period between the start of a process/task and the end of its execution/result returning back to a customer/user. According to IMP/CMP management policy at RC, TAT is finished after SaaS is available for the users and tested, customer impact has ended and IMP/JIRA ticket is closed that is equivalent to a sum of TTD, TTC and TTC, or their average values:

$$\text{TAT} = \text{TTD} + \text{TTC} + \text{TTR}, \tag{4}$$

$$\text{TAT} = \text{TTD} + \max(\text{TTC}, \text{TTR}), \tag{5}$$

Expression (5) is applicable when a predefined remediation procedure is fully automated where TTC is usually much less than TTR. In this case the TTC and TTR phases are performed in parallel without manual involvement to IMP escalation process.

**Table 3.** Sample Zabbix daily report for CMP KPI at RC Company.

| CMP KPI metric name | Latest daily value |
| --- | --- |
| % changes caused by incidents | 4% |
| % changes implemented within target time | 84% |
| % unplanned emergency changes | 12% |
| Average change implementation time, minutes | 67 |
| Total changes implemented | 25 |
| Total changes successfully tested | 24 |
| Total changes cancelled/rollback | 1 |

## 4    Key Performance Indicators of Change Management

Analysis of IMP DB statistics shows many cases when the incidents are caused by certain changes to production infrastructure, either hot fix or planned upgrade. Therefore the change management and CMP KPI computations are of key importance to meet high SLA level of 99.99%. At RC, every Change Management Request (CMR) is always planned as a part of a bigger or smaller Roadmap project with a predefined lifecycle, including all necessary phases implementation, debug and stress tests on stage environment, handover and release notes meetings, deployment to production IaaS and monitoring, post-mortem analysis of possible rollback.

All CMR changes are tracked in CMP DB in real time. For better efficiency of CMP processes, new CMP KPI metrics are introduced to Zabbix monitoring

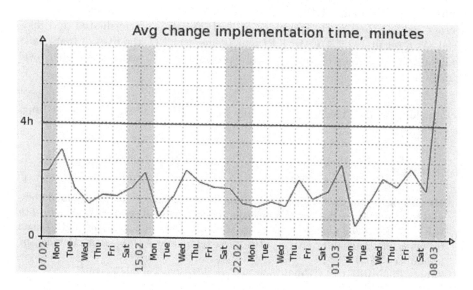

**Fig. 2.** Monthly CMP KPI statistics of CMR implementation time at RC Company.

system similar to IMP KPI described in Sect. 3. Table 3 shows sample report generated in Zabbix with some CMP KPI metrics and daily values at RC.

Daily KPI reports are expanded to weekly, monthly, quarterly, etc. Zabbix functionality allows defining the aggregated hosts/metrics and triggering the alerts for specified thresholds in case overall degradation of IMP/CMP management process. Example in Fig. 2 shows monthly average statistics of CMR implementation time. This CMP KPI metric is useful for analysis of the performance and workload of daily planned maintenance. According to the graph, one value has exceeded the 4 h limit that is not an incident and is allowed during non-business weekend. On the other hand, the overall usage of daily maintenance window is less than 50 % at RC Company and could be improved.

Paper [13] presents more examples of KPI evaluation specifically for big dynamic enterprise environment with continuous delivery of SaaS updates to the customers.

## 5   Conclusions

Integration of separate tracking systems of different vendors into a centralized management portal is the effective way to improve world-class KPIs of a big IT company. At RC with distributed IaaS/SaaS provided globally in the Internet cloud, Zabbix enterprise-class monitoring solution is considered for the purpose of integration with incident and change management, JIRA documenting, Roadmap project planning and other information systems. Zabbix DB stands for integrated storage where both real time and historical data from remote hosts and management systems is collected and analyzed.

New IMP/CMP/JIRA metrics are implemented in Zabbix to automatically compute the world-class KPIs, such as TTD, TTC, TTR, TAT, etc., and their average values. Real world examples of KPI statistics are evaluated and analyzed. As a result of IMP/CMP KPI management at RC, the SLA of IaaS/SaaS availability for business customers improved to 99.998 %, which is a higher level than IT industry standard.

**Acknowledgments.** This work is partially supported by and implemented at Ring-Central Inc., USA [3].

## References

1. Mescheryakov, S., Shchemelinin, D., Efimov, V.: Adaptive control of cloud computing resources in the internet telecommunication multiservice system. In: 6th IEEE International Congress on Ultra Modern Telecommunications and Control Systems, St. Petersburg (2014). http://ieeexplore.ieee.org/xpl/articleDetails.jsp?arnumber=7002117
2. Zabbix Enterprise-Class Monitoring System. http://www.zabbix.com
3. RingCentral Cloud Services. http://www.ringcentral.com

4. Volkov, A., Efimov, V., Mescheryakov, S., Shchemelinin, D.: Integrated data model for managing a multi-service dynamic infrastructure. In: Proceedings of the 3rd International Conference on Computer Modeling and Simulation. Polytechnic University Publishing House, St. Petersburg (2014). http://dcn.icc.spbstu.ru/index.php?id=344&L=2

5. Atlassian JIRA. https://www.atlassian.com/software/jira

6. Mescheryakov, S., Shchemelinin, D.: Capacity management of java-based business applications running on virtualized environment. In: Proceedings of the 39th International Conference for the Performance and Capacity by CMG, La Jolla, CA, USA (2013). http://www.cmg.org/conference/cmg2013/

7. Fault Detection and Isolation. http://en.wikipedia.org/wiki/Fault_detection_and_isolation

8. Nonverbal Communication. http://en.wikipedia.org/wiki/Nonverbal_communication

9. Mean Time to Repair. http://en.wikipedia.org/wiki/Mean_time_to_repair

10. Service Level Agreement. http://en.wikipedia.org/wiki/Service-level_agreement

11. Performance Engineering. http://en.wikipedia.org/wiki/Performance_engineering

12. Turnaround Time, Performance Management. http://en.wikipedia.org/wiki/Turnaround_time

13. Mescheryakov, S., Shchemelinin, D.: Big software deployments in a big enterprise environment. In: 2nd ASE International Conference on Big Data Science and Computing, Stanford University, CA, USA (2014). http://www.ase360.org/handle/123456789/135/

# Heuristic Solution for the Optimal Thresholds in a Controllable Multi-server Heterogeneous Queueing System Without Preemption

Dmitry Efrosinin[1,2](✉) and Vladimir Rykov[3]

[1] Johannes Kepler University Linz, Altenbergerstrasse 69, 4040 Linz, Austria
dmitry.efrosinin@jku.at
[2] Institute of Control Sciences, RAS,
Profsoyuznaya street 65, 117997 Moscow, Russia
[3] Gubkin Russian State University of Oil and Gas,
Leninskiy prospect 65, 119991 Moscow, Russia
vladimir_rykov@mail.ru
http://www.jku.at, http://www.ipu.ru, http://www.gubkin.ru

**Abstract.** As it known the optimal policy which minimizes the long-run average cost per unit of time in a multi-server queueing system with heterogeneous servers without preemption has a threshold structure. It means that the slower server must be activated whenever all faster servers are busy and the number of customers in the queue exceeds some specified for this server threshold level. The optimal thresholds can be evaluated using the Howard iteration algorithm or by minimizing the function of the average cost which can be obtained in closed form as a function of unknown threshold levels. The both cases have sufficient restrictions on dimensionality of the model. In present paper we provide a heuristic method to derive expressions for the optimal threshold levels in explicit form as functions of system parameters like service intensities, usage and holding costs for an arbitrary number of servers. The proposed method is based on the fitting of the boundary planes between the areas where the optimal threshold takes a certain value.

**Keywords:** Controllable queueing system · Heterogeneous servers · Long-run average cost · Threshold policy · Optimal allocation

## 1 Introduction

Controllable queueing systems with heterogeneous servers and single queue find wide application in various fields of human activity including resource allocation in telecommunication and computer networks, control problems in production lines and so on. For detailed review of the literature on heterogeneous queues the reader is referred to [2]. The queues with preemption or priority interruption of

D. Efrosinin—This work was funded by the Russian Foundation for Basic Research (RFBR), Project No 15-08-08677-a.

V. Vishnevsky and D. Kozyrev (Eds.): DCCN 2015, CCIS 601, pp. 238–252, 2016.
DOI: 10.1007/978-3-319-30843-2_25

service, i.e. when the customer can change the server during the service process, are well studied, see e.g. [6]. The optimal allocation mechanism for customers in such systems is defined normally through a simple Fastest Free Server policy, when a new customer is transfered to the fastest available server even if it is already under service on some slower server. In this case the system can be exhaustively studied independently of the number of servers in the system. In contrast for the queueing systems without preemption the Fastest Free Server allocation policy occurs to be not the best one. As it was shown in papers [8,9] the optimal policy, which minimizes the long-run average cost per unit of time, is of threshold type, i.e. for any server exists some threshold level, which specifies the number of customers in the queue when the server must be activated. There are a number of methods how to calculate the optimal threshold levels of the control policy. They can be evaluated numerically by means of the Howard iteration algorithm [5] or by numerical minimization of the average cost function evaluated in closed form, see e.g. [3,4]. In both cases we have problems with dimensionality of the model if the number of servers is relative high. Therefore a natural question arises: Is there a possibility to get some heuristic solution which will be valid for an arbitrary number of servers.

In framework of the paper we solve such a problem and construct formulas for the optimal threshold levels which seems to be quite appropriate for the studied systems. The kernel element of the proposed heuristic method consists in evaluation of the average cost function in closed form as a function of unknown threshold levels. It is performed by the method of the difference equations. Then we analyze the boundaries between the optimality regions where the fixed threshold level takes a certain value. It can be done at least for two and three server models but obtained results can be generalized to the case of an arbitrary number of servers.

The rest of the paper is organized as follows. Section 2 describes the mathematical model based on a homogeneous Markov process under fixed threshold policy. In Sect. 3 the stationary state probabilities are evaluated by means of the difference equations approach. Section 4 deals with optimization problem. Heuristic method for the optimal threshold level estimation is presented in Sect. 5. Finally, some numerical examples are illustrated in Sect. 6.

In further sections we will use the notations $e_j$ for the vector with 1 in the $j$-th (beginning from 0-th) position and 0 elsewhere, $1_{\{A\}}$ for the indicator function, where $1_{\{A\}} = 1$ if the condition $A$ holds, and 0 otherwise, $\overline{a,b}$ for the elements of the set $[a, b] \cap \mathbb{N}_0$.

## 2    Mathematical Model

Consider a queueing system where the customers arrive according to the Poisson process with intensity $\lambda$ and $K$ heterogeneous servers have exponentially distributed service times with intensities $\mu_1, \mu_2, \ldots, \mu_K$. The service of the customers is assumed to be without preemption, i.e. the customer being served on some server can not change this server. The interarrival and service times are assumed to be mutually independent. Define the control policy

$$f = (q_1, q_2, \ldots, q_K),$$

which prescribes how to allocate the customers between the servers and is of threshold type defined as a sequence of threshold levels

$$1 = q_1 \leq q_2 \leq \cdots \leq q_K < \infty.$$

According to this policy the first $k$ servers must be occupied whenever there are $q$ customers in the queue and $q = \overline{q_k, q_{k+1} - 1}$. The cost structure consists of the following components:

$c_0$ – the holding cost per unit of time for any waiting customer in the queue,

$c_j$ – the usage cost per unit of time for any busy server $j$.

The servers are enumerated in such a way that

$$0 < c_1 \mu_1^{-1} \leq c_2 \mu_2^{-1} \leq \cdots \leq c_K \mu_K^{-1}, \tag{1}$$
$$0 < \mu_1^{-1} \leq \mu_2^{-1} \leq \cdots \leq \mu_K^{-1},$$

where the ratio $c_j \mu_j^{-1}$ stands for the average usage cost of the $j$th server. The states of the system at time $t$ are described by the vector $\{Q(t), D(t)\}_{t\geq 0}$, where $Q(t)$ – the number of customers in the queue and $D(t) = \{D_1(t), \ldots, D_K(t)\}$ – the states of the servers,

$$D_j(t) = \begin{cases} 0, & \text{the server } j \text{ is idle,} \\ 1, & \text{the server } j \text{ is busy,} \end{cases} \quad j = \overline{1, K}.$$

The multidimensional random process

$$\{X(t)\}_{t\geq 0} = \{Q(t), D(t)\}_{t\geq 0} \tag{2}$$

is a homogeneous Markov process. The state space of the process $\{X(t)\}_{t\geq 0}$ is

$$E = \{x = (q, d); q \in \mathbb{N}_0, d \in D(q)\}. \tag{3}$$

Note that the state space of the servers $E_D(q)$ depends on the queue length $q$,

$$E_D(q) = \left\{ \begin{array}{l} d_j \in \{0,1\}, j = \overline{1, K}, q = 0 \\ d; d_j = 1, d_i \in \{0,1\}, 1 \leq j \leq k \leq i - 1 \leq K - 1, q = \overline{q_k, q_{k+1} - 1} \\ d_j = 1, j = \overline{1, K}, q \geq q_K \end{array} \right\},$$

and the number of states $|E_D(q)|$ is

$$|E_D(q)| = \begin{cases} 2^K, & q = 0, \\ 2^{K-k}, & q = \overline{q_k, q_{k+1} - 1}, k = \overline{1, K-1}, \\ 1, & q \geq q_K. \end{cases}$$

Denote by $A_{q_k}(x)$ and $\overline{A}_{q_k}(x)$, $k = \overline{1,K}$, the following events and their complements,

$$A_{q_k}(x) = \{q(x) \geq q_k,\, d_i(x) = 1,\, i = \overline{1,k-1},\, d_k(x) = 0\}, \tag{4}$$

$$\overline{A}_{q_k}(x) = \{q(x) \leq q_k - 1,\, d_i(x) = 1,\, i = \overline{1,k-1},\, d_k(x) = 0\}.$$

Analyzing the transitions of the Markov process $\{X(t)\}_{t \geq 0}$ and using the notations (4) we get the system of balance equations for the stationary state probabilities $\pi_x$, $x = (q, d) \in E$,

$$\pi_x = \lim_{t \to \infty} \mathbb{P}[X(t) = x]$$

in the form

$$\left(\lambda + \sum_{k=1}^{K} d_k(x)\mu_k\right)\pi_x = \lambda\left(\sum_{k=1}^{K} \pi_{x-e_k} 1_{\{A_{q_k-1}(x-e_k)\}} + \pi_{x-e_0} 1_{\{\overline{A}_{q_k-1}(x-e_0)\}}\right)$$

$$+ \sum_{k=1}^{K} d_k(x)\mu_k \pi_{x+e_0} 1_{\{A_{q_k}(x+e_0)\}} + \sum_{k=1}^{K} (1 - d_k(x))\mu_k \pi_{x+e_k} 1_{\{\overline{A}_{q_k}(x+e_k)\}}. \tag{5}$$

To get a solution of this system for the fixed threshold policy in closed form we intend to apply the method of the difference equations.

## 3   Evaluation of the Stationary State Probabilities

**Theorem 1.** *For the two-server queueing system $M/M/2$ with a threshold policy $f = (1, q_2)$ the stationary state probabilities satisfy the relations,*

$$\pi_{(0,0,0)} = [R_{q_2+1}^{-1}B_{q_2+1} - 1]\pi_{(0,0,1)}, \tag{6}$$

$$\pi_{(q,1,0)} = [R_{k-q_2}B_{q_2+1} - B_{q+1}]\pi_{(0,0,1)},\ k = \overline{0, q_2 - 1},$$

$$\pi_{(q,1,1)} = B_{q+1}\pi_{(0,0,1)},\ q = \overline{0, q_2 - 1},$$

$$\pi_{(q,1,1)} = N_{q-q_2+1}\pi_{(q_2-1,1,1)},\ q \geq q_2$$

$$\pi_{(0,0,1)} = \left[R_{q_2+1}B_{q_2+1} + \frac{R_{q_2-1}B_{q_2+1}}{R_{q_2+1} - R_{q_2}} + \frac{N_1 B_{q_2+1}}{1 - N_1}\right]^{-1},$$

*where*

$$R_n = \left(\frac{\lambda}{\mu_1}\right)^n,\ N_n = \left(\frac{\lambda}{\mu_1 + \mu_2}\right)^n,\ B_n = b_1\beta_1^n + b_2\beta_2^n,$$

$$\beta_{1,2} = \frac{(\lambda + \mu_1 + \mu_2) \pm \sqrt{(\lambda + \mu_1 + \mu_2)^2 - 4\lambda\mu_1}}{2\mu_1},$$

$$b_1 = \frac{1 - \beta_1}{\beta_2 - \beta_1},\ b_2 = \frac{\beta_2 - 1}{\beta_2 - \beta_1}.$$

*Proof.* Due to threshold structure of the control policy the system of balance Eq. (5) are divided into separate subsystems which can be treated as homogeneous difference equations.

*Subsystem 1.*

$$(\lambda + \mu_2)\pi_{(0,0,1)} = \mu_1\pi_{(0,1,1)}, \tag{7}$$
$$(\lambda + \mu_1 + \mu_2)\pi_{(0,1,1)} = \lambda\pi_{(0,0,1)} + \mu_1\pi_{(1,1,1)},$$
$$(\lambda + \mu_1 + \mu_2)\pi_{(q,1,1)} = \lambda\pi_{(q-1,1,1)} + \mu_1\pi_{(q+1,1,1)}, \ q = \overline{1, q_2 - 2}.$$

For this system the solution is assumed to be in form

$$\pi_{(q,1,1)} = \beta^{q+1}\pi_{(0,0,1)}, \ q = \overline{0, q_2 - 1}. \tag{8}$$

The substitution of this solution to the equation of the previous system for the arbitrary $q$ and subsequent division by $\beta^{q-1}$ lead to the equation

$$\mu_1\beta^2 - (\lambda + \mu_1 + \mu_2)\beta + \lambda = 0.$$

This quadratic equation has two real roots,

$$\beta_{1,2} = \frac{(\lambda + \mu_1 + \mu_2) \pm \sqrt{(\lambda + \mu_1 + \mu_2)^2 - 4\mu_1}}{2\mu_1}.$$

Therefore the probabilities $\pi_{(k,1,1)}$ as solutions of the difference equations must satisfy the relation

$$\pi_{(q,1,1)} = B_{q+1}\pi_{(0,0,1)}, \ q = \overline{0, q_2 - 1}, \tag{9}$$

where

$$B_n = (b_1\beta_1^n + b_2\beta_2^n).$$

From the first equation of the subsystem (6) for the probability $\pi_{(0,1,1)}$ we get

$$b_1\beta_1 + b_2\beta_2 = \beta_1 + \beta_2 - 1,$$
$$b_1 + b_2 = 1,$$

whence it appears the relations for the constants $b_1$ $b_2$.

*Subsystem 2.* For the states $x \in E$, where the queue length exceeds the level $q_2$, we have

$$(\lambda + \mu_1 + \mu_2)\pi_{(q,1,1)} = \lambda\pi_{(q-1,1,1)} + (\mu_1 + \mu_2)\pi_{(q+1,1,1)}, \ q \ge q_2. \tag{10}$$

Assuming the solution of these difference equations in the form

$$\pi_{(q,1,1)} = \eta^{q-q_2+1}\pi_{(q_2-1,1,1)}$$

and substituting of this result to the equality (10), it is easy to show, that there is exist the only one non-trivial solution of the quadratic equation, $\eta = \frac{\lambda}{\mu_1+\mu_2}$, i.e.

$$\pi_{(q,1,1)} = N_{q-q_2+1}\pi_{(q_2-1,1,1)}, \ q \ge q_2, \tag{11}$$

where

$$N_n = \eta^n = \left(\frac{\lambda}{\mu_1 + \mu_2}\right)^n.$$

*Subsystem 3.* Consider the following subsystem of the system of balance equations,

$$(\lambda + \mu_1)\pi_{(0,1,0)} = \lambda\pi_{(0,0,0)} + \mu_1\pi_{(1,1,0)} + \mu_2\pi_{(0,1,1)}, \tag{12}$$
$$(\lambda + \mu_1)\pi_{(q,1,0)} = \lambda\pi_{(q-1,1,0)} + \mu_1\pi_{(q+1,1,0)} + \mu_2\pi_{(q,1,1)}, \ q = \overline{1, q_2 - 2},$$
$$(\lambda + \mu_1)\pi_{(q_2-1,1,0)} = \lambda\pi_{(q_2-2,1,0)} + \mu_2\pi_{(q_2-1,1,1)},$$
$$(\lambda + \mu_1 + \mu_2)\pi_{(q_2-1,1,1)} = \lambda\pi_{(q_2-2,1,1)} + \lambda\pi_{(q_2-1,1,0)} + (\mu_1 + \mu_2)\pi_{(q_2,1,1)}.$$

To solve the system the corresponding equations are summed up for the probabilities $\pi_{(q,1,0)}$ and $\pi_{(q,1,1)}$. As a result we obtain the following equations

$$\pi_{(q,1,1)} + \pi_{(q,1,0)} = R_{q+1-q_2}(\pi_{(q_2-1,1,1)} + \pi_{(q_2-1,1,0)}), \ q = \overline{0, q_2 - 1},$$

where $R_n = \left(\frac{\lambda}{\mu_1}\right)^n$. Now we rewrite the last equation of the subsystem (12),

$$\lambda(\pi_{(q_2-1,1,1)} + \pi_{(q_2-1,1,0)}) = (\lambda + \mu_1 + \mu_2)\pi_{(q_2-1,1,1)} - \lambda\pi_{(q_2-2,1,1)}.$$

But from the system (7) it follows,

$$(\lambda + \mu_1 + \mu_2)\pi_{(q_2-1,1,1)} - \pi_{(q_2-2,1,1)} = \frac{\mu_1}{\lambda}(b_1\beta_1^{q_2+1} + b_2\beta_2^{q_2+1}),$$

that in turn leads to the relation

$$\pi_{(q,1,0)} = [R_{q-q_2}(b_1\beta_1^{q_2+1} + b_2\beta_2^{q_2+1}) - (b_1\beta_1^{q+1} + b_2\beta_2^{q+1})]\pi_{(0,0,1)} \tag{13}$$
$$= [R_{q-q_2}B_{q_2+1} - B_{q+1}]\pi_{(0,0,1)}, \ q = \overline{0, q_2 - 1}.$$

Finally from

$$\lambda\pi_{(0,0,0)} = \mu_1\pi_{(0,1,0)} + \mu_2\pi_{(0,0,1)}$$

we get

$$\pi_{(0,0,0)} = [\left(\frac{\mu_1}{\lambda}\right)^{q_2+1}(b_1\beta_1^{q_2+1} + b_2\beta_2^{q_2+1}) - 1]\pi_{(0,0,1)} \tag{14}$$
$$= [R_{q_2+1}^{-1}B_{q_2+1} - 1]\pi_{(0,0,1)}.$$

**Theorem 2.** *For the three-server queueing system M/M/3 with a threshold policy $f = (1, q_2, q_3)$ the stationary state probabilities satisfy the following relations,*

$$\pi_{(0,0,1,1)} + \pi_{(0,0,1,0)} = \left(\frac{(R_{q_2+1} - 1)B_{q_2+1}}{R_{q_2+2} - R_{q_2+1}} + \frac{(N_{q_3-q_2+1} - N_1)B_{q_2}}{N_1 - 1}\right.$$
$$\left. + \frac{\Phi_1 H_{q_3} N_{q_3-q_2+1}B_{q_2}}{(1 - \Phi_1)H_{q_3+1}}\right)^{-1}, \tag{15}$$

$$\pi_{(0,0,1,1)} = \frac{N_{q_3-q_2+1}B_{q_2}}{H_{q_3+1}A_{q_2}}[\pi_{(0,0,1,1)} + \pi_{(0,0,1,0)}],$$

$$\pi_{(0,0,0,1)} = (G_{q_2}^{-1}[A_{q_2} + R_1^{-1}A_{q_2+1} - N_1^{-1}H_{q_2+1}A_{q_2}] - 1)\pi_{(0,0,1,1)},$$

$$\pi_{(0,0,0,0)} = (R_{1+q_2}^{-1}B_{q_2+1} - 1)[\pi_{(0,0,1,1)} + \pi_{(0,0,1,0)}] - \pi_{(0,0,0,1)}$$

$$\pi_{(q,1,0,0)} = (R_{q_2-q}^{-1}B_{q_2+1} - B_{q+1})[\pi_{(0,0,1,1)} + \pi_{(0,0,1,0)}]$$
$$\qquad - G_{q+1}[\pi_{(0,0,1,1)} + \pi_{(0,0,0,1)}] + A_{q+1}\pi_{(0,0,1,1)}, \quad q = \overline{0, q_2 - 1},$$

$$\pi_{(q,1,0,1)} = G_{q+1}[\pi_{(0,0,1,1)} + \pi_{(0,0,0,1)}] - A_{q+1}\pi_{(0,0,1,1)}, \quad q = \overline{0, q_2 - 1},$$

$$\pi_{(q,1,1,0)} = B_{q+1}[\pi_{(0,0,1,1)} + \pi_{(0,0,1,0)}] - A_{q+1}\pi_{(0,0,1,1)}, \quad q = \overline{0, q_2 - 1},$$

$$\pi_{(q,1,1,0)} = N_{q+1-q_2}B_{q_2}\pi_[\pi_{(0,0,1,1)} + \pi_{(0,0,1,0)}] - H_{q+1}A_{q_2}\pi_{(0,0,1,1)}, \quad q = \overline{q_2, q_3 - 1},$$

$$\pi_{(q,1,1,1)} = A_{q+1}\pi_{(0,0,1,1)}, \quad q = \overline{0, q_2 - 1},$$

$$\pi_{(q,1,1,1)} = H_{q+1}A_{q_2}\pi_{(0,0,1,1)}, \quad q = \overline{q_2, q_3 - 1},$$

$$\pi_{(q,1,1,1)} = \Phi_{q-q_3+1}H_{q_3}A_{q_2}\pi_{(0,0,1,1)}, \quad q \geq q_3,$$

*where*

$$R_n = \left(\frac{\lambda}{\mu_1}\right)^n, \; N_n = \left(\frac{\lambda}{\mu_1+\mu_2}\right)^n, \; \Phi_n = \left(\frac{\lambda}{\mu_1+\mu_2+\mu_3}\right)^n,$$

$$A_n = a_1\alpha_1^n + a_2\alpha_2^n, \; B_n = b_1\beta_1^n + b_2\beta_2^n, \; G_n = g_1\gamma_1^n + g_2\gamma_2^n, \; H_n = h_1\xi_1^n + h_2\xi_2^n,$$

$$\alpha_{1,2} = \frac{(\lambda+\mu_1+\mu_2+\mu_3) \pm \sqrt{(\lambda+\mu_1+\mu_2+\mu_3)^2 - 4\lambda\mu_1}}{2\mu_1},$$

$$\beta_{1,2} = \frac{(\lambda+\mu_1+\mu_2) \pm \sqrt{(\lambda+\mu_1+\mu_2)^2 - 4\lambda\mu_1}}{2\mu_1},$$

$$\gamma_{1,2} = \frac{(\lambda+\mu_1+\mu_3) \pm \sqrt{(\lambda+\mu_1+\mu_3)^2 - 4\lambda\mu_1}}{2\mu_1},$$

$$\xi_{1,2} = \frac{(\lambda+\mu_1+\mu_2+\mu_3) \pm \sqrt{(\lambda+\mu_1+\mu_2+\mu_3)^2 - 4\lambda(\mu_1+\mu_2)}}{2(\mu_1+\mu_2)},$$

$$a_1 = \frac{1-\alpha_1}{\alpha_2-\alpha_1}, \; a_2 = \frac{\alpha_2-1}{\alpha_2-\beta_1}, \; b_1 = \frac{1-\beta_1}{\beta_2-\beta_1}, \; b_2 = \frac{\beta_2-1}{\beta_2-\beta_1},$$

$$g_1 = \frac{1-\gamma_1}{\gamma_2-\gamma_1}, \; g_2 = \frac{\gamma_2-1}{\gamma_2-\gamma_1}, \; h_1 = \frac{L_2-L_1}{K_1-K_2}, \; h_2 = 1-h_1,$$

$$K_1 = \left[\left(\frac{1}{\xi_2}-1\right)\xi_1^{q_3-q_2} - \left(\frac{1}{\xi_1}-1\right)\xi_2^{q_3-q_2}\right]A_{q_2},$$

$$L_1 = \left(\frac{1}{\xi_1}-1\right)\xi_2^{q_3-q_2}A_{q_2},$$

$$K_2 = \frac{\mu_3}{\lambda}A_{q_2}\left(\frac{\xi_1-\xi_1^{q_3-q_2+1}}{1-\xi_1} - \frac{\xi_2-\xi_2^{q_3-q_2+1}}{1-\xi_2} - G_{q_2}^{-1}\frac{\gamma_2^{q_2+1}-\gamma_1^{q_2+1}}{\gamma_2-\gamma_1}\left(\frac{1}{\xi_1}-\frac{1}{\xi_2}\right)\right),$$

$$L_2 = \frac{\mu_3}{\lambda}G_{q_2}^{-1}\frac{\gamma_2^{q_2+1}-\gamma_1^{q_2+1}}{\gamma_2-\gamma_1}\left(\left(1-\frac{1}{\xi_1}\right)A_{q_2} + R_1^{-1}A_{q_2+1}\right) + \frac{\mu_3}{\lambda}\frac{\xi_2-\xi_2^{q_3-q_2+1}}{1-\xi_2}A_{q_2}.$$

*Proof.* The proof can be performed in a similar way as before by dividing the system of balance equations into subsystems, which are solved as homogeneous difference equations.

## 4    Optimization Problem for Performance Characteristics

For every fixed threshold policy $f$ we wish to guarantee that the process $\{X(t)\}_{t\geq 0}$ with a state space $E$ is an irreducible, positive recurrent Markov process defined through its infinitesimal matrix $\Lambda = [\lambda_{xy}(q_2, \ldots, q_K)]$, which depends on the threshold policy. The immediate cost $c(x)$ of the specified controllable Markov model is defined as

$$c(x) = c_0 q_0(x) + \sum_{j=1}^{K} c_j d_j(x),$$

where the notations $q(x)$ and $d_j(x)$ stand for the elements of the vector state $x \in E$. As it is known [11], for ergodic Markov process with costs the long-run average cost per unit of time for the policy $f$ coincides with corresponding assemble average,

$$g^f = \lim_{t \to \infty} \frac{1}{t} V^f(x, t) = \sum_{y \in E} c(y) \pi_y^f, \tag{16}$$

where

$$V^f(x, t) = \int_0^t \sum_{y \in E} \mathbb{P}^f[X(u) = y | X(0) = x] c(y) du \tag{17}$$

denotes the total average cost up to time $t$ when the process starts in state $x$ and $\pi_y^f = \mathbb{P}^f[X(t) = y]$ denotes a stationary probability of the process given policy $f$. The policy $f^*$ is said to be optimal when for any available policy $f$

$$g^{f^*} = \min_f g^f. \tag{18}$$

For existence of optimal policies we refer to Aviv and Federgruen [1], Puterman [7], Sennott [10]. Obviously the optimal policy exists if $\sum_{y \in E} |c(y) \pi_y^f| < \infty$. For the model under study it is always the case, since the infinite part of this sum is definitely finite,

$$\sum_{q=0}^{\infty} \left[ c_0(q + q_K - 1) + \sum_{j=1}^{K} c_j \right] \rho^q \pi_{(q_K-1, d)} = c_0 \frac{q_K - 1 - \rho(q_K - 2)}{(1 - \rho)^2} + \sum_{j=1}^{K} c_j \frac{1}{1 - \rho} < \infty,$$

if

$$\rho = \frac{\lambda}{\sum_{j=1}^{K} \mu_j} < 1,$$

which coincides with a stability condition of the system. Hence we can obtain the explicit solution for the function $g^f := g(q_2, \ldots, q_K)$ at least up to the case of three servers. These formulas are used to get a heuristic solution.

*Remark 1.* The method based on a solution of difference equations can be applied theoretically to the system with a higher number of servers as well, although it requires to solve a larger number of subsystems. The number of rows in the system of balance equations is equal to $\sum_{k=0}^{K}(q_{k+1} - q_k)2^{K-k}$, where $q_0 = 0$ and $q_{K+1} = N, N$ - maximal possible number of customers in the system. It is clear that the calculation of the stationary state probabilities and minimization of the function $g$ over $K$ parameters is computationally not feasible for large $K$, what really motivates us to look for some heuristic solution.

## 5   Heuristic Solution for the Optimal Thresholds

As it was shown in [2] the optimal thresholds $q_k, k = \overline{2, K}$, can be calculated by means of the Howard iteration algorithm. But it has sufficient restrictions on dimensionality of the model and number of states. In this section we propose a heuristic method for optimal thresholds estimation, which provide us with explicit formulas depending on the system parameters. Two types of model are discussed: With $\lambda = 0$ and $\lambda > 0$. In the first case the model reduces to the equivalent scheduling problem. It is assumed here that there are a number of customers in the system and no new customers join the system. The optimization problem consists in allocation of the customers between the heterogeneous servers with the aim to minimize the total average cost function until the system becomes empty. For the scheduling problem the optimal thresholds $q_k$ are evaluated exactly. For the original problem with new arrivals the optimal thresholds can be only estimated. Since we are not able to give the rigorous proofs to all our results, they are summarized in the following conjecture. The details and main principles of the used approach are given afterwards.

*Conjecture 1.* The optimal thresholds $q_k, \; k = \overline{2, K}$, are defined by

$$q_k \approx \hat{q}_k = \left\lfloor \frac{1}{c_0} \left[ \frac{c_k}{\mu_k} F_k - \sum_{j=1}^{k-1} c_j \right] \right\rfloor, \quad \text{where} \tag{19}$$

$$F_k = \begin{cases} \sum_{j=1}^{k-1} \mu_j, & \lambda = 0, \\ \frac{1}{2}\left[ \sum_{j=1}^{k-1} \mu_j - \lambda + \sqrt{\left( \sum_{j=1}^{k-1} \mu_j - \lambda \right)^2 + 4(k-1)\mu_k \lambda} \right], & \lambda > 0. \end{cases} \tag{20}$$

*Proof.* Consider the case $\lambda = 0$. Due to our assumption about threshold structure of the control policy $f$, we can calculate the total average cost until the system becomes empty recursively. Assuming the known values of thresholds $q_2, \ldots, q_{k-1}$ we obtain the value of $q_k$. Obviously for the state $x = (0, \ldots, 0)$ we have

$$V(x) = 0,$$
$$V(x + e_1) = \frac{c_1}{\mu_1},$$
$$V(x + e_0 + e_1) = \frac{c_0 + c_1}{\mu_1} + V(x + e_1) = \frac{c_0}{\mu_1} + \frac{2c_1}{\mu_1},$$
$$\cdots$$
$$V(x + (q_2 - 1)e_0 + e_1) = \frac{(q_2 - 1)c_0 + c_1}{\mu_1} + V(x + (q_2 - 2)e_0 + e_1) = \frac{q_2(q_2 - 1)c_0}{2\mu_1} + \frac{q_2 c_1}{\mu_1}.$$

When the queue length has reached the level $q_2$, it becomes optimal to use the second server. Then we have

$$V(x + (q_2 - 1)e_0 + e_1 + e_2) = \frac{c_2}{\mu_2} + \frac{q_2(q_2 - 1)c_0}{2\mu_1} + \frac{q_2 c_1}{\mu_1}.$$

Repeating the procedure up to the level $q_k$ one can obtain

$$V\left(x + (q_k - 1)e_0 + \sum_{j=1}^{k-1} e_j\right) = \frac{(q_k - q_{k-1})\sum_{j=1}^{k-1} c_j}{\sum_{j=1}^{k-1} \mu_j} + \frac{(q_k - q_{k-1})(q_{k-1} + q_k - 1)c_0}{2\sum_{j=1}^{k-1} \mu_j}.$$

Again, since $q_k$ is an optimal threshold, the condition

$$V\left(x + (q_k - 1)e_0 + \sum_{j=1}^{k} e_j\right) = \frac{c_k}{\mu_k} + V\left(x + (q_k - 1)e_0 + \sum_{j=1}^{k-1} e_j\right) < v\left(x + q_k e_0 + \sum_{j=1}^{k-1} e_j\right)$$

implies

$$\frac{c_k}{\mu_k} + \frac{(q_k - q_{k-1})\sum_{j=1}^{k-1} c_j}{\sum_{j=1}^{k-1} \mu_j} + \frac{(q_k - q_{k-1})(q_{k-1} + q_k - 1)c_0}{2\sum_{j=1}^{k-1} \mu_j}$$
$$< \frac{(q_k - q_{k-1} + 1)\sum_{j=1}^{k-1} c_j}{\sum_{j=1}^{k-1} \mu_j} + \frac{(q_k - q_{k-1} + 1)(q_{k-1} + q_k)c_0}{2\sum_{j=1}^{k-1} \mu_j}.$$

From the last inequality we get

$$q_k = \frac{1}{c_0}\left[\frac{c_k}{\mu_k}\sum_{j=1}^{k-1} \mu_j - \sum_{j=1}^{k-1} c_j\right].$$

*Remark 2.* Threshold levels defined by (19) satisfy the inequalities

$$\frac{1}{c_0}\left[\frac{c_k}{\mu_k}\left(\sum_{j=1}^{k-1} \mu_j - \lambda\right) - \sum_{j=1}^{k-1} c_j\right] \leq \hat{q}_k^* \leq \frac{1}{c_0}\left[\frac{c_k}{\mu_k}\sum_{j=1}^{k-1} \mu_j - \sum_{j=1}^{k-1} c_j\right]. \qquad (21)$$

*Proof.* The left inequality of (21) follows directly from

$$F_k \geq \frac{1}{2}\left[\sum_{j=1}^{k-1} \mu_j - \lambda + \sqrt{\left(\sum_{j=1}^{k-1} \mu_j - \lambda\right)^2}\right] = \sum_{j=1}^{k-1} \mu_j - \lambda.$$

To prove the inequality at the right hand side it is necessary to show that

$$F_k \leq \sum_{j=1}^{k-1} \mu_j.$$

By solving this inequality using simple algebraic manipulations we get

$$(k - 1)\mu_k \leq \sum_{j=1}^{k-1} \mu_j,$$

which is true due to the ordering (1).

To get the formulas for the case $\lambda > 0$ consider first the problem of the mean number of customers minimization, i.e. when $c_j = 1, j = \overline{0,K}$. To reduce the number of system parameters in the system (5) we divide the right and the left hand side by $\lambda$ and introduce the notations $r_j = \frac{\mu_j}{\lambda}, j = \overline{1,K}$. The proposed Algorithm 3 is based on a functional estimation of the boundaries between the areas where the optimal threshold $q_k, k = \overline{1,K}$, takes a certain value. Numerical analysis confirms our expectations that the optimal threshold $q_k$ depends only on parameters of the first $k$ servers and the boundaries between the regions of optimality have a linear structure, i.e. they can be represented as hyperplanes.

**Algorithm 3.**
**Step 1.** *Calculation of the optimal thresholds $q_k, k = \overline{1,K}$, for all possible values of $(r_1, \ldots, r_K)$ by means of a function*

$$\bar{N} := \bar{N}(r_1, \ldots, r_K, q_2^*, \ldots, q_K^*) = \sum_{x \in E} \left( q(x) + \sum_{j=1}^{K} d_j(x) \right) \pi_x,$$

*derived in closed form as discussed in previous section.*

**Step 2.** *Labeling of the regions, where the optimal thresholds $(q_2, \ldots, q_K)$ take certain values.*

**Step 3.** *Identification of the points laying on the boundaries between the regions, where*

$$q_k^* = i \ \ and \ \ q_k = i + 1, \ i \geq q_{k-1}.$$

**Step 4.** *Application of the least-squares method to estimate the unknown coefficients $a_{kj}$ of the hyperplanes for each threshold $q_k, k = \overline{1,K}$,*

$$\sum_{j=1}^{k} a_{kj} r_j + a_{kk+1} = 0,$$

*and their representation as functions of $q_k$, $a_{kj} := a_{kj}(q_k^*)$.*

**Step 5.** *Express $q_k, k = \overline{2,K}$, through $r_1, \ldots, r_k$.*

If $K = 2$ and $K = 3$, then the geometric visualization of the algorithm is possible. For the two-server model the regions of optimality of the level $q_2$ and asymptotic lines for the boundaries are illustrated in Figs. 1(a,b). One can easily identify the boundary lines between the areas,

$$a_{21} r_1 + a_{22} r_2 + a_{23} = 0,$$

where the unknown coefficients are evaluated through coordinates of the two points $(x_1, y_1)$ and $(x_2, y_2)$: $a_{21} = y_2 - y_1, a_{22} = x_1 - x_2, a_{23} = x_2 y_1 - x_1 y_2$. By selecting appropriate coordinates we get the coefficients summarized in the next table:

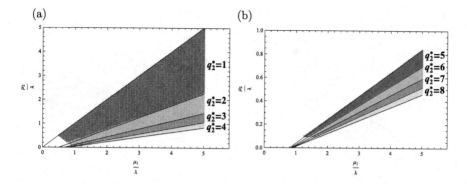

**Fig. 1.** The regions of optimality for $q_2 = i, i = \overline{1,8}$

| $i$ | 1 | 2 | 3 | 4 | 5 | 6 | 7 | 8 |
|---|---|---|---|---|---|---|---|---|
| $x_1$ | 6.47 | 9.65 | 12.74 | 15.79 | 18.83 | 21.85 | 24.87 | 27.90 |
| $y_1$ | 3.00 | 3.00 | 3.00 | 3.00 | 3.00 | 3.00 | 3.00 | 3.00 |
| $x_2$ | 10.48 | 18.66 | 28.74 | 40.79 | 54.83 | 70.87 | 88.87 | 108.90 |
| $y_2$ | 5.00 | 6.00 | 7.00 | 8.00 | 9.00 | 10.00 | 11.00 | 12.00 |
| $a_{21}$ | 2.00 | 3.00 | 4.00 | 5.00 | 6.00 | 7.00 | 8.00 | 9.00 |
| $a_{22}$ | −4.01 | −9.01 | −16.00 | −25.00 | −36.00 | −49.01 | −64.00 | −81.00 |
| $a_{23}$ | −0.91 | −1.92 | −2.96 | −3.95 | −4.98 | −5.89 | −6.96 | −8.09 |

From the table one can see that the coefficients of the line can be represented as a function of threshold estimation $\hat{q}_2$,

$$a_{21}(\hat{q}_2) = \hat{q}_2 + 1, \; a_{22}(\hat{q}_2) = -(\hat{q}_2 + 1)^2, \; a_{23}(\hat{q}_2) = -\hat{q}_2.$$

Conjecture 1 follows from solution of the quadratic equation

$$(\hat{q}_2 + 1)r_1 - (\hat{q}_2 + 1)^2 r_2 - \hat{q}_2 = r_1(\hat{q}_2)^2 - (r_1 - 2r_2 - 1)\hat{q}_2 + r_2 - r_1 = 0,$$

with two roots

$$(\hat{q}_2)_{1,2} = \frac{r_1 - 1 \pm \sqrt{(r_1 - 1)^2 + 4r_2}}{2r_2} - 1.$$

Eliminating the negative root due to condition $\hat{q}_2 \geq 1$ and taking into account the notations $r_j, j = \overline{1,K}$ for the integer-valued level $q_2$ we get an estimation in the form

$$\hat{q}_2 = \left\lfloor \frac{\mu_2 - \lambda + \sqrt{(\mu_2 - \lambda)^2 + 4\mu_2\lambda}}{2\mu_2} - 1 \right\rfloor.$$

To get the formulas for the model with costs we assume that it should have the same structure as (19), where $\lambda = 0$. Hence we get the relation (19) for $\lambda > 0$. In the same way one can derive the formulas for the case $K = 3$. For

(a)                                    (b)

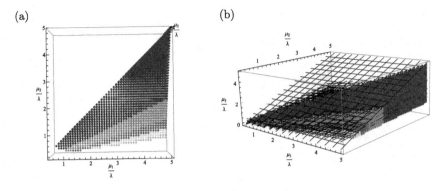

**Fig. 2.** The regions of optimality for $q_k = i, i = \overline{q_{k-1}, 5}, k = \overline{2,3}$

three server model the regions of optimality for the levels $q_2$ and $q_3$ are the planes as illustrated in Fig. 2. These planes are defined as

$$a_{21}r_1 + a_{22}r_2 + a_{23} = 0, \ a_{31}r_1 + a_{32}r_2 + a_{33}r_3 + a_{34} = 0,$$

where the coefficients are expressed through the estimations $\hat{q}_2$ and $\hat{q}_3$,

$$a_{21}(\hat{q}_2) = \hat{q}_2 + 1, \ a_{22}(\hat{q}_2) = -(\hat{q}_2 + 1)^2, \ a_{23}(\hat{q}_2) = -\hat{q}_2,$$
$$a_{31}(\hat{q}_3) = a_{32}(\hat{q}_3) = \hat{q}_3 + 2, \ a_{33}(\hat{q}_3) = -(\hat{q}_3 + 2)^2, \ a_{34}(\hat{q}_3) = -\hat{q}_3.$$

As before, the solution of quadratic equations leads to the formulas corresponding to the problem of the mean number of customers minimization. These formulas can be rewritten for the scheduling model ($\lambda = 0$) with costs as in (19). The form of the recently derived formulas for the queues with $K = 2$ and $K = 3$ reveals a possible general expression for the estimations of threshold levels with arbitrary $K$, as was represented in Conjecture 1.

## 6 Numerical Examples

*Example 1.* Consider the queueing system $M/M/5$ arrival intensity $\lambda = 0.9$. Other parameters take the following values:

| $j$ | 0 | 1 | 2 | 3 | 4 | 5 |
|---|---|---|---|---|---|---|
| $c_j$ | 1.50 | 3.00 | 2.80 | 2.60 | 2.40 | 2.00 |
| $\mu_j$ | – | 2.00 | 0.70 | 0.50 | 0.40 | 0.30 |
| $c_j\mu_j^{-1}$ | – | 1.50 | 4.00 | 5.20 | 6.00 | 6.66 |

In the following table we list the results of calculations for the levels $q_k^*$ and $\hat{q}_k^*, k = \overline{2, K}$ as well as the results for $g$ in case of different control policies and arrival intensity $\lambda$

| $\lambda$ | $g^{OTP}$ | $(q_2^*,q_3^*,q_4^*,q_5^*)$ | $g^{HTP}$ | $(\hat{q}_2^*,\hat{q}_3^*,\hat{q}_4^*,\hat{q}_5^*)$ | $g^{FFS}$ | $g^{RSS}$ | $g^{HS}$ |
|---|---|---|---|---|---|---|---|
| 0.1 | 0.154 | (4,6,7,9) | 0.154 | (4,6,7,9) | 0.162 | 0.459 | 0.328 |
| 0.9 | 1.812 | (3,4,5,6) | 1.812 | (3,4,6,7) | 2.197 | 3.707 | 2.957 |
| 1.5 | 3.687 | (2,3,4,5) | 3.690 | (2,3,5,6) | 4.379 | 5.832 | 4.972 |
| 2.5 | 8.294 | (2,2,3,3) | 8.296 | (2,2,3,4) | 8.917 | 15.502 | 8.982 |
| 3.5 | 19.739 | (2,2,2,3) | 19.857 | (2,2,3,3) | 19.972 | 116.274 | 21.419 |
| 3.8 | 54.689 | (2,2,2,2) | 54.689 | (2,2,2,2) | 54.707 | 586.186 | 65.852 |

Here $OTP$ means *Optimal Threshold Policy, HTP – Heuristic Threshold Policy, FFS – Fastest Free Server* policy, $RSS$ – *Random Server Selection* policy and $HS$ – *Homogeneous System*, where the total service rate $K\mu$ of the homogeneous system is equal to the total service rate $\sum_{j=1}^{K}\mu_j$ of the heterogeneous system.

*Example 2.* Consider the queueing system $M/M/5$ from the previous example. Let $c_j = 1, j = \overline{0,K}$. The next table includes the results of calculations for the levels $q_k^*$ and $\hat{q}_k^*, k = \overline{2,K}$, as well as the values of the mean number of customers in the system $\bar{N}$ for different control policies and $\lambda$.

| $\lambda$ | $\bar{N}^{OTP}$ | $(q_2^*,q_3^*,q_4^*,q_5^*)$ | $\bar{N}^{HTP}$ | $(\hat{q}_2^*,\hat{q}_3^*,\hat{q}_4^*,\hat{q}_5^*)$ | $\bar{N}^{FFS}$ | $\bar{N}^{RSS}$ | $\bar{N}^{HS}$ |
|---|---|---|---|---|---|---|---|
| 0.1 | 0.053 | (2,4,5,8) | 0.053 | (2,4,5,8) | 0.055 | 0.191 | 0.128 |
| 0.9 | 0.692 | (2,3,4,6) | 0.692 | (2,3,4,7) | 0.788 | 1.509 | 1.156 |
| 1.5 | 1.459 | (2,2,3,5) | 1.463 | (2,2,4,6) | 1.649 | 2.222 | 1.955 |
| 2.5 | 3.454 | (1,2,2,3) | 3.510 | (2,2,3,4) | 3.674 | 4.506 | 3.723 |
| 3.5 | 11.181 | (1,1,2,3) | 11.185 | (2,2,2,3) | 11.219 | 28.944 | 11.109 |
| 3.8 | 29.815 | (1,2,2,2) | 29.815 | (2,2,2,3) | 33.029 | 144.798 | 40.458 |

The results show that the difference in performance between the OTP and HTP does not exceed 1.5 % and these policies can be more than 25 % superior in performance comparing to the mostly used in practice allocation policy $FFS$.

# 7   Conclusion

In this paper we have obtained heuristic relations for the optimal threshold levels which minimize the long-run average cost per unit of time or, in particular, the mean number of customers in the queue system with heterogeneous servers. These formulas give satisfactory values for the levels $q_k, k = \overline{2,K}$, with a relative error which was less than 1.5 % in all realized examples. Also it was proved that this policy is superior in performance comparing to some alternative allocation policies. Therefore this policy can be treated at least as a quasi-optimal one. The formulas were verified for different number of servers with quite equal relative errors so we may expect the validity of these expressions to an arbitrary $K$. These approach was successfully applied to the retrial heterogeneous queueing systems as well, the results will be published shortly.

# References

1. Aviv, Y., Federgruen, A.: The value-iteration method for countable state Markov decision processes. Oper. Res. Lett. **24**(5), 223–234 (1999)
2. Efrosinin, D.: Controlled Queueing Systems with Heterogeneous Servers. Dynamic Optimization and Monotonicity Properties. VDM Verlag, Saarbrücken (2008)
3. Efrosinin, D.: Queueing model of a hybrid channel with faster link subject to partial and complete failures. Ann. Oper. Res. **202**(1), 75–102 (2013)
4. Efrosinin, D., Rykov, V.: On performance characteristics for queueing systems with heterogeneous servers. Autom. Remote Control **69**(1), 61–75 (2008)
5. Howard, R.: Dynamic Programming and Markov Processes. Wiley Series, New York (1960)
6. Kumar, B.K., Arivudainambi, D.: Transient solution of an $M/M/c$ queue with heterogeneous servers and balking. Inf. Manage. Sci. **12**(3), 15–27 (2001)
7. Puterman, M.L.: Markov Decision Process. Wiley Series in Probability and Mathematical Statistics. Wiley, New York (1994)
8. Rykov, V., Efrosinin, D.: Optimal control of queueing systems with heterogeneous servers. Queueing Syst. **46**, 389–407 (2004)
9. Rykov, V., Efrosinin, D.: On the slow server problem. Autom. Remote Control **70**(12), 2013–2023 (2009)
10. Sennott, L.I.: Stochastic Dynamic Programming and the Control of Queueing Systems. Wiley, New York (1999)
11. Tijms, H.C.: Stochastic Models. An Algorithmic Approach. Wiley, New York (1994)

# Evaluation of Functionality's Efficiency in Priority Telecommunication Networks with Heterogeneous Traffic

I.V. Kalinin and L.A. Muravyeva-Vitkovskaya$^{(\boxtimes)}$

ITMO University, Sablinskaya str. 14, 197101 Saint-Petersburg, Russia
{kalinin,muravyeva-vitkovskaya}@cs.ifmo.ru

**Abstract.** Analytic dependencies for message delivery time computation from sender node to recipient node in priority telecommunication networks with heterogeneous traffic are described in this article.

**Keywords:** Computer network · Calculation of intensity · Message delivery time · Message flow · Queuing system

## 1 Introduction

Telecommunication network's (TCN) specificity is conditioned on the following factors: variety of network technologies and architectures, variety of QoS requirements for different data types (e.g. the main important text file transmitting index is delivery reliability, guarantees the data loss and distortion absence in received file, and as for audio and video data the main important index is delay jitter for received data packets); traffic heterogeneity, where traffic is controlled by different methods (LAN accessing methods, routing algorithms, establishing connections methods and etc.), aimed at preserving overload and blocking in networks and provide required QoS.

Contemporary TCNs are characterized by variety of provided services, by increasing number of users and amount of transmitted data; by raising level of user's QoS requirements. TCN's requirements can be achieved by selecting structural and functional organization, including such issues as choosing specific technology of data transmitting and processing, defining the most rational communication network topology, choosing network equipment, traffic management mechanisms and etc.

Now days due to intensive growing number of users and applications multi-server networks become more and more common with it's dozens of traffic types [1], caused by new information technology's implementation and using various type applications: Internet, VoIP (Voice over IP), videoconferencing, enterprise resource planning (ERP), client relations management (CRM) and etc. Thus the traffic's heterogeneity [1,2] is one of the TCN's accents. Traffic's heterogeneity lies in transmitting different types of data packets over TCN (video and audio

V. Vishnevsky and D. Kozyrev (Eds.): DCCN 2015, CCIS 601, pp. 253–259, 2016.
DOI: 10.1007/978-3-319-30843-2_26

data packets, speech data packets, text data packets and etc.), where different delivery requirements exist [3]. This circumstance must be kept in network administrator's mind for increasing using TCN's resources efficiency. Traffic's prioritizing is one of the ways to distribute network resources according to existing priorities.

Building networks using switches allows enabling network technology independent traffic's prioritization for increasing user traffic's servicing quality. This new capability in comparison with networks built only upon hubs is consequence of buffering data frames in switches before sending them to another port. Usually switch keeps not one but some queues for every input and output ports, where every queue has it's own processing priority.

Queue processing algorithms are the one of the basic methods of QoS in network elements.

Solving TCN design problems assumes using efficient models and mathematical methods, which give an opportunity to make qualitative and quantitative analysis of TCN's functioning characteristics depending on structural, functional and load parameters. In the TCN's analyzing process one of the major characteristics being determined is delivery time from sender to recipient. Results, represented in [4], give on opportunity to calculate only average delivery time for various message types. But in practice not average time but probability of prompt delivery from sender to recipient for different message types is more interesting (e.g. operative, service, dialog, file message types). Moreover, it's necessary to keep in mind a possibility of using priority techniques for information flows management, where priority techniques are based on general discipline with mixed priorities.

To solve a given task an open queue network with heterogeneous request flow can be used. Let's illustrate method of analyzing and getting relatively simple results assuming nodes having non-priority information flows management methods.

## 2    Method

**Problem Definition.** Let messages of $K$ types circulate in TCN with $n$ nodes. Let us assume that abonent, connected to sender node $h$, generates markovian exponential flow of $k$-type messages with $\lambda_k(0 \,|h, l)$ intensity to destination recipient node $l$, this kind of messages will be called "$(h, \ l)$-messages of $k$-type" for short. Also let us assume that probability $\pi_k(i, j \,|h, l)$ of transferring $(h, \ l)$-messages of $k$-type from node $i$ to connected node $j$ is defined based on chosen routing algorithm. For each $(h, \ l)$-direction these probabilities $\pi_k(i, j \,|h, l)$ form transfer probability matrix which describes possible transfer routes from node $h$ to node $l$ $(i, \ j, \ h, \ l = \overline{1, \ n}; \ k = \overline{1, \ K})$.

Message processing duration in node $i$, represents processing time in node and transferring time to neighbor node, will be assumed exponentially distributed with central tendency equals to $b_k(i)$. Different type messages are serviced in nodes in incoming order. It is necessary to determine transferring time distribution law for $k$-type messages routed from node $h$ to node $l$ ($k = \overline{1, K}$; $i, j, h, l = \overline{1, n}$).

TCN's analysis method is based on network decomposition and led to computation of separate open queue network's nodes like computation of queue system of $M_K/M_K/1$ type with non-priority request processing. This method gives exact results in case of $b_h(i) = b(i)$ ($i = \overline{1,n}$) for all $k = \overline{1, K}$ and approximate otherwise. In addition, measure of result's inaccuracy is decreasing with increasing number of routes and number of types of messages, circulating in TCN by these routes, and also with decreasing difference in message processing duration for different message types in node. It is necessary to determine message flow intensity in every network's node to decompose open queue network.

**Message Flow Intensity Computation.** Message flow intensity for (h, l)-messages of k-type in node $j$ is defined by the following set of linear equations:

$$\lambda_k(j|h,l) = \sum_{i=0}^{n} \lambda_k(i|h,l)\, \pi_k(i,j|h,l) \quad (j = \overline{0,n}), \tag{1}$$

where for all $i,j,h,l = \overline{1,n}$   $\pi_k(0,0|h,l) = 0$;

$$\pi_k(0,j|h,l) = \begin{cases} 1, & j = h; \\ 0, & j \neq h; \end{cases} \qquad \pi_k(i,0|h,l) = \begin{cases} 1, & i = l; \\ 0, & i \neq l. \end{cases}$$

Thus summary intensity of k-type message flow into node $j$ is

$$\lambda_k(j) = \sum_{h=1}^{n}\sum_{l=1}^{n} \lambda_k(j|h,l) \quad (j = \overline{0,n}), \tag{2}$$

where $\lambda_k(0)$ - is intensity of k-type messages incoming into the network.

Transfer probabilities in the open queue network can be calculated in terms of obtained intensities:

$$p_k(0,0) = 0;$$

$$p_k(i,0) = \sum_{h=1}^{n} \lambda_k(i|h,i)/\lambda_k(i) \quad (i = \overline{1,n});$$

$$p_k(0,j) = \sum_{l=1}^{n} \lambda_k(0|j,l)/\lambda_k(0) \quad (j = \overline{1,n}); \tag{3}$$

$$p_k(i,j) = \sum_{h=1}^{n}\sum_{l=1}^{n} \lambda_k(i|h,l)\, \pi_k(i,j|h,l)/\lambda_k(i)\lambda \quad (i,j = \overline{1,n}).$$

Message flow intensities $\lambda_k(j)$ are connected by the obvious dependency:

$$\lambda_k(j) = \sum_{i=0}^{n} \lambda_k(i)\cdot p_k(i,j) \quad (j = \overline{0,n}). \tag{4}$$

The open queue network's node $j=0$ in expressions (1)–(4) corresponds to outer environment - the source of incoming and the sink of returning messages.

**Message Delivery Time Computation.** First, let's determine stay duration of k-type message in data transmission node $i$, considering this node as queue system of $M_K/M_K/1$ type, which is receiving $K$ exponential message flows with intensities $\lambda_k(i)$ $(k = \overline{1,K};\ i = \overline{1,n})$. In case of non-priority message processing in node $i$ the Laplace transformation for probability density of stay duration of k-type message is defining in the following way [5]:

$$U_k^*(i,s) = \frac{[1 - R(i)][1 + s\,b_k(i)]^{-1}\,s}{s - \Lambda(i) + \sum_{r=1}^{K}\lambda_r(i)/[1 + s\,b_r(i)]}, \tag{5}$$

where $\Lambda(i) = \sum_{k=1}^{K}\lambda_k(i)$; $R(i) = \sum_{k=1}^{K}\lambda_k(i)b_k(i)$; $i = \overline{1,n}$; $k = \overline{1,K}$; $s > 0$.

The Laplace transformation $V_k^*(h,l,s)$ for probability density of delivery time of k-type messages from node $h$ to node $l$ defines by the following set of equations:

$$V_k^*(h,l,s) = U_k^*(h,s) \sum_{j=1}^{n} \pi_k(h,j|h,l)\, V_k^*(j,l,s) \quad (h,l = \overline{1,n}), \tag{6}$$

where $V_k^*(h,h,s) = U_k^*(h,s)$.

Two first initial moments are determined from the following set of equations:

$$V_k(h,l) = U_k(h) + \sum_{j=1}^{n} \pi_k(h,j|h,l)\, V_k(j,l) \quad (h,l = \overline{1,n}); \tag{7}$$

$$V_k^{(2)}(h,l) = U_k^{(2)}(h) + 2[V_k(h,l) - U_k(h)]\,U_k(h)$$
$$+ \sum_{j=1}^{n} \pi_k(h,j|h,l)\, V_k^{(2)}(j,l) \quad (h,l = \overline{1,n}), \tag{8}$$

where $U_k(h)$ and $U_k^{(2)}(h)$ — respectively first and second initial moments of stay duration of k-type messages in node $h$. These moments are defined by derivation (5) by $s$ at point $s=0$.

## 3    Results

It's possible to determine various probabilistic and pertaining to time telecommunication system's characteristics, specifically, probability of message's prompt delivery [6], using Laplace transformations $V_k^*(h,l,s)$ or moments $V_k(h,l)$ and $V_k^{(2)}(h,l)$. Prompt delivery probability equals value of delivery time Laplace transformation calculated with $s = s_k$ if message's aging function is exponential and average aging time for k-type messages equals $1/s$.

For $(i,j|h,l) = (1,2|1,2);\ (1,3|1,3);\ (3,4|1,4)\ (2,3|2,3);\ (3,1|2,1);$

$(3,4|2,4); (2,1|3,1); (2,4|3,4); (3,2|3,2); (2,1|4,1); (3,4|2,4); (3,1|4,1);$

$(4,2|4,2); (4,3|4,3)$ **and** $h = \overline{1,2}:\ \pi_k(i,j|h,l) = 1.$

**Table 1.** Probabilities "pi" k(i, j | h, l).

| Direction (h, l) | Route (i, j) | $p_k(i, j \mid h, l)$ | |
|---|---|---|---|
| | | $h = 1$ | $h = 2$ |
| 1,4 | 1,2 | 0,9 | 0,3 |
| | 1,3 | 0,1 | 0,7 |
| | 2,3 | 0,1 | 0,7 |
| | 2,4 | 0,9 | 0,3 |
| 2,1 | 2,1 | 0,9 | 0,3 |
| | 2,3 | 0,1 | 0,7 |
| 2,4 | 2,3 | 0,1 | 0,7 |
| | 2,4 | 0,9 | 0,3 |
| 3,1 | 3,1 | 0,9 | 0,3 |
| | 3,2 | 0,1 | 0,7 |
| 3,4 | 3,2 | 0,1 | 0,7 |
| | 3,4 | 0,9 | 0,3 |
| 4,1 | 4,2 | 0,1 | 0,7 |
| | 4,3 | 0,9 | 0,3 |

**Example.** Let's examine TCN, containing four nodes and two types of messages circulating in it. Markovian exponential flow of k-type messages generates with $\lambda_k(0|h, l)$ intensity from abonents, connected to sender node $h$, to destination recipient node $l$.

$$\lambda_1(0|h, l) = \begin{cases} 0,02 \; if \; h < l \\ 0,04 \; if \; h > l \end{cases}; \lambda_2(0|h, l) = \begin{cases} 0,03 \; if \; h < l \\ 0,01 \; if \; h > l \end{cases}; \quad h, l = \overline{1,4}.$$

Processing durations are the same and equals 2 s. for all messages in every node. Delivery times for k-type messages from sender node $h$ to destination node $l$ are constrained by $\hat{V}_k(h, l) = 50$ s for all $k = \overline{1,2}$; $h, l = \overline{1,4}$, where $h \neq l$. Average aging time for k-type message equals 10 s ($k = \overline{1,2}$). Probabilities $\pi_k(i, j|h, l)$, shown in Table 1, are defined based upon chosen routing algorithms where $\pi_k(0, h|h, l) = \pi_k(l, 0|h, l) = 1$ ($i, j = \overline{1,4}$; $h, l = \overline{1,4}$; $k = \overline{1,2}$). For this TCN we have message's prompt delivery probabilities $P(V_k(h, l) < \hat{V}_k(h, l))$ ($h, l = \overline{1,4}$; $k = \overline{1,2}$), shown in Table 2, and prompt delivery probabilities, shown in Table 3, granting aging functioning.

TCN's characteristic's computation method, based on decomposition, also can be used in case of priority message management methods used in TCN's nodes. Stay durations for nodes can be determined as described in [7] in case of using mixed priorities message management methods. Final results, computed by this method, are approximate, because message flows of different types at output, and thus at input, differ from exponential in case of priority management methods. However result's inaccuracies lie in acceptable for engineering computations limits as was discovered in research of flow's characters and their influencing on results in wide range of parameters corresponding to real systems.

Table 2. Message's prompt delivery probabilities.

| K | H | l | | | |
|---|---|---|---|---|---|
| | | 1 | 2 | 3 | 4 |
| 1 | 1 | 1,000 | 0,996 | 0,989 | 0,984 |
| | 2 | 0,994 | 1,000 | 0,973 | 0,994 |
| | 3 | 0,985 | 0,973 | 1,000 | 0,985 |
| | 4 | 0,973 | 0,996 | 0,989 | 1,000 |
| 2 | 1 | 1,000 | 0,996 | 0,989 | 0,704 |
| | 2 | 0,764 | 1,000 | 0,073 | 0,764 |
| | 3 | 0,749 | 0,973 | 1,000 | 0,749 |
| | 4 | 0,712 | 0,996 | 0,989 | 1,000 |

Table 3. Prompt delivery probabilities granting aging function.

| K | h | L | | | |
|---|---|---|---|---|---|
| | | 1 | 2 | 3 | 4 |
| 1 | 1 | 0,667 | 0,374 | 0,327 | 0,233 |
| | 2 | 0,354 | 0,561 | 0,275 | 0,354 |
| | 3 | 0,312 | 0,275 | 0,490 | 0,312 |
| | 4 | 0,211 | 0,374 | 0,327 | 0,667 |
| 2 | 1 | 0,667 | 0,374 | 0,327 | 0,109 |
| | 2 | 0,174 | 0,561 | 0,275 | 0,174 |
| | 3 | 0,158 | 0,275 | 0,490 | 0,158 |
| | 4 | 0,113 | 0,374 | 0,327 | 0,667 |

## 4    Conclusions

Found results can be used for solving TCN's optimization problem, which lies in routing algorithm determination (transfer probabilities $\pi_k(i,j|h,l)$) and in assigning priorities to messages of different types, providing specified message delivery time.

Described method of TCN's computation is implemented in program system.

## References

1. Jakubovich, D.: Network traffic optimization. Commun. Netw. Syst. **10**, 92–97 (2001)
2. Aliev, T.I.: Characteristics of mixed priorities message servicing disciplines. Izv. vuzov. Priborostroenie **4**(57), 30–35 (2014)
3. Goldstein B.S., Pinchuk A.V., Suhovickiy, A.L. VoIP. – M.: Radio and communication (2001)

4. Basharin, G.P., Tolmachev, A.L.: Queue network's theory and its application to analysis of information-calculating systems. Science and Engineering Summaries. Probability Theory. Mathematical Statistics. Theoretical Cybernetics. – M.: VINITI T. 21, p. 3–119 (1983)

5. Kleinrock, L.: Computation systems with queues: Russian translation. – M.: Mir (1979)

6. Zaharov, G.P.: Research methods of data networks. – M.: Radio and communication (1982)

7. Aliev, T.I., Muravyeva-Vitkovskaya, L.A.: Prioritetnye strategii upravleniya trafikom v multiservisnykh komp'yuternykh setyakh [Priority-based strategies of traffic management in multiservice computer networks]. Izv. vuzov. Priborostroenie **54**(6), 44–48 (2011)

# Estimation Quality Parameters of Transferring Image and Voice Data over ZigBee in Transparent Mode

R. Kirichek$^{(\boxtimes)}$, M. Makolkina, J. Sene, and V. Takhtuev

The Bonch-Bruevich State University of Telecommunications,
St. Petersburg, Russia
kirichek@sut.ru, makolkina@list.ru, senejeanvalery@yahoo.fr,
sbst@yandex.ru

**Abstract.** Wireless Sensor Networks, operating on the basis of IEEE 802.15.4, have gained more popularity in all sphere of life. The new direction of using ZigBee Networks is Flying Ubiquitous Sensor Networks. Data communication with terrestrial segment of network has become more effective than the other technologies, due to self-organizing function and low power consumption.

**Keywords:** Ubiquitous sensor networks · Multimedia data · ZigBee · QoE · Flying Ubiquitous Sensor Networks (FUSN)

## 1 Introduction

Today, Ubiquitous Sensor Networks (USN) are deeply integrated in the everyday life of an ordinary person. Research in the field of capability, network security, traffic characteristics are actively conducted by large number of scientists in different countries of the world [1–3].

Currently, in the interaction of the FUSN, within the concept of the Internet of Things [4,5], there are a number of unresolved issues. One of them is the delivery of various types of traffic based on the USN.

Recently a new direction of development in USN is Flying Ubiquitous Sensor Networks (FUSN) [6] has become popular. The scope of FUSN is beneficial for industry, agriculture, transport and monitoring of the functioning over various objects.

FUSN is a part of USN, thereby it has a similar principles of organization and architecture. Additionally, this networks have peer-to-peer or hierarchical (cluster) models. Also, there are 2 segments of FUSN: terrestrial and flying.

Being a part of the FUSN USN, it has a similar organization principles and architecture. Such networks can be peer or hierarchical (cluster). Also allocate 2 segments FUSN: ground and flying, which in turn can be peer or hierarchical.

FUSNs are realized by ZigBee specification, 6 LoWPAN, RPL, Bluetooth Low Energy (BLE) and so on [7,8], which enables self-organizing networks with

© Springer International Publishing Switzerland 2016
V. Vishnevsky and D. Kozyrev (Eds.): DCCN 2015, CCIS 601, pp. 260–267, 2016.
DOI: 10.1007/978-3-319-30843-2_27

clusters topologies, and low cost realization, transferring small amounts of information and which is characterized as a low power consumption.

The self-organizition in Flying Ubiquitous Sensor Networks (FUSN) is an automatic creation of network topology, self-acting connection of new nodes, automatic choice of routes the packet network without human intervention and this is the one of typical sides of ZigBee specification. The main feature of FUSN is a large coverage of sensors fields for terrestrial segment. Therefore, there is a possibility of clustering of the terrestrial segment and the using of UAVs as a main node.

The public unmanned aerial vehicles (UAV-P) can be used for FUSN creation and FUSN-P is the network abbreviation in this case.

The voice and video transmission from terrestrial segment to UAV-P is the next investigation task because often it may be the only chance to pass the necessary information to the area of terrestrial sensor fields. There is a positive experience of transfer of voice data over the protocol ZigBee [9]. In connection with this problem the comparison of voice quality and video quality of experience during their transmission from the terrestrial segment for years using different protocols (ZigBee, 6LoWPAN, RPL), and different languages is of great interest for FUSN-P.

## 2   Goal of Investigation

To assess the quality of voice in different languages and videos from the terrestrial segment, we have chosen the ZigBee protocol, whereas it is commonly known, has many applications in various fields. It is planned to conduct researches and compare with the others protocols.

ZigBee networks are created on the base of nodes of 3 types: coordinators, routers and end points [10]. Coordinator is a generating network, forming and functioning as the control center and network trust center - setting security policy defining the settings in the process of joining devices to the network, responsible for security keys. Router is transferring packets of data, realizing dynamic routing, restoring routes on network congestion or failure of any device. Routers are connected to the coordinator or others routers on the forming of network, and can be attached to child devices — routers or end points. Router works in continuous mode, has permanent power consumption and can provide up to 32 end devices. End point can send and receive packets, but does not translate and route. End points can connect to the coordinator or router, but cannot have child devices. Additionally, end points can be converted into sleep mode to conserve battery power. Developers of this specification have documented of transferring small packets of data, mostly text and packets insensitive for delays.

Despite on this, the goal was to analyze the opportunities of transferring multimedia data over ZigBee with appropriate quality. Investigation possibilities of transmitting data such as voice, video, image will expand the range of services to end users on the basis of WPAN networks.

## 3    Experimental Evaluation

To achieve the goal, the following tasks were allocated: investigation of existing algorithms and approaches to transferring data via FUSN; development laboratory test bed of transferring image data and voice; the experiment of broadcast image and voice data; evaluation of the obtained results quality over the method of assessing the quality of perception, according ITU recommendations [11,12].

It is known, that developed solutions for voice channel through ZigBee have already existed and studied, but the commercial development which is performed for specific hardware and software technology are not universal, and the solutions do not involve an assessment of transmission quality, according to the existing standards.

For practical implementation and experimentation laboratory test bed was assembled based on debugging kits by company Silicon Labs, which was based on ZigBee modules by Telegesis. Two devices of ETRX3 were selected, satisfying the research conditions, that enable managing the network through the AT command and quickly establishing a connection via asynchronous receiver-transmitter (UART). Since the task is to translate the voice and the image, test beds have to include different components, besides the main part of research — the ZigBee network. The wireless network was installed between two computers, which have simulated a transparent channel (Transparent mode). Scheme of laboratory test bed shown in Fig. 1.

The initial data was selected as an image size of 37 KB, it is obvious that the resolution of the image isn't large, but it is more than enough to recognize objects and analyze their actions. After event from software information is input to the ZigBee ETRX3. The image then is transmitted to the coordinator via the wireless network. The next step is transferring bytes to the computer through a UART interface that can operate at a speed of 115200 bits per second, using the hardware flow control. ETRX3 modules can transmit data in two modes. The first approach is "AT+DMODE" which simulates transparent mode working. This method is transmitted data without acknowledgment, thus a higher data rate is 12.7 Kbit per second. The second method allows transferring numbered packets and confirms the correct data sending. It helps to recover lost packets; this method is "AT+SCAST". Thereby, for analyze network possibilities we needed to take into account different parameters, such as indicators of quality perception, delays and losses, thus a second mode have selected for laboratory test bed —"AT+SCAST". If the loss had occurred, the resulting image would

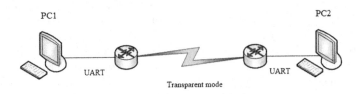

**Fig. 1.** Laboratory test bed for image transferring.

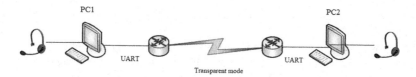

**Fig. 2.** Laboratory test bed for image transferring.

have damaged and as a result, would have not displayed. "AT+SCAST" allows broadcasting data from the router to the coordinator, which is marked in the network as a Sink. This module is part of the network ZigBee, which performs a function of receiving all transmitted data over the network. "AT+SCAST" provides correct transferring and receiving image without distortion, but the speed of this process is much slower than the first mode. The speed was 7.5 Kbit per second. The original image was transmitted in 39 s. The next experiment has a direction on voice over ZigBee network; the scheme of laboratory test bed is shown on Fig. 2.

The goal of this experiment was to transmit voice in real time. The parameters of the experiment are the same as in the previous case, but this test bed has included a microphone for inputting information and a speaker for outputting information. Each experiment was performed at 15 s, the experts were speaking into a microphone in different languages, and the voice is transmitted to an analog-digital converter and then is converted into the PCM format. A method for transmitting data via network has selected "AT+DMODE", because this type of traffic is not sensitive to losses, and is more dependent on the speed of transmission and delays. The sampling rate was 8000 Hz with 8 quantization bits, and was sending the data to the transmission buffer every 10 ms. Therefore, one packet contained the 80 bits. These characteristics ensured low sound quality; moreover collection on the receiving side for full audio output was failed. Thereby, test bed was using voice compression codec. Lossy compression algorithm A-Law was chosen, compliant with G.711 [13]. Due to the compression of traffic was able to increase quantization up to 16-bit for filling each packet of 80 bits. After transferring data to the PCM format at the transmitting side of test bed traffic, it was compressed and then was passed via transparent channel. The receiver side is a decompressed data and output to the speaker, where Assistant assessed results. One of the important cases, connecting with realization network solutions, is provisioning quality of service for each service. Additionally, the requirements for network transmission parameters are specific to different applications and types of traffic. Recommendation ITU-T Y.1540 [14] has identified the following characteristics of the network as the most important in terms of their impact on the quality of service: throughput, network reliability, delays and delay variability, loss ratio, network survivability asses.

This research has analyzed the influence of network characteristics on the quality of voice and image transmission over a wireless sensor network, according to the existing recommendations. This will determine the further expediency of using the ZigBee specification for the transmission of multimedia traffic in similar technologies.

The most interesting thing is the transfer of the image, because today most services are focused on the visualization of information. But the widespread usage of Ubiquitous Sensor Networks suggests that image transmission over ZigBee becomes ordinary technology. For example, if there is a sensor movement in the house or in a protected area, you will be interested to get the picture of what or who caused the alarm. If it is the hare or other animal, there is no necessity to call special services.

Obviously, the existing Sensor Networks are problematic to transmit image or video with acceptable levels of quality of service. But for the implementation of applications such as "smart infrastructure", the values of some parameters will be enough.

For the quality assessment of the image and voice transmission a methodology was chosen proposed in the ITU-T Recommendation R.913. As a subjective evaluation method, method of Absolute Category Rating (ACR) was chosen. This method uses a categorical estimation. The test sequence is evaluated according to the established scale of assessments. The advantage of the method is the ability to evaluate the ACR which only receives test sequence on the receive side without the etalons, so it gets you closer to the real conditions of the network and the estimated of end-users.

The experiment evaluated such parameters as transmission speed, the number of losses, the quality of speech recognition in multiple languages. Evaluations were conducted by four experts. The all duration was 15 s. The five experiments for each language were provided. The decision to perform experiments on the transmission of speech using fragments in different languages was made on the basis of the fact that the quality of speech in each language demands different requirements. Subjective assessment may vary from one language to another, due to the fact that languages differ in sound, some are more melodic, some are more clear and easy to recognize and understand, etc. That is why the experiments were conducted: Russian, English, French, Arabic and Belarusian languages. To reduce the chance of exhibiting lower valuation of the expert to foreign speech, at least one native speaker of each language is included in the expert group. The expert estimates are determined according to the following five-point scale: 5 — excellent; 4 — good; 3 — acceptable; 2 — bad; 1 — unacceptable.

The first stage of the results analysis was calculation of the average rating for each demonstration by the formula (1).

$$\overline{u}_{jkr} = \frac{1}{N} \sum_{i=1}^{N} u_{ijkr} \tag{1}$$

Table 1 illustrates an example of evaluation of the first fragment of speech in English language.

As it can be seen, the results of fragment evaluation show that the English higher than the Russian at approximately equal transmission conditions.

**Table 1.** Results in English.

| Experiment | Losses, % | Speed, bps | Expert 1 | Expert 2 | Expert 3 | Expert 4 |
|---|---|---|---|---|---|---|
| 1 | 11,4 | 353 | 3,0 | 3,0 | 3,5 | 3,0 |
| 2 | 12,0 | 312 | 2,5 | 3,0 | 3,0 | 3,0 |
| 3 | 10,1 | 412 | 3,5 | 4,0 | 3,5 | 4,0 |
| 4 | 11,3 | 368 | 3,5 | 3,5 | 3,5 | 3,5 |
| 5 | 12,2 | 302 | 3,0 | 3,0 | 2,5 | 3,0 |
| **Average** | 11,4 | 349 | 3,1 | 3,3 | 3,2 | 3,3 |

This is related to the subjective perception of information by the experts. The group of experts consists of foreign students who take English familiar, thus there was some tendency of inflated estimates, caused by human characteristics.

For further processing of the initial data the confidence interval was calculated by the formula (2), which is derived from the standard deviation of the estimates (1) and the size of each sample, according to ITU BT.500-13. For example, it is calculated for the English and Russian language.

$$\delta_{jkr} = 1,96 \frac{S_j kr}{\sqrt{N}} \tag{2}$$

$$S_{jkr} = \sqrt{\sum_{i=1}^{N} \frac{(\overline{u}_j kr - u_{ij} kr)^2}{N-1}} \tag{3}$$

Similarly, the confidence intervals were calculated for the other languages. The results are combined in Table 2.

Evaluation of subjective methods performed to research the connection between objective indicators of the network and the subjective perception of information by users based on changes during transmission. For displaying the results the logistic curve approximation based on the function (4) was used. The result is shown in Fig. 3.

$$\sigma = \frac{5}{1 + e^{\frac{-t-t_0}{B}}} \tag{4}$$

**Table 2.** Evaluation MOS for five languages.

| Language | Rating | Losses, % | Speed, bps | Confidence interval |
|---|---|---|---|---|
| Russian | 3,2 | 11,4 | 383 | (3,06; 3,38) |
| English | 3,4 | 11,4 | 349 | (3,26; 3,54) |
| French | 3,3 | 11,4 | 348 | (3,22; 3,48) |
| Arabic | 3,0 | 11,1 | 331 | (2,81; 3,19) |
| Belarussian | 3,2 | 11,2 | 352 | (3,04; 3,36) |

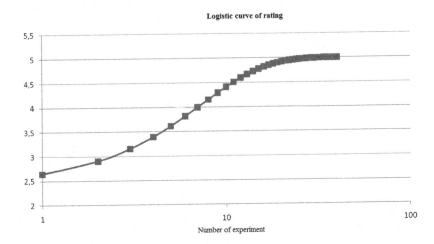

**Fig. 3.** Logistic curve approximation.

The next step in the analysis was to determine the relationship between the results. To determine the correlation of subjective assessments transmission quality, lossy and speed of transmission were calculated by the correlation coefficient. Based on the fact, that the coefficient of correlation between subjective assessment and loss - 0.9, and between the speed of transmission - 0.3, it can be concluded that the estimates are more dependent on the number of lost packets, rather than on the speed of transmission over the network ZigBee.

## 4    Conclusions

According to the results of the study, conclusions are following:

- In case of low speed and high losses, quality of voice transmission is significantly inferior of traditional communication networks. The number of losses provides a large impact of transferring data via ZigBee.
- At that moment transferring video over ZigBee is problematic in the case of low bandwidth of channel. The transmission of low resolution images makes rational requirements for latency and jitter.
- On the one hand, the compression of voice by codecs with losses of significant information increases probability of packet delivery. On the other hand, if a packet is lost in the network, user will lose a significant number of data. The choice of codec with enhanced compression algorithm will improve the results.

Thus, the scope of the using ZigBee networks for transmitting multimedia information is specific. These networks are useful where requirements to the quality of voice and image transmission rate are low. The main advantage of using a networking standard ZigBee is a low cost, high autonomy, simple creation and survivability of these networks, which allows us to consider the networks for the transmission of multimedia data in the future. Video and voice are the only one way to communicate over FUSN in some cases. Especially, if it is realized in the countryside or outback. Whereas quality of transferring voice and image is satisfactory for all tested languages, this direction can take important place in FUSN.

**Acknowledgments.** The reported study was supported by RFBR, research project No 15 07-09431a "Development of the principles of construction and methods of self-organization for Flying Ubiquitous Sensor Networks".

# References

1. Koucheryavy, A., Prokopiev, A.: Ubiquitous sensor networks traffic models for telemetry applications. In: Balandin, S., Koucheryavy, Y., Hu, H. (eds.) NEW2AN 2011 and ruSMART 2011. LNCS, vol. 6869, pp. 287–294. Springer, Heidelberg (2011)
2. Vybornova, A., Koucheryavy, A.: Ubiquitous sensor networks traffic models for medical and tracking applications. In: Andreev, S., Balandin, S., Koucheryavy, Y. (eds.) NEW2AN/ruSMART 2012. LNCS, vol. 7469, pp. 338–346. Springer, Heidelberg (2012)
3. ITU-T Recommendation Y.2060: Overview of Internet of Things. Geneva, February 2012
4. Lera, A., Floerkemeier, C., Mitsugi, J., Morabito, G.: The internet of things. IEEE Wirel. Commun. **17**(6), 44–51 (2010)
5. Kirichek, R., Paramonov, A., Koucheryavy, A.: Flying ubiquitous sensor networks as a quening system. In: Proceedings of International Conference on Advanced Communication Technology (ICACT 2015), Phoenix Park, Korea (2015)
6. IEEE Standard for Local and metropolitan area networks–Part 15.4: Low-Rate Wireless Personal Area Networks (LR-WPANs):IEEE 802.15.4 (2011)
7. RPL: IPv6 Routing Protocol for Low-Power and Lossy Networks: RFC 6550– 03. 2012
8. Touloupis, A., Meliones, S.: Apostolacos speech codes for high-quality voice over ZigBee applications: evaluation and implementation challenges. IEEE Commun. Mag. **50**(4), 122–128 (2012)
9. ZigBee Alliance, ZigBee Specification, 17 January, 2008
10. ITU-R Recommendation BT.500-13: Methodology for the Subjective Assessment of the Quality of Television Pictures. Geneva, January 2012
11. ITU-T Recommendation P.910: Subjective video quality assessment methods for multimedia applications. Geneva, April 2008
12. ITU-T Recommendation G.711: Pulse code modulation (PCM) of voice frequencies. Amendment 1: New Annex A on lossless encoding of PCM frames (1998) Amd.1 (08/2009)
13. ITU-T. Recommendation Y.1540: Internet protocol data communication service– IP packet transfer and availability performance parameters. Geneva, November 2007
14. ITU-T Recommendation P.913: Methods for the subjective assessment of video quality, audio quality and audiovisual quality of Internet video and distribution quality television in any environment. Geneva, January 2014

# Low-Priority Queue Fluctuations in Tandem of Queuing Systems Under Cyclic Control with Prolongations

Victor Kocheganov$^{(\boxtimes)}$ and Andrei V. Zorine

N. I. Lobachevsky State University of Nizhny Novgorod,
Gagarina Avenue 23, 603950 Nizhny Novgorod, Russia
kocheganov@gmail.com, zoav1602@gmail.com

**Abstract.** A tandem of queuing systems is considered. Each system has a high-priority input flow and a low-priority input flow which are conflicting. In the first system, the customers are serviced in the class of cyclic algorithms. The serviced high-priority customers are transferred from the first system to the second one with random delays and become the high-priority input flow of the second system. In the second system, customers are serviced in the class of cyclic algorithms with prolongations. Low-priority customers are serviced when their number exceeds a threshold. A mathematical model is constructed in form of a multidimensional denumerable discrete-time Markov chain. The recurrent relations for partial probability generating functions for the low-priority queue in the second system are found.

**Keywords:** Tandem of controlling queuing systems · Cyclic algorithm with prolongations · Conflicting flows · Multidimensional denumerable discrete-time markov chain

## 1 Introduction

Conflicting traffic flows control at a crossroad is one of classical problems in queuing theory. In the literature several algorithms were investigated: fixed duration cyclic algorithm, cyclic algorithm with a loop, cyclic algorithm with changing regimes, etc [1–6]. However, several (two in our case) consecutive crossroads are of great interest, because in a real-life situation a vehicle having passed one highway intersection finds itself at another one. In other words, an output flow from the first intersection forms an input flow of the second intersection. Hence, the second input flow no longer has an *a priori* known simple probabilistic structure (for example, that of a non-ordinary Poison flow), and knowledge about the service algorithm should be taken into account to deduce formation conditions of the first output flow.

Tandems of intersections were considered by a few authors. In [7] a computer-aided simulation of adjacent intersection was carried out. In [8] a mathematical

© Springer International Publishing Switzerland 2016
V. Vishnevsky and D. Kozyrev (Eds.): DCCN 2015, CCIS 601, pp. 268–279, 2016.
DOI: 10.1007/978-3-319-30843-2_28

model of two intersection in tandem governed by cyclic algorithms was investigates and stability conditions were found. In this paper we assume that the first intersection is governed by a cyclic algorithm while the econd intersection is governed by a cyclic algorithm with prolongations. In particular, we pay attention to the low-priority queue in the second intersection.

## 2 The Problem Settings

Consider a queuing system with a scheme shown Fig. 1. There are four input flows of customers $\Pi_1$, $\Pi_2$, $\Pi_3$, and $\Pi_4$ entering the single server queueing system. Customers in the input flow $\Pi_j$, $j \in \{1, 2, 3, 4\}$ join a queue $O_j$ with an unlimited capacity. For $j \in \{1, 2, 3\}$ the discipline of the queue $O_j$ is FIFO (First In First Out). Discipline of the queue $O_4$ will be described later. The input flows $\Pi_1$ and $\Pi_3$ are generated by an external environment, which has only one state. Each of these flows is a nonordinary Poisson flow. Denote by $\lambda_1$ and $\lambda_3$ the intensities of bulk arrivals for the flows $\Pi_1$ and $\Pi_3$ respectively. The probability generating function of number of customers in a bulk in the flow $\Pi_j$ is

$$f_j(z) = \sum_{\nu=1}^{\infty} p_\nu^{(j)} z^\nu, \quad j \in \{1, 3\}. \tag{1}$$

We assume that $f_j(z)$ converges for any $z \in \mathbb{C}$ such that $|z| < (1+\varepsilon)$, $\varepsilon > 0$. Here $p_\nu^{(j)}$ is the probability of a bulk size in flow $\Pi_j$ being exactly $\nu = 0, 1, \ldots$. Having been serviced the customers from $O_1$ come back to the system as the $\Pi_4$ customers. The $\Pi_4$ customers in turn after service enter the system as the $\Pi_2$ ones. The flows $\Pi_2$ and $\Pi_3$ are conflicting in the sense that their customers can't be serviced simultaneously. This implies that the problem can't be reduced to a problem with fewer input flows by merging the flows together.

In order to describe the server behavior we fix positive integers $d, n_0, n_1, \ldots, n_d$ and we introduce a finite set $\Gamma = \{\Gamma^{(k,r)} : k = 0, 1, \ldots, d; r = 1, 2, \ldots n_k\}$ of states server can reside in. At the state $\Gamma^{(k,r)}$ sever stays during constant time $T^{(k,r)}$. Define disjoint subsets $\Gamma^{\mathrm{I}}$, $\Gamma^{\mathrm{II}}$, $\Gamma^{\mathrm{III}}$, and $\Gamma^{\mathrm{IV}}$ of $\Gamma$ as follows. In the state $\gamma \in \Gamma^{\mathrm{I}}$ only customers from the queues $O_1$, $O_2$ and $O_4$ are serviced. In the state $\gamma \in \Gamma^{\mathrm{II}}$ only customers from the queues $O_2$ and $O_4$ are serviced. In the state $\gamma \in \Gamma^{\mathrm{III}}$ only customers from queues $O_1, O_3$, and $O_4$ are serviced. In the state $\gamma \in \Gamma^{\mathrm{IV}}$ only customers from queues $O_3$ and $O_4$ are serviced. We assume that $\Gamma = \Gamma^{\mathrm{I}} \cup \Gamma^{\mathrm{II}} \cup \Gamma^{\mathrm{III}} \cup \Gamma^{\mathrm{IV}}$. Set also $^1\Gamma = \Gamma^{\mathrm{I}} \cup \Gamma^{\mathrm{III}}$, $^2\Gamma = \Gamma^{\mathrm{I}} \cup \Gamma^{\mathrm{II}}$, $^3\Gamma = \Gamma^{\mathrm{III}} \cup \Gamma^{\mathrm{IV}}$.

The server changes its state according to the following rules. We call a set $C_k = \{\Gamma^{(k,r)} : r = 1, 2, \ldots n_k\}$ the $k$-th cycle, $k = 1, 2, \ldots, d$. For $k = 0$ the state $\Gamma^{(0,r)}$ with $r = 1, 2, \ldots, n_0$ is called a prolongation state. Put $r \oplus_k 1 = r + 1$ for $r < n_k$, and $r \oplus_k 1 = 1$ for $r = n_k$ ($k = 0, 1, \ldots, d$). In the cycle $C_k$ we select a subset $C_k^{\mathrm{I}}$ of input states, a subset $C_k^{\mathrm{O}}$ of output states, and a subset $C_k^{\mathrm{N}} = C_k \setminus (C_k^{\mathrm{O}} \cup C_k^{\mathrm{I}})$ of neutral states. After the state $\Gamma^{(k,r)} \in C_k \setminus C_k^{\mathrm{O}}$ the server switches to the state $\Gamma^{(k,r \oplus_k 1)}$ within the same cycle $C_k$. After the state $\Gamma^{(k,r)}$ in $C_k^{\mathrm{O}}$ the server switches to the state $\Gamma^{(k,r \oplus_k 1)}$ if number of customers

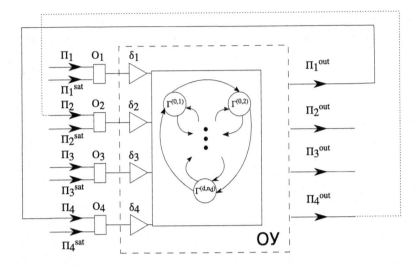

**Fig. 1.** Scheme of the queuing system as a cybernetic control system

in the queue $O_3$ at switching instant is greater than a predetermined threshold $L$. Otherwise, if the number of customers in the queue $O_3$ is less than or equal to $L$ then the new state is a prolongation one $\Gamma^{(0,r_1)}$ where $r_1 = h_1(\Gamma^{(k,r)})$ and $h_1(\cdot)$ is a given mapping of $\bigcup_{k=1}^{d} C_k^O$ into $\{1, 2, \ldots, n_0\}$. After the state $\Gamma^{(0,r)}$ if the number of customers in $O_3$ is not above $L$ the state of the same type $\Gamma^{(0,r_2)}$ is chosen where $r_2 = h_2(r)$ and $h_2(\cdot)$ is a given mapping of the set $\{1, 2, \ldots, n_0\}$ into itself; in the other case the new state is $\Gamma^{(k,r_3)} \in C_k^I$ where $\Gamma^{(k,r_3)} = h_3(r)$ and $h_3(\cdot)$ is a given mapping of $\{1, 2, \ldots, n_0\}$ to $\bigcup_{k=1}^{d} C_k^I$. We assume that each prolongation state $\Gamma^{(0,r)}$ belongs to the set ${}^2\Gamma$ and that relations $C_k^O \subset {}^2\Gamma$ and $C_k^I \subset {}^3\Gamma$ hold. We also assume that all the cycles have exactly one input and output state. Finally, we assume that all the prolongation states make a cycle, that is $h_2(r) = r \oplus_0 1$. Putting all together, we introduce a function which formalizes the server state changes:

$$h(\Gamma^{(k,r)}, y) = \begin{cases} \Gamma^{(k,r\oplus_k 1)} & \text{if } \Gamma^{(k,r)} \in C_k \setminus C_k^O \text{ or } (\Gamma^{(k,r)} \in C_k^O) \wedge (y > L); \\ \Gamma^{(0,h_1(\Gamma^{(k,r)}))} & \text{if } \Gamma^{(k,r)} \in C_k^O \text{ and } y \leqslant L; \\ \Gamma^{(0,r\oplus_0 1)} & \text{if } k = 0 \text{ and } y \leqslant L; \\ h_3(r) & \text{if } k = 0 \text{ and } y > L. \end{cases} \quad (2)$$

In general, service durations of different customers can be dependent and may have different laws of probability distributions. So, saturation flows will be used to define the service process. A saturation flow $\Pi_j^{\text{sat}}$, $j \in \{1, 2, 3, 4\}$, is defined as a virtual output flow under the maximum usage of the server and unlimited number of customer in the queue $O_j$. The saturation flow $\Pi_j^{\text{sat}}$, $j \in \{1, 2, 3\}$ contains a non-random number $\ell(k, r, j) \geqslant 0$ of customers in the server state $\Gamma^{(k,r)}$. In particular, $\ell(k, r, j) \geqslant 1$ for $\Gamma^{(k,r)} \in {}^j\Gamma$ and $\ell(k, r, j) = 0$

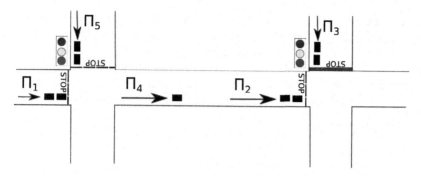

**Fig. 2.** A tandem of crossroads, the physical interpretation of the queuing system under study

for $\Gamma^{(k,r)} \notin {}^j\Gamma$. Let $\mathbb{Z}_+$ be the set of non-negative integer numbers. If the queue $O_4$ contains $x \in \mathbb{Z}_+$ customers the saturation flow $\Pi_4^{\text{sat}}$ also contains $x$ customers. Finally, in the state $\Gamma^{(k,r)}$ every customer from queue $O_4$ with probability $p_{k,r}$ and independently of others ends servicing and joins $\Pi_2$ to go to $O_2$. With the complementary probability $1 - p_{k,r}$ the customer stays in $O_4$ until the next time slot. In the next time slot it repeats its attempt to join $\Pi_2$ with a proper probability.

A real-life example of just described queuing system is a tandem of two consecutive crossroads (Fig. 2). The input flows are flows of vehicles. The flows $\Pi_1$ and $\Pi_5$ at the first crossroad are conflicting; $\Pi_2$ and $\Pi_3$ at the second crossroad are also conflicting. Every vehicle from the flow $\Pi_1$ after passing first road intersection joint the flow $\Pi_4$ and enters the queue $O_4$. After some random time interval the vehicle arrives to the next road intersection. Such a pair of crossroads is an instance of a more general queuing model described above.

## 3    Mathematical Model

The queuing system under investigation can be regarded as a cybernetic control system what helps to rigorously construct a formal stochastic model [8]. The scheme of the control system is shown in Fig. 1. There are following blocks present in the scheme: (1) the external environment with one state; (2) input poles of the first type — the input flows $\Pi_1$, $\Pi_2$, $\Pi_3$, and $\Pi_4$; (3) input poles of the second type — the saturation flows $\Pi_1^{\text{sat}}$, $\Pi_2^{\text{sat}}$, $\Pi_3^{\text{sat}}$, and $\Pi_4^{\text{sat}}$; (4) an external memory — the queues $O_1$, $O_2$, $O_3$, and $O_4$; (5) an information processing device for the external memory — the queue discipline units $\delta_1$, $\delta_2$, $\delta_3$, and $\delta_4$; (6) an internal memory — the server (OY); (7) an information processing device for internal memory — the graph of server state transitions; (8) output poles — the output flows $\Pi_1^{\text{out}}$, $\Pi_2^{\text{out}}$, $\Pi_3^{\text{out}}$, and $\Pi_4^{\text{out}}$. The coordinate of a block is its number on the scheme.

Let us introduce the following variables and elements along with their value ranges. To fix a discrete time scale consider the epochs $\tau_0 = 0$, $\tau_1$, $\tau_2$, ... when the server changes its state. Let $\Gamma_i \in \Gamma$ be the server state during the interval $(\tau_{i-1}; \tau_i]$, $\varkappa_{j,i} \in \mathbb{Z}_+$ be the number of customers in the queue $O_j$ at the instant $\tau_i$, $\eta_{j,i} \in \mathbb{Z}_+$ be the number of customers arrived into the queue $O_j$ from the flow $\Pi_j$ during the interval $(\tau_i; \tau_{i+1}]$, $\xi_{j,i} \in \mathbb{Z}_+$ be the number of customers in the saturation flow $\Pi_j^{\mathrm{sat}}$ during the interval $(\tau_i; \tau_{i+1}]$, $\overline{\xi}_{j,i} \in \mathbb{Z}_+$ be the actual number of serviced customers from the queue $O_j$ during the interval $(\tau_i; \tau_{i+1}]$, $j \in \{1, 2, 3, 4\}$.

The server changes its state according to the following rule:

$$\Gamma_{i+1} = h(\Gamma_i, \varkappa_{3,i}), \tag{3}$$

where the mapping $h(\cdot, \cdot)$ is defined by formula (2). To determine the duration $T_{i+1}$ of the next time slot it useful to introduce a mapping $h_T(\cdot, \cdot)$ by

$$T_{i+1} = h_T(\Gamma_i, \varkappa_{3,i}) = T^{(k,r)}, \qquad \text{where } \Gamma^{(k,r)} = \Gamma_{i+1} = h(\Gamma_i, \varkappa_{3,i}).$$

A functional relation

$$\overline{\xi}_{j,i} = \min\{\varkappa_{j,i} + \eta_{j,i}, \xi_{j,i}\}, \quad j \in \{1, 2, 3\}, \tag{4}$$

between $\overline{\xi}_{j,i}$ and $\varkappa_{j,i}$, $\eta_{j,i}$, $\xi_{j,i}$ describes the service strategy. Further, since

$$\varkappa_{j,i+1} = \varkappa_{j,i} + \eta_{j,i} - \overline{\xi}_{j,i}, \quad j \in \{1, 2, 3\},$$

and due to (4) it follows that

$$\varkappa_{j,i+1} = \max\{0, \varkappa_{j,i} + \eta_{j,i} - \xi_{j,i}\}, \quad j \in \{1, 2, 3\}. \tag{5}$$

We also have from the problem settings the following relations for the flow $\Pi_4$:

$$\eta_{4,i} = \min\{\xi_{1,i}, \varkappa_{1,i} + \eta_{1,i}\}, \quad \varkappa_{4,i+1} = \varkappa_{4,i} + \eta_{4,i} - \eta_{2,i}, \quad \xi_{4,i} = \varkappa_{4,i}. \tag{6}$$

Put $\varkappa_i = (\varkappa_{1,i}, \varkappa_{2,i}, \varkappa_{3,i}, \varkappa_{4,i})$. The non-local description of the input and saturation flows consists of specifying particular features of the conditional probability distribution of selected discrete components $\eta_i = (\eta_{1,i}, \eta_{2,i}, \eta_{3,i}, \eta_{4,i})$ and $\xi_i = (\xi_{1,i}, \xi_{2,i}, \xi_{3,i}, \xi_{4,i})$ of marked point processes $\{(\tau_i, \nu_i, \eta_i); i \geqslant 0\}$ and $\{(\tau_i, \nu_i, \xi_i); i \geqslant 0\}$ with marks $\nu_i = (\Gamma_i; \varkappa_i)$. Let $\varphi_1(\cdot, \cdot)$ and $\varphi_3(\cdot, \cdot)$ be defined by series expansions

$$\sum_{\nu=0}^{\infty} z^\nu \varphi_j(\nu, t) = \exp\{\lambda_j t (f_j(z) - 1)\}$$

with functions $f_j(z)$ defined by (1), $j \in \{1, 3\}$. The function $\varphi_j(\nu, t)$ equals the probability of $\nu = 0, 1, \ldots$ arrivals in the flow $\Pi_j$ during time $t \geqslant 0$. If $\nu < 0$ the value of $\varphi_j(\nu, t)$ is set to zero. Define function $\psi(\cdot, \cdot, \cdot)$ by

$$\psi(k; y, u) = C_y^k u^k (1 - u)^{y-k}.$$

Then $\psi(k; y, p_{k,r})$ is the probability of $k$ arrival from flow $\Pi_2$ given the queue $O_4$ contains $y$ customers and the server state is $\Gamma^{(k,r)}$. For values $k \notin \{0, 1, \dots, y\}$ the value of $\psi(k; y, u)$ is set to zero.

Let $a = (a_1, a_2, a_3, a_4) \in \mathbb{Z}_+^4$ and $x = (x_1, x_2, x_3, x_4) \in \mathbb{Z}_+^4$. If the mark value is $\nu_i = (\Gamma^{(k,r)}; x)$ then the probability $\varphi(a, k, r, x)$ of simultaneous equalities $\eta_{1,i} = a_1$, $\eta_{2,i} = a_2$, $\eta_{3,i} = a_3$, $\eta_{4,i} = a_4$ according the problem statement is

$$\varphi_1(a_1, h_T(\Gamma^{(k,r)}, x_3)) \cdot \psi(a_2, x_4, p_{\tilde{k},\tilde{r}}) \cdot \varphi_3(a_3, h_T(\Gamma^{(k,r)}, x_3)) \cdot \delta_{a_4, \min\{\ell(\tilde{k}, \tilde{r}, 1), x_1 + a_1\}}$$

where $\Gamma^{(\tilde{k}, \tilde{r})} = h(\Gamma^{(k,r)}, x_3)$ and $\delta_{i,j}$ is the Kroneker's delta:

$$\delta_{i,j} = \begin{cases} 1 & \text{if } i = j, \\ 0 & \text{if } i \neq j. \end{cases}$$

Let $b = (b_1, b_2, b_3, b_4) \in \mathbb{Z}_+^4$. The probability $\zeta(b, k, r, x)$ of simultaneous equalities $\xi_{1,i} = b_1$, $\xi_{2,i} = b_2$, $\xi_{3,i} = b_3$, $\xi_{4,i} = b_4$ given the fixed label value $\nu_i = (\Gamma^{(k,r)}; x)$ is

$$\delta_{b_1, \ell(\tilde{k}, \tilde{r}, 1)} \cdot \delta_{b_2, \ell(\tilde{k}, \tilde{r}, 2)} \cdot \delta_{b_3, \ell(\tilde{k}, \tilde{r}, 3)} \cdot \delta_{b_4, x_4}.$$

The assumptions on statistical properties of some blocks and function relations between blocks are not contradicting and sufficient to construct a formal probability model, as the following theorem first proven in [9] demonstrates.

**Theorem 1.** *Choose $\gamma_0 = \Gamma^{(k_0, r_0)} \in \Gamma$ and $x^0 = (x_{1,0}, x_{2,0}, x_{3,0}, x_{4,0}) \in \mathbb{Z}_+^4$. There exists a probability space $(\Omega, \mathcal{F}, \mathbf{P}(\cdot))$, random variables $\eta_{j,i} = \eta_{j,i}(\omega)$, $\xi_{j,i} = \xi_{j,i}(\omega)$, $\varkappa_{j,i} = \varkappa_{j,i}(\omega)$, and random elements $\Gamma_i = \Gamma_i(\omega)$, $i \geq 0$, $j \in \{1, 2, 3, 4\}$ defined on this space, such that: (1) equalities $\Gamma_0(\omega) = \gamma_0$ and $\varkappa_0(\omega) = x^0$ hold; (2) relations (3), (5), (6) hold; (3) for any $a$, $b$, $x^t = (x_{1,t}, x_{2,t}, x_{3,t}, x_{4,t}) \in \mathbb{Z}_+^4$, $\Gamma^{(k_t, r_t)} \in \Gamma$, $t = 1, 2, \dots$ the joint conditional probability distribution of vectors $\eta_i$ and $\xi_i$ has the form*

$$\mathbf{P}\left(\{\omega : \eta_i(\omega) = a, \xi_i(\omega) = b\} \,\Big|\, \bigcap_{t=0}^{i} \{\omega : \Gamma_t(\omega) = \Gamma^{(k_t, r_t)}, \varkappa_t(\omega) = x^t\}\right)$$
$$= \varphi(a, k_i, r_i, x^i) \cdot \zeta(b, k_i, r_i, x^i).$$

From now on we focus on low-priority customers in the queue $O_3$.

## 4   The Low-Priority Queue

Here we will consider the stochastic sequence

$$\{(\Gamma_i(\omega), \varkappa_{3,i}(\omega)); i = 0, 1, \dots\}, \tag{7}$$

which includes the number of low-priority customers $\varkappa_{3,i}(\omega)$ in the queue $O_3$. In this section we will report several results concerning this stochastic sequence.

**Theorem 2.** *Let* $\Gamma_0(\omega) = \Gamma^{(k,r)} \in \Gamma$ *and* $\varkappa_{3,0}(\omega) = x_{3,0} \in \mathbb{Z}_+$ *be fixed. Then the stochastic sequence* (7) *is a homogeneous denumerable Markov chain.*

**Theorem 3.** *Let* $x_3, \tilde{x}_3 \in \mathbb{Z}_+$ *and* $\Gamma^{(k,r)} \in \Gamma$, $\Gamma^{(\tilde{k},\tilde{r})} = h(\Gamma^{(k,r)}, x_3) \in \Gamma$. *Then the transition probabilities of the Markov chain* (7) *are*

$$\mathbf{P}(\{\omega\colon \Gamma_{i+1}(\omega) = \Gamma^{(\tilde{k},\tilde{r})}, \varkappa_{3,i+1}(\omega) = \tilde{x}\} \mid \{\omega\colon \Gamma_i(\omega) = \Gamma^{(k,r)}, \varkappa_{3,i}(\omega) = x\})$$

$$= (1 - \delta_{\tilde{x}_3,0}) \cdot \varphi\Big(\tilde{x}_3 + \ell(\tilde{k}, \tilde{r}, 3) - x_3, h_T\big(\Gamma^{(k,r)}, x_3\big)\Big)$$

$$+ \delta_{\tilde{x}_3,0} \sum_{a=0}^{\ell(\tilde{k},\tilde{r},3)-x_3} \varphi_3\Big(a, h_T\big(\Gamma^{(k,r)}, x_3\big)\Big).$$

The next theorem clarifies which states of the Markov chain $\{(\Gamma_i, \varkappa_{3,i}); i \geqslant 0\}$ are essential. To make a complete classification we introduce sets

$$S_{0,r}^3 = \Big\{(\Gamma^{(0,r)}, x_3)\colon x_3 \in \mathbb{Z}_+, \ x_3 > L - \max_{k=1,2,\ldots,d}\Big\{\sum_{t=0}^{n_k}\ell(k,t,3)\Big\}\Big\}, \quad 1 \leqslant r \leqslant n_0,$$

$$S_{k,r}^3 = \Big\{(\Gamma^{(k,r)}, x_3)\colon x_3 \in \mathbb{Z}_+, \ x_3 > L - \sum_{t=0}^{r}\ell(k,t,3)\Big\}, \quad 1 \leqslant k \leqslant d, \quad 1 \leqslant r \leqslant n_k.$$

**Theorem 4.** *The set of essential states of the Markov chain* $\{(\Gamma_i, \varkappa_{3,i}); i \geqslant 0\}$ *consists of sets* $\bigcup_{1 \leqslant r \leqslant n_0} S_{0,r}^3$ *and* $\bigcup_{\substack{1 \leqslant k \leqslant d \\ 1 \leqslant r \leqslant n_k}} S_{k,r}^3$.

As before, let $\Gamma^{(k,r)} \in \Gamma$ and $x_3 \in \mathbb{Z}_+$. Denote by $H_{-1}(\Gamma^{(k,r)}, x_3)$ the set of all server states $\gamma \in \Gamma$ such that $h(\gamma, x_3) = \Gamma^{(k,r)}$ and put $r \ominus_k 1 = r - 1$ for $n_k \geqslant r > 1$, and $r \ominus_k 1 = n_k$ for $r = 1$ $(k = 0, 1, \ldots, d)$. Then formula (2) makes it possible to define the mapping $H_{-1}(\Gamma^{(k,r)}, x_3)$ explicitly:

$$H_{-1}(\Gamma^{(k,r)}, x_3) = \begin{cases} \{\Gamma^{(k_1,r_1)}, \Gamma^{(0,r\ominus_0 1)}\} & \text{if } (k = 0) \wedge (x_3 \leqslant L), \\ \{\Gamma^{(k,r\ominus_k 1)}, \Gamma^{(0,r_2)}\} & \text{if } (\Gamma^{(k,r)} \in C_k^{\mathrm{I}}) \wedge (x_3 > L), \\ \{\Gamma^{(k,r\ominus_k 1)}\} & \text{if } (\Gamma^{(k,r)} \in C_k^{\mathrm{O}}) \vee (\Gamma^{(k,r)} \in C_k^{\mathrm{N}}), \\ \varnothing & \text{if } (k = 0) \wedge (x_3 > L) \\ & \quad \text{or } (\Gamma^{(k,r)} \in C_k^{\mathrm{I}}) \wedge (x_3 \leqslant L), \end{cases}$$

where $h_1(\Gamma^{(k_1,r_1)}) = r$ and $h_3(r_2) = \Gamma^{(k,r)}$.

Let's define for $\gamma \in \Gamma$ and $x_3 \in \mathbb{Z}_+$ values

$$Q_{3,i}(\gamma, x) = \mathbf{P}(\{\omega\colon \Gamma_i(\omega) = \gamma, \varkappa_{3,i}(\omega) = x_3\}).$$

**Lemma 1.** *Let* $\tilde{\gamma} = \Gamma^{(\tilde{k},\tilde{r})} \in \Gamma$ *and* $\tilde{x}_3 \in \mathbb{Z}_+$. *Then following recurrent w.r.t.* $i \geqslant i$ *relations take place for transition probabilities* $\{Q_{3,i}(\cdot,\cdot)\}_{i\geqslant 0}$ *of Markov chain* $\{(\Gamma_i, \varkappa_{3,i}); i \geqslant 0\}$:

$$Q_{3,i+1}(\tilde{\gamma}, \tilde{x}_3) = (1 - \delta_{\tilde{x}_3,0}) \times \sum_{x_3=0}^{\tilde{x}_3 + \ell(\tilde{k}, \tilde{r}, 3)} \sum_{\gamma \in \mathbb{H}_{-1}(\tilde{\gamma}, x_3)} Q_{3,i}(\gamma, x_3) \times \varphi_3(\tilde{x}_3 + \ell(\tilde{k}, \tilde{r}, 3)$$

$$- x_3, T^{(\tilde{k}, \tilde{r})}) + \delta_{\tilde{x}_3,0} \sum_{x_3=0}^{\ell(\tilde{k}, \tilde{r}, 3)} \sum_{\gamma \in \mathbb{H}_{-1}(\tilde{\gamma}, x_3)} Q_{3,i}(\gamma, x_3) \sum_{a=0}^{\ell(\tilde{k}, \tilde{r}, 3) - x_3} \varphi_3(a, T^{(\tilde{k}, \tilde{r})})$$

$$(8)$$

Suppose $k$ and $r$ are such that $\Gamma^{(k,r)} \in \Gamma$. Let's define partial probability generating functions

$$\mathfrak{M}^{(i)}(k, r, v) = \sum_{w=0}^{\infty} Q_{3,i}(\Gamma^{(k,r)}, w)v^w,$$

$$q_{k,r}(v) = v^{-\ell(k,r,3)} \sum_{w=0}^{\infty} \varphi_3(w, T^{(k,r)})v^w.$$

and auxilary functions

$$\tilde{\alpha}_i(\tilde{k}, \tilde{r}, v) = \sum_{x_3=0}^{\ell(\tilde{k}, \tilde{r}, 3)} \sum_{\gamma \in H_{-1}(\tilde{\gamma}, x_3)} Q_{3,i}(\gamma, x_3) \sum_{a=0}^{\ell(\tilde{k}, \tilde{r}, 3) - x_3} \varphi_3(a, T^{(\tilde{k}, \tilde{r})})$$

$$- \sum_{x_3=0}^{\ell(\tilde{k}, \tilde{r}, 3)} \sum_{\gamma \in H_{-1}(\tilde{\gamma}, x_3)} Q_{3,i}(\gamma, x_3)v^{x_3 - \ell(\tilde{k}, \tilde{r}, 3)} \sum_{w=0}^{\ell(\tilde{k}, \tilde{r}, 3) - x_3} \varphi_3(w, T^{(\tilde{k}, \tilde{r})})v^w$$

$$\alpha_i(0, \tilde{r}, v) = \tilde{\alpha}_i(0, \tilde{r}, v) + q_{0,\tilde{r}}(v) \times \sum_{x_3=0}^{L} \Big[ Q_{3,i}(\Gamma^{(k_1, r_1)}, x_3)$$

$$+ Q_{3,i}(\Gamma^{(0, r \ominus_0 1)}, x_3) \Big] v^{x_3} - q_{\tilde{k}, \tilde{r}}(v) \sum_{x_3=0}^{L} Q_{3,i}(\Gamma^{(0, r \ominus_0 1)}, x_3), \quad \Gamma^{(0,\tilde{r})} \in \Gamma.$$

$$\alpha_i(\tilde{k}, \tilde{r}, v) = \tilde{\alpha}_i(\tilde{k}, \tilde{r}, v) - q_{\tilde{k}, \tilde{r}}(v) \sum_{x_3=0}^{L} \Big[ Q_{3,i}(\Gamma^{(k, r \ominus_k 1)}, x_3)$$

$$+ Q_{3,i}(\Gamma^{(0, r_2)}, x_3) \Big] v^{x_3}, \quad \Gamma^{(\tilde{k}, \tilde{r})} \in C_{\tilde{k}}^{\mathrm{I}}$$

$$\alpha_i(\tilde{k}, \tilde{r}, v) = \tilde{\alpha}_i(\tilde{k}, \tilde{r}, v) - q_{\tilde{k}, \tilde{r}}(v) \sum_{x_3=0}^{L} Q_{3,i}(\Gamma^{(k, r \ominus_k 1)}, x_3)v^{x_3}, \quad \Gamma^{(\tilde{k}, \tilde{r})} \in C_{\tilde{k}}^{\mathrm{O}} \cup C_{\tilde{k}}^{\mathrm{N}}$$

**Theorem 5.** *Following recurrent w.r.t. $i \geqslant 0$ relations take place for the partial probability generating functions:*

*1. $\Gamma^{(0,\tilde{r})} \in \Gamma$, $\tilde{r} = \overline{1, n_0}$*

$$\mathfrak{M}^{(i+1)}(0, \tilde{r}, v) = \alpha_i(0, \tilde{r}, v);$$

2. $\Gamma^{(\tilde{k},\tilde{r})} \in \Gamma$, $\tilde{k} = \overline{1,d}$, $\tilde{r} = \overline{1,n_{\tilde{k}}}$

$$\mathfrak{M}^{(i+1)}(\tilde{k},\tilde{r},v) = q_{\tilde{k},\tilde{r}}(v) \times \mathfrak{M}^{(i)}(\tilde{k},\tilde{r} \ominus_{\tilde{k}} 1, v) + \alpha_i(\tilde{k},\tilde{r},v);$$

*Proof.* Using (8) following can be deduced

$$\mathfrak{M}^{(i+1)}(\tilde{k},\tilde{r},v) = \sum_{w=0}^{\infty} Q_{3,i+1}(\Gamma^{(k,r)},w)v^w$$

$$= Q_{3,i+1}(\Gamma^{(k,r)},0) + \sum_{w=1}^{\infty} Q_{3,i+1}(\Gamma^{(k,r)},w)v^w$$

$$= \sum_{x_3=0}^{\ell(\tilde{k},\tilde{r},3)} \sum_{\gamma \in \mathbb{H}_{-1}(\tilde{\gamma},x_3)} Q_{3,i}(\gamma,x_3) \sum_{a=0}^{\ell(\tilde{k},\tilde{r},3)-x_3} \varphi_3(a,T^{(\tilde{k},\tilde{r})})$$

$$+ \sum_{w=1}^{\infty} \sum_{x_3=0}^{w+\ell(\tilde{k},\tilde{r},3)} \sum_{\gamma \in \mathbb{H}_{-1}(\tilde{\gamma},x_3)} Q_{3,i}(\gamma,x_3) \times \varphi_3(w + \ell(\tilde{k},\tilde{r},3) - x_3, T^{(\tilde{k},\tilde{r})})v^w.$$

(9)

After changing summation order by $x_3$ and $w$ second summand splits into two

$$\sum_{w=1}^{\infty} \sum_{x_3=0}^{w+\ell(\tilde{k},\tilde{r},3)} \sum_{\gamma \in \mathbb{H}_{-1}(\tilde{\gamma},x_3)} Q_{3,i}(\gamma,x_3) \times \varphi_3(w + \ell(\tilde{k},\tilde{r},3) - x_3, T^{(\tilde{k},\tilde{r})})v^w$$

$$= \sum_{x_3=0}^{\ell(\tilde{k},\tilde{r},3)} \sum_{w=1}^{\infty} \sum_{\gamma \in \mathbb{H}_{-1}(\tilde{\gamma},x_3)} Q_{3,i}(\gamma,x_3) \times \varphi_3(w + \ell(\tilde{k},\tilde{r},3) - x_3, T^{(\tilde{k},\tilde{r})})v^w$$

$$+ \sum_{x_3=\ell(\tilde{k},\tilde{r},3)+1}^{\infty} \sum_{w=x_3-\ell(\tilde{k},\tilde{r},3)}^{\infty} \sum_{\gamma \in \mathbb{H}_{-1}(\tilde{\gamma},x_3)} Q_{3,i}(\gamma,x_3) \times \varphi_3(w + \ell(\tilde{k},\tilde{r},3)$$

$$- x_3, T^{(\tilde{k},\tilde{r})})v^w.$$

(10)

Lets transform them a little to get more convinient form.

$$\sum_{x_3=0}^{\ell(\tilde{k},\tilde{r},3)} \sum_{w=1}^{\infty} \sum_{\gamma \in \mathbb{H}_{-1}(\tilde{\gamma},x_3)} Q_{3,i}(\gamma,x_3) \times \varphi_3(w + \ell(\tilde{k},\tilde{r},3) - x_3, T^{(\tilde{k},\tilde{r})})v^w$$

$$= \sum_{x_3=0}^{\ell(\tilde{k},\tilde{r},3)} \sum_{\gamma \in \mathbb{H}_{-1}(\tilde{\gamma},x_3)} Q_{3,i}(\gamma,x_3) \sum_{w=1}^{\infty} \varphi_3(w + \ell(\tilde{k},\tilde{r},3) - x_3, T^{(\tilde{k},\tilde{r})})v^w$$

$$= \sum_{x_3=0}^{\ell(\tilde{k},\tilde{r},3)} \sum_{\gamma \in \mathbb{H}_{-1}(\tilde{\gamma},x_3)} Q_{3,i}(\gamma,x_3)v^{x_3-\ell(\tilde{k},\tilde{r},3)} \sum_{w=\ell(\tilde{k},\tilde{r},3)+1-x_3}^{\infty} \varphi_3(w, T^{(\tilde{k},\tilde{r})})v^w.$$

(11)

Using the same calculations for the second summand one can derive

$$
\sum_{x_3=\ell(\tilde{k},\tilde{r},3)+1}^{\infty} \sum_{w=x_3-\ell(\tilde{k},\tilde{r},3)}^{\infty} \sum_{\gamma\in\mathbb{H}_{-1}(\tilde{\gamma},x_3)} Q_{3,i}(\gamma,x_3)
$$

$$
\times \varphi_3(w+\ell(\tilde{k},\tilde{r},3)-x_3, T^{(\tilde{k},\tilde{r})})v^w
$$

$$
= \sum_{x_3=\ell(\tilde{k},\tilde{r},3)+1}^{\infty} \sum_{\gamma\in\mathbb{H}_{-1}(\tilde{\gamma},x_3)} Q_{3,i}(\gamma,x_3)v^{x_3-\ell(\tilde{k},\tilde{r},3)} \sum_{w=0}^{\infty} \varphi_3(w,T^{(\tilde{k},\tilde{r})})v^w
$$

$$
= \sum_{x_3=0}^{\infty} \sum_{\gamma\in\mathbb{H}_{-1}(\tilde{\gamma},x_3)} Q_{3,i}(\gamma,x_3)v^{x_3-\ell(\tilde{k},\tilde{r},3)} \sum_{w=0}^{\infty} \varphi_3(w,T^{(\tilde{k},\tilde{r})})v^w
$$

$$
- \sum_{x_3=0}^{\ell(\tilde{k},\tilde{r},3)} \sum_{\gamma\in\mathbb{H}_{-1}(\tilde{\gamma},x_3)} Q_{3,i}(\gamma,x_3)v^{x_3-\ell(\tilde{k},\tilde{r},3)} \sum_{w=0}^{\infty} \varphi_3(w,T^{(\tilde{k},\tilde{r})})v^w \tag{12}
$$

Substitute (11) and (12) to (10) and after that (10) to (9), we get:

$$
\mathfrak{M}^{(i+1)}(\tilde{k},\tilde{r},v) = \sum_{x_3=0}^{\ell(\tilde{k},\tilde{r},3)} \sum_{\gamma\in\mathbb{H}_{-1}(\tilde{\gamma},x_3)} Q_{3,i}(\gamma,x_3) \sum_{a=0}^{\ell(\tilde{k},\tilde{r},3)-x_3} \varphi_3(a,T^{(\tilde{k},\tilde{r})})
$$

$$
+ \sum_{x_3=0}^{\ell(\tilde{k},\tilde{r},3)} \sum_{\gamma\in\mathbb{H}_{-1}(\tilde{\gamma},x_3)} Q_{3,i}(\gamma,x_3)v^{x_3-\ell(\tilde{k},\tilde{r},3)} \sum_{w=\ell(\tilde{k},\tilde{r},3)+1-x_3}^{\infty} \varphi_3(w,T^{(\tilde{k},\tilde{r})})v^w
$$

$$
+ \sum_{x_3=0}^{\infty} \sum_{\gamma\in\mathbb{H}_{-1}(\tilde{\gamma},x_3)} Q_{3,i}(\gamma,x_3)v^{x_3-\ell(\tilde{k},\tilde{r},3)} \sum_{w=0}^{\infty} \varphi_3(w,T^{(\tilde{k},\tilde{r})})v^w
$$

$$
- \sum_{x_3=0}^{\ell(\tilde{k},\tilde{r},3)} \sum_{\gamma\in\mathbb{H}_{-1}(\tilde{\gamma},x_3)} Q_{3,i}(\gamma,x_3)v^{x_3-\ell(\tilde{k},\tilde{r},3)} \sum_{w=0}^{\infty} \varphi_3(w,T^{(\tilde{k},\tilde{r})})v^w .
$$

Group the second and the fourth summand:

$$
\mathfrak{M}^{(i+1)}(\tilde{k},\tilde{r},v) = \sum_{x_3=0}^{\ell(\tilde{k},\tilde{r},3)} \sum_{\gamma\in\mathbb{H}_{-1}(\tilde{\gamma},x_3)} Q_{3,i}(\gamma,x_3) \sum_{a=0}^{\ell(\tilde{k},\tilde{r},3)-x_3} \varphi_3(a,T^{(\tilde{k},\tilde{r})})
$$

$$
+ \sum_{x_3=0}^{\ell(\tilde{k},\tilde{r},3)} \sum_{\gamma\in\mathbb{H}_{-1}(\tilde{\gamma},x_3)} Q_{3,i}(\gamma,x_3)v^{x_3-\ell(\tilde{k},\tilde{r},3)}[\sum_{w=\ell(\tilde{k},\tilde{r},3)+1-x_3}^{\infty} \varphi_3(w,T^{(\tilde{k},\tilde{r})})v^w
$$

$$
- \sum_{w=0}^{\infty} \varphi_3(w,T^{(\tilde{k},\tilde{r})})v^w] + q_{\tilde{k},\tilde{r}}(v) \sum_{x_3=0}^{\infty} \sum_{\gamma\in\mathbb{H}_{-1}(\tilde{\gamma},x_3)} Q_{3,i}(\gamma,x_3)v^{x_3}
$$

and consequently

$$\mathfrak{M}^{(i+1)}(\tilde{k}, \tilde{r}, v) = \sum_{x_3=0}^{\ell(\tilde{k}, \tilde{r}, 3)} \sum_{\gamma \in \mathbb{H}_{-1}(\tilde{\gamma}, x_3)} Q_{3,i}(\gamma, x_3) \sum_{a=0}^{\ell(\tilde{k}, \tilde{r}, 3) - x_3} \varphi_3(a, T^{(\tilde{k}, \tilde{r})})$$

$$- \sum_{x_3=0}^{\ell(\tilde{k}, \tilde{r}, 3)} \sum_{\gamma \in \mathbb{H}_{-1}(\tilde{\gamma}, x_3)} Q_{3,i}(\gamma, x_3) v^{x_3 - \ell(\tilde{k}, \tilde{r}, 3)} \sum_{w=0}^{\ell(\tilde{k}, \tilde{r}, 3) - x_3} \varphi_3(w, T^{(\tilde{k}, \tilde{r})}) v^w$$

$$+ q_{\tilde{k}, \tilde{r}}(v) \sum_{x_3=0}^{\infty} \sum_{\gamma \in \mathbb{H}_{-1}(\tilde{\gamma}, x_3)} Q_{3,i}(\gamma, x_3) v^{x_3}$$

$$= \tilde{\alpha}_i(\tilde{k}, \tilde{r}, v) + q_{\tilde{k}, \tilde{r}}(v) \sum_{x_3=0}^{\infty} \sum_{\gamma \in \mathbb{H}_{-1}(\tilde{\gamma}, x_3)} Q_{3,i}(\gamma, x_3) v^{x_3}. \qquad (13)$$

Consider sum $\sum_{x_3=0}^{\infty} \sum_{\gamma \in \mathbb{H}_{-1}(\tilde{\gamma}, x_3)} Q_{3,i}(\gamma, x_3) v^{x_3}$ in more details depending on $\tilde{\gamma}$. In case $\tilde{\gamma} = \Gamma^{(0,\tilde{r})}$ from definition $\mathbb{H}_{-1}(\cdot, \cdot)$ one can derive $\mathbb{H}_{-1}(\tilde{\gamma}, x_3) = \{\Gamma^{(k_1,r_1)}, \Gamma^{(0,\tilde{r}\ominus_0 1)}\}$ for $x_3 \leqslant L$ and $\mathbb{H}_{-1}(\tilde{\gamma}, x_3) = \emptyset$ for $x_3 > L$. Here pair $(k_1, r_1)$ such that $h_1(\Gamma^{(k_1,r_1)}) = \tilde{r}$. The sum is transformed to

$$\sum_{x_3=0}^{\infty} \sum_{\gamma \in \mathbb{H}_{-1}(\tilde{\gamma}, x_3)} Q_{3,i}(\gamma, x_3) v^{x_3} = \sum_{x_3=0}^{L} \left[ Q_{3,i}(\Gamma^{(k_1,r_1)}, x_3) + Q_{3,i}(\Gamma^{(0,r\ominus_0 1)}, x_3) \right] v^{x_3}.$$
$$(14)$$

In case $\tilde{\gamma} \in C_{\tilde{k}}^{\mathrm{I}}$ from definition $\mathbb{H}_{-1}(\cdot, \cdot)$ one finds that $\mathbb{H}_{-1}(\tilde{\gamma}, x_3) = \emptyset$ for $x_3 \leqslant L$ and $\mathbb{H}_{-1}(\tilde{\gamma}, x_3) = \{\Gamma^{(\tilde{k}, \tilde{r}\ominus_{\tilde{k}} 1)}, \Gamma^{(0,r_2)}\}$ for $x_3 > L$. Here $r_2$ such that $h_3(r_2) = \Gamma^{(\tilde{k}, \tilde{r})}$. The sum is transformed to

$$\sum_{x_3=0}^{\infty} \sum_{\gamma \in \mathbb{H}_{-1}(\tilde{\gamma}, x_3)} Q_{3,i}(\gamma, x_3) v^{x_3} = \sum_{x_3=L+1}^{\infty} \left[ Q_{3,i}(\Gamma^{(k,r\ominus_k 1)}, x_3) + Q_{3,i}(\Gamma^{(0,r_2)}, x_3) \right] v^{x_3}$$

$$= \sum_{x_3=0}^{\infty} \left[ Q_{3,i}(\Gamma^{(k,r\ominus_k 1)}, x_3) + Q_{3,i}(\Gamma^{(0,r_2)}, x_3) \right] v^{x_3}$$

$$- \sum_{x_3=0}^{L} \left[ Q_{3,i}(\Gamma^{(k,r\ominus_k 1)}, x_3) + Q_{3,i}(\Gamma^{(0,r_2)}, x_3) \right] v^{x_3}.$$
$$(15)$$

In the last case $\tilde{\gamma} \in C_{\tilde{k}}^{\mathrm{O}} \cup C_{\tilde{k}}^{\mathrm{N}}$, one can derive $\mathbb{H}_{-1}(\tilde{\gamma}, x_3) = \{\Gamma^{(\tilde{k}, \tilde{r}\ominus_{\tilde{k}} 1)}\}$ for all $x_3 \geqslant 0$. The sum takes following form

$$\sum_{x_3=0}^{\infty} \sum_{\gamma \in \mathbb{H}_{-1}(\tilde{\gamma}, x_3)} Q_{3,i}(\gamma, x_3) v^{x_3} = \sum_{x_3=L+1}^{\infty} Q_{3,i}(\Gamma^{(k,r\ominus_k 1)}, x_3) v^{x_3}$$

$$= \sum_{x_3=0}^{\infty} Q_{3,i}(\Gamma^{(k,r\ominus_k 1)}, x_3) v^{x_3} - \sum_{x_3=0}^{L} Q_{3,i}(\Gamma^{(k,r\ominus_k 1)}, x_3) v^{x_3}.$$
$$(16)$$

Substituting (14), (15) and (16) to the (13) one gets theorem statement.

**Acknowledgments.** This work was fulfilled as a part of State Budget Research and Development program No. 01201456585 "Mathematical modeling and analysis of stochastic evolutionary systems and decision processes" of N.I. Lobachevsky State University of Nizhny Novgorod and was supported by State Program "Promoting the competitiveness among world's leading research and educational centers"

# References

1. Neimark, Y., Fedotkin, M.A., Preobrazhenskaja, A.M.: Operation of an automate with feedback controlling traffic at an intersection. Izvestija of USSR Academy of Sciences, Technical Cybernetic (5), 129–141 (1968)
2. Fedotkin, M.A.: On a class of stable algorithms for control of conflicting flows or arriving airplanes. Prob. Control Inf. Theor. **6**(1), 13–22 (1977)
3. Fedotkin, M.A.: Construction of a model and investigation of nonlinear algorithms for control of intense conflict flows in a system with variable structure of servicing demands. Lith. Math. J. **17**(1), 129–137 (1977)
4. Litvak, N.V., Fedotkin, M.A.: A probabilistic model for the adaptive control of conflict flows. Autom. Remote Control **61**(5), 777–784 (2000)
5. Proidakova, E.V., Fedotkin, M.A.: Control of output flows in the system with cyclic servicing and readjustments. Autom. Remote Control **69**(6), 993–1002 (2008)
6. Afanasyeva, L.G., Bulinskaya, E.V.: Mathematical models of transport systems based on queueing theory. Trudy of Moscow Institute of Physocs and Technology (4), 6–21 (2010)
7. Yamada, K., Lam, T.N.: Simulation analysis of two adjacent traffic signals. In: Proceedings of the 17th Winter Simulation Conference, pp. 454–464. ACM, New York (1985)
8. Zorin, A.V.: Stability of a tandem of queueing systems with Bernoulli noninstantaneous transfer of customers. Theor. Probab. Math. Stat. **84**, 173–188 (2012)
9. Kocheganov, V.M., Zorine, A.V.: Probabilistic model of tandem of queuing systems under cyclic control with prolongations. In: Proceedings of Internation Conference Probability Theory, Stochastic Processes, Mathematical Statistics and Applications, Minsk, 23–26 Feburary 2015, pp. 94–99 (2015)

# On Conditional Monte Carlo Estimation of Busy Period Probabilities in Gaussian Queues

Oleg Lukashenko[1,2(✉)], Evsey Morozov[1,2], and Michele Pagano[3]

[1] Institute of Applied Mathematical Research of the Karelian Research Centre RAS,
Petrozavodsk, Russia
`lukashenko-oleg@mail.ru`, `emorozov@karelia.ru`
[2] Petrozavodsk State University, Petrozavodsk, Russia
[3] University of Pisa, Pisa, Italy
`m.pagano@iet.unipi.it`

**Abstract.** Due to the self-similar nature of broadband traffic, the arrival rate can persist on relatively high values for a considerable amount of time. Such a behavior, closely related to the duration of busy periods, has a deep impact on queueing performance in terms of loss probability and distribution of losses. In the paper we consider the probability that the normalized cumulative workload grows at least as the length $T$ of the considered interval in case of Gaussian input traffic. As $T$ increases, the event becomes rare and standard Monte Carlo simulation would require a large number of generated sample paths to get an accurate estimate. To cope with this problem, we propose a variant of the well-known conditional Monte Carlo method, in which conditioning is expressed in terms of the bridge process. We derive the analytical expression of the estimator and verify its effectiveness through simulations.

**Keywords:** Gaussian processes · Conditional Monte Carlo · Bridge process · Rare events · Variance reduction

## 1 Introduction

The self-similar nature of broadband traffic [7] has a deep impact in terms of network dimensioning and Quality of Service (QoS) issues [3]. Indeed, persistence phenomena (known in the literature as Noah effect) imply that the arrival rate can remain on relatively high values for a considerable amount of time. Such a behavior, closely related to the duration of busy periods in the underlying queueing system, negatively affects QoS performance in terms of loss probability and distribution of losses.

In traffic modelling, Gaussian processes have emerged as a flexible and powerful tool, able to take into account the long memory properties of real traffic,

This work is supported by Russian Foundation for Basic research, projects 15–07–02341 A, 15–07–02354 A, 15–07–02360 A, and also by the Program of strategic development of Petrozavodsk State University.

V. Vishnevsky and D. Kozyrev (Eds.): DCCN 2015, CCIS 601, pp. 280–288, 2016.
DOI: 10.1007/978-3-319-30843-2_29

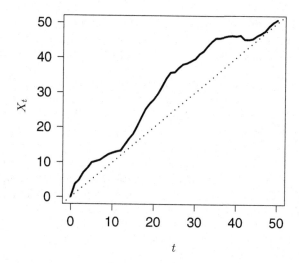

**Fig. 1.** The sample path of the process $X$ lies above the diagonal

while keeping a relatively simple and elegant description. The best known model is Fractional Brownian Motion (FBM), originally proposed by Norros [10], but our method is more general and includes FBM as a special case.

In more detail, in this work we consider a centered Gaussian process with stationary increments $\{X_t, t \in \mathbb{R}_+\}$. Let us denote by $v_t := Var X_t$ the variance of $X_t$; then the covariance function has the following expression:

$$\Gamma_{s,t} = \frac{1}{2} \left( v_t + v_s - v_{|t-s|} \right). \tag{1}$$

We are interested in the estimation of the following probability

$$\pi(\mathbb{T}) := \mathbb{P}\left( \forall t \in \mathbb{T} : X_t > t \right), \tag{2}$$

where $\mathbb{T} = [0, T] \subseteq \mathbb{R}_+$ (see Fig. 1). Such probability is closely related to the duration of busy periods and plays an important role in the study of QoS indexes since it takes into account bursts of losses, see [9,11] for more details.

It is worth mentioning that our approach only requires that $v_t$ is an increasing function. Such condition is quite general and holds for the following processes, well-known in the literature:

1. Fractional Brownian Motion (FBM). In this case $v_t = t^{2H}$, with Hurst parameter $H \in (0, 1)$; in the teletraffic framework usually $H \in (0.5, 1)$, corresponding to traffic processes with long-range dependence. It has been shown in [13] that FBM arises as the scaled limit process when the cumulative workload is a superposition of on-off sources with mutually independent heavy-tailed on and/or off periods.
2. Sum of independent FBMs with $v_t = \sum_i t^{2H_i}$. The use of this model is also motivated by the fundamental result in [13] in case of heterogeneous on-off sources.

3. Integrated Ornstein-Uhlenbeck process (IOU) with $v_t = t + e^{-t} - 1$. IOU is the Gaussian counterpart of the well-known Anick-Mitra-Sondi fluid model [1], and its relevance in the framework of teletraffic is also motivated in [6].

The key contribution of this work is the application of a variant of the conditional Monte Carlo method for variance reduction, to estimate the target probability (2) when $T \to \infty$. Indeed in this case the event $\{\forall t \in \mathbb{T} : X_t > t\}$ becomes rare and hence standard Monte Carlo requires an unacceptable large number of generated sample paths.

Note that the analytical expression of the target probability is not known in explicit form for general Gaussian input (including considered examples). Indeed, there are only a few asymptotic results available based on the large deviation theory. For instance in case of FBM input

$$\lim_{T \to \infty} \frac{1}{T^{2-2H}} \log \mathbb{P}\left(\forall t \in (0,T] : X_t > t\right)$$

$$= \lim_{n \to \infty} \frac{1}{n} \log \mathbb{P}\left(\forall t \in (0,1] : \frac{X_t}{\sqrt{n}} > t\right) = -\inf_{f \in \mathscr{B}} I(f), \quad (3)$$

where

$$\mathscr{B} := \{f \in \mathcal{R} : f(r) > r, \, \forall r \in (0,1]\},$$

$I$ denotes the rate function and $\mathcal{R}$ is the reproducing kernel Hilbert space (RKHS) associated with the distribution of FBM (see [2] for more details).

## 2    Conditional Monte Carlo

Let $X$ be a random process. Consider estimating the probability

$$\pi = \mathbb{P}(X \in A) = \mathbb{E}I(X \in A)$$

for some Borel set $A$ of the paths of the process X, where $I$ denotes the indicator function. To estimate $\pi$ by standard Monte Carlo simulation, we should generate $N$ replications $X_1, ..., X_N$ of the process $X$ and calculate the sample mean

$$\tilde{\pi}_N = \frac{1}{N} \sum_{n=1}^{N} I_n.$$

Then the relative error (RE) of the estimate $\tilde{\pi}_N$ behaves as

$$\mathrm{RE}(\tilde{\pi}_N) := \frac{\sqrt{\mathbb{V}ar\tilde{\pi}_N}}{\mathbb{E}\tilde{\pi}_N} \sim \frac{1}{\sqrt{\pi N}} \text{ as } \pi \to 0.$$

Therefore the relative error of the standard Monte Carlo estimation is unbounded when the event becomes rare. That is why for the rare event simulation it is crucially important to use the modified estimators in order to reduce variance (and, as a result, relative error).

Let us briefly describe the method of simulation by conditioning, known in the literature as conditional Monte Carlo [12]. Denote

$$Z = I(X \in A)$$

and assume we have an auxiliary random variable $Y$ correlated with $Z$ such that $\mathbb{E}[Z|Y]$ is available in explicit form. Let $Y_1, ..., Y_N$ be the samples of $Y$, then the corresponding unbiased estimator is defined as

$$\widehat{\pi}_N = \frac{1}{N} \sum_{n=1}^{N} \mathbb{E}[Z|Y_n]. \tag{4}$$

Note that the variance of this estimator is always less than the variance of the standard Monte Carlo one since

$$\mathbb{V}ar Z = \mathbb{E}[\mathbb{V}ar[Z|Y]] + \mathbb{V}ar[\mathbb{E}[Z|Y]]. \tag{5}$$

## 3   BMC Estimator

The Bridge Monte Carlo (BMC) is a special case of the conditional Monte Carlo method, particularly suitable for the estimation of the rare event probabilities in a queueing system with Gaussian input.

Originally proposed by some of the authors in [4,5], BMC is based on the idea of expressing the overflow probability as the expectation of a function of the *Bridge* $Y := \{Y_t\}$ of the Gaussian input process $X$, i.e., the process obtained by conditioning $X$ to reach a certain level at some prefixed (deterministic) time $\tau$:

$$Y_t = X_t - \psi_t X_\tau, \tag{6}$$

where function $\psi_t$ is expressed via covariance function as

$$\psi_t := \frac{\Gamma_{t,\tau}}{\Gamma_{\tau,\tau}}.$$

Because the variance of the input is an increasing function of $t$ in all models we consider, it is easy to see that $\psi_t > 0$ for all $t \in T$. Moreover, we note that the process $Y$ is independent of $X_\tau$ since

$$\mathbb{E}[X_\tau Y_t] = \Gamma_{\tau,t} - \frac{\Gamma_{t,\tau}}{\Gamma_{\tau,\tau}} \Gamma_{\tau,\tau} = 0.$$

The BMC approach, originally proposed for the estimation of the overflow probability [8], can also be adopted for estimating our target probability $\pi(\mathbb{T})$. Indeed,

$$\begin{aligned}
\pi(\mathbb{T}) &:= \mathbb{P}\left(\forall t \in \mathbb{T} : \ X_t > t\right) \\
&= \mathbb{P}\left(\forall t \in \mathbb{T} : \ Y_t + \psi_t X_\tau > t\right) \\
&= \mathbb{P}\left(\forall t \in \mathbb{T} : \ X_\tau > \frac{t - Y_t}{\psi_t}\right) \\
&= \mathbb{P}\left(X_\tau \geq \sup_{t \in \mathbb{T}} \frac{t - Y_t}{\psi_t}\right) = \mathbb{P}\left(X_\tau \geq \overline{Y}\right),
\end{aligned}$$

where $\mathbb{T}$ denotes the set of interest and

$$\overline{Y} := \sup_{t \in \mathbb{T}} \frac{t - Y_t}{\psi_t}. \tag{7}$$

Recall that

$$X_t =_d N(0, \Gamma_{t,t}) =_d \sqrt{\Gamma_{t,t}} N(0, 1),$$

where $=_d$ stands for stochastic equivalence. Then the considered probability can be rewritten as follows

$$\pi = \mathbb{P}\left(X_\tau \geq \overline{Y}\right) = \int_R \mathbb{P}(X_\tau \geq u)\mathbb{P}(\overline{Y} \in du)$$

$$= \int_R \mathbb{P}\left(N(0, 1) \geq \frac{u}{\sqrt{\Gamma_{\tau,\tau}}}\right)\mathbb{P}(\overline{Y} \in du)$$

$$= \mathbb{E}\left[\Phi\left(\frac{\overline{Y}}{\sqrt{\Gamma_{\tau,\tau}}}\right)\right],$$

where independence between $\overline{Y}$ and $X_\tau$ is used and $\Phi$ denotes the tail distribution of standard normal variable, that is

$$\Phi(x) = \frac{1}{\sqrt{2\pi}} \int_x^\infty e^{-y^2/2} dy.$$

Hence, given an i.i.d. sequence $\{\overline{Y}^{(i)}, i = 1, ..., N\}$ distributed as $\overline{Y}$, the estimator of $\pi(\mathbb{T})$, by (9), is defined as follows:

$$\widehat{\pi}_N^{BMC} = \frac{1}{N} \sum_{i=1}^N \Phi\left(\frac{\overline{Y}^{(i)}}{\sqrt{\Gamma_{\tau,\tau}}}\right). \tag{8}$$

Note that

$$\Phi\left(\frac{\overline{Y}}{\sqrt{\Gamma_{\tau,\tau}}}\right) = \mathbb{E}\left[I(X_\tau > \overline{Y})|\overline{Y}\right], \tag{9}$$

therefore the BMC approach is actually a special case of the conditional Monte Carlo method. By (5), $\mathbb{V}arZ \geq \mathbb{V}ar[\mathbb{E}[Z|\overline{Y}]]$, so we can expect that the BMC estimator implies variance reduction (comparing to the Crude Monte Carlo simulation) in the estimation of the target probability $\pi(\mathbb{T})$.

## 4 Simulation Results

In this section, we point out through simulation results the accuracy of the BMC estimator and the dependence of its performance on different parameters. For sake of brevity, we only present results for FBM input. Moreover, in the following we consider $N = 10000$ replications and a discrete lattice $\mathbb{T} = \{0, 1, ..., T\}$.

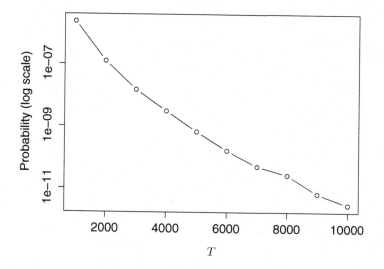

**Fig. 2.** Dependence of $\pi(\mathbb{T})$ on parameter $T$

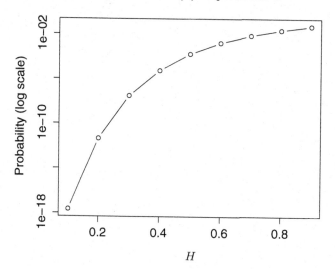

**Fig. 3.** Dependence $\pi(\mathbb{T})$ on Hurst parameter $H$

Figure 2 shows the dependence of the target probability on the time duration $T$ in case of $H = 0.8$. The probability $\pi(T)$ exhibits an exponential decay in agreement with the known large deviations asymptotic results, see formula (3).

Figure 3 highlight that $\pi(T)$ strongly depends on the Hurst parameter $H$. It is quite natural that in the short-range dependent case ($H < 1/2$) the probability the process $X$ is above the diagonal for some fixed time duration $T$ (in this set of simulation $T = 10$) is extremely small, while in presence of long-range dependence (that is $H \in (1/2, 1)$) $\pi(\mathbb{T})$ is much higher.

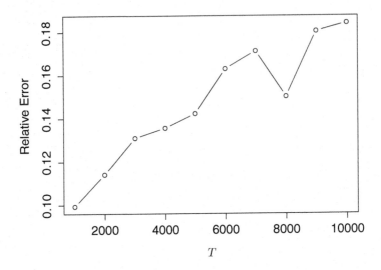

**Fig. 4.** Dependence of the relative error on $T$

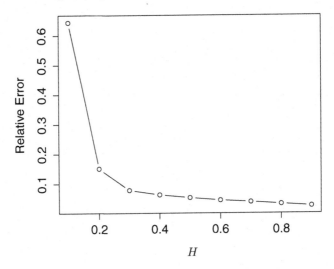

**Fig. 5.** Dependence of the relative error on $H$

In order to verify the goodness of our estimator, Figs. 4 and 5 show the dependence of the relative error on parameters $T$ and $H$ respectively. Although the relative error is not bounded (indeed, it is even unknown whether BMC is asymptotically efficient), it grows slowly, and for the target probabilities of the order of $10^{-11}$ is still less than 18 % (compare for example the values in Figs. 2 and 4).

Finally, Fig. 6 highlights the dependence of the relative error on the conditioning time $\tau$, which was used for constructing the Bridge process $Y$.

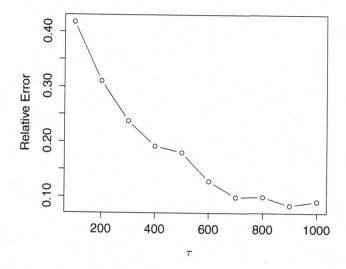

**Fig. 6.** Dependence of relative error on $\tau$

Empirical results highlight that it is better to choose larger values of free parameter $\tau$ in order to reduce the relative error.

## 5   Conclusions

In this paper, we addressed the busy period duration in Gaussian queue and estimated the related probabilities by means of the BMC approach, a special case of conditional Monte Carlo estimator. The BMC simulation approach exploits the Gaussian nature of the input process (independence is equivalent to uncorrelatedness) of the input process and relies on the properties of bridges.

Several simulation experiments have been carried out in order to study the properties of the proposed estimator in case of FBM traffic for different values of the relevant parameters (duration of the interval, value of the Hurst parameter and choice of the conditioning point).

As a further research step, it is possible to study the influence of so-called discretization step on simulation results and the asymptotic efficiency/inefficiency of BMC estimator.

## References

1. Addie, R., Mannersalo, P., Norros, I.: Most probable paths and performance formulae for buffers with Gaussian input traffic. Eur. Trans. Telecommun. **13**, 183–196 (2002)
2. Deuschel, J.D., Stroock, D.W.: Large Deviations. Academic Press, San Diego (1989)

3. Erramilli, A., Narayan, O., Willinger, W.: Experimental queueing analysis with long-range dependent packet traffic. IEEE/ACM Trans. Netw. **4**(2), 209–223 (1996)
4. Giordano, S., Gubinelli, M., Pagano, M.: Bridge Monte-Carlo: a novel approach to rare events of Gaussian processes. In: Proceedings of the 5th St. Petersburg Workshop on Simulation, pp. 281–286, St. Petersburg, Russia (2005)
5. Giordano, S., Gubinelli, M., Pagano, M.: Rare events of gaussian processes: a performance comparison between Bridge Monte-Carlo and importance sampling. In: Koucheryavy, Y., Harju, J., Sayenko, A. (eds.) NEW2AN 2007. LNCS, vol. 4712, pp. 269–280. Springer, Heidelberg (2007)
6. Kulkarni, V., Rolski, T.: Fluid model driven by an Ornstein-Uhlenbeck process. Probab. Eng. Info. Sci. **8**, 403–417 (1994)
7. Leland, W.E., Taqqu, M.S., Willinger, W., Wilson, D.V.: On the self-similar nature of Ethernet traffic (extended version). IEEE/ACM Trans. Netw. **2**(1), 1–15 (1994)
8. Lukashenko, O.V., Morozov, E.V., Pagano, M.: Performance analysis of Bridge Monte-Carlo estimator. Trans. KarRC RAS **3**, 54–60 (2012)
9. Mandjes, M., Norros, I., Glynn, P.: On convergence to stationarity of fractional Brownian storage. Ann. Appl. Probab. **19**, 1385–1403 (2009)
10. Norros, I.: On the use of fractional Brownian motion in the theory of connectionless networks. IEEE J. Sel. Areas Commun. **13**(6), 953–962 (1995)
11. Norros, I.: Busy periods for fractional Brownian storage: a large deviation approach. Adv. in Perf. Anal. **2**, 1–19 (1999)
12. Ross, S.M.: Simulation. Elsevier, Amsterdam (2006)
13. Taqqu, M.S., Willinger, W., Sherman, R.: Proof of a fundamental result in self-similar traffic modeling. Comput. Commun. Rev. **27**, 5–23 (1997)

# Stability and Admissible Densities in Transportation Flow Models

A.A. Lykov$^{(\boxtimes)}$, V.A. Malyshev, and M.V. Melikian

Faculty of Mechanics and Mathematics, Moscow State University,
Moscow, Russia
alekslyk@yandex.ru, 2malyshev@mail.ru, magaarm@list.ru

**Abstract.** One-way road traffic model with a local control is considered.
For given density of transportation units we discover phase transitions in
the control parameter space when there exists or not safe transportation.
Also we discuss possible densities and instability in several road networks
with crosses, if no control is imposed.

## 1 Introduction

Theoretical modelling and computer simulation of transportation systems is a
very popular field, see very impressive review [3]. There are two main directions
in this research - macro and micro models. Macro approach does not distinguish
individual transportation units and uses analogy with the notion of flow in hydro-
dynamics, see [2]. Stochastic micro models are most popular and use almost all
types of stochastic processes: mean field, queueing type and local interaction
models. We consider here deterministic transportation flows. Although not as
popular as stochastic traffic, there is also a big activity in this field, see [1,4–8].

In Sect. 2 we do not pursue maximal generality but rather consider the simplest flow, that is the following one-way road traffic model. Namely, at any time
$t \geq 0$ there are many (even infinite number) of point particles (may be called also
cars, units etc.) with coordinates $z_k(t)$ on the real line, enumerated as follows

$$... < z_N(t) < ... < z_1(t) < z_0(t) \tag{1}$$

For this infinite chain of cars we try to find control mechanism which garanties
that the distance between any pair of neighbouring cars is greater or equal (on
all time interval $(0, \infty)$) to some fixed number $d$, called safe distance, but at
the same time is not too far from it (then we say that the density of cars is
admissible).

This control mechanism is assumed to be local - any car has information only
about the previous car. Moreover, this mechanism is of physical nature, like forces
between molecules in crystals but our "forces" are not symmetric. The safety
(stability) conditions appear to be similar to the dynamical phase transition in
the model of the molecular chain rapture under the action of external force.
However here we do not need the double scaling limit, used in [9].

© Springer International Publishing Switzerland 2016
V. Vishnevsky and D. Kozyrev (Eds.): DCCN 2015, CCIS 601, pp. 289–295, 2016.
DOI: 10.1007/978-3-319-30843-2_30

In Sect. 3 we consider idealistic models of transportation network as the collection of roads with crosses (intersections). Along each road the units (cars) move deterministically. We get admissible density of units on the given transportation network and show the dependence of this density on the number and multiplicity of crosses and number of cycles in the network.

## 2 Local Flow Control

In this section we consider the simplest flow (1), where it is assumed that the rightmost unit moves "as it wants". More exactly, the trajectory $z_0(t)$ is only assumed to be sufficiently smooth with positive velocity $v_0(t) = \dot{z}_0(t)$ and natural upper bounds on the velocity and acceleration

$$\sup_{t>0} v_0(t) \leqslant v_{max}, \quad \sup_{t>0} |\ddot{z}_0(t)| \leqslant a_{max} \tag{2}$$

We would like to organize such traffic so that for any $t$ and $k$ the distances $r_k(t) = z_{k-1}(t) - z_k(t)$ were greater or equal to some number $d > 0$, which is chosen to avoid collisions and keep maximal possible density of traffic. Moreover, the organization should use only (maximally) local control. More exactly, the $k$-th driver at any time $t$ knows only its own coordinate and velocity and the coordinate $z_{k-1}(t)$ of the previous car. Thus, for any $k \geq 1$ the trajectory $z_k(t)$ is uniquely defined by the trajectory $z_{k-1}(t)$ of the previous particle.

Using physical terminology one could say that if, for example, $r_k(t)$ becomes larger than $d$, then some positive force $F_k$ increases acceleration of the particle $k$. We will see that, besides $F_k$, for such stability, also friction force $-\alpha v_k(t)$, which, on the contrary, restrains the growth of the velocity $v_k(t)$, is necessary. The constant $\alpha > 0$ should be chosen appropriately.

Thus the trajectories are uniquely defined by the system of equations for $k \geq 1$

$$z_k''(t) = F_k(t) - \alpha \frac{dz_k}{dt} = \omega^2(z_{k-1}(t) - z_k(t) - d) - \alpha \frac{dz_k}{dt} \tag{3}$$

where $F_k$ is taken to be simplest possible

$$F_k(t) = \omega^2(z_{k-1}(t) - z_k(t) - d)$$

*Stability Conditions.* For given $\alpha, \omega, d, z_0(t)$ and initial conditions the trajectories are uniquely defined and we can denote

$$I = \inf_{k \geqslant 1} \inf_{t \geqslant 0} r_k(t), \quad S = \sup_{k \geqslant 1} \sup_{t \geqslant 0} r_k(t)$$

Put also

$$d^* = d^*(v_{max}, a_{max}) = \frac{1}{\omega^2}(a_{max} + \alpha v_{max})$$

Consider firstly the simplest initial conditions

$$z_k(0) = -kd, \quad \frac{dz_k}{dt}(0) = v, k \geq 0 \tag{4}$$

**Theorem 1.** *Assume (4) and $\alpha > 2\omega > 0$. Then for any chosen "safe distance" parameter $d > d^*$ in the Eq. (3) the following bounds hold*

$$I > (d - d^*) > 0, \quad S < 2d.$$

Now consider more general initial conditions satisfying

$$z_0(0) = 0, z_0'(0) = v, 0 < a \leqslant r_k(0) \leqslant b, \quad |\frac{dr_k}{dt}(0)| \leqslant c, \qquad (5)$$

for any $k > 0$ and some non-negative $a$, $b$. Denote also

$$D^* = \max\{A, d^*\},$$

where

$$A = \frac{\alpha a' + 2c}{2\gamma}, \gamma = \sqrt{\frac{\alpha^2}{4} - \omega^2}, a' = \max\{|a - d|, |b - d|\}.$$

**Theorem 2.** *If $\alpha > 2\omega$, $\frac{\alpha a - 2c}{\alpha - 2\gamma} > \frac{a+b}{2}$, $\frac{a+b}{2} < d^* < \frac{\alpha a - 2c}{\alpha - 2\gamma}$ then for any initial conditions (5) and any smooth function $z_0(t)$, satisfying (2), there exists open subset $D \subset \mathbb{R}$ such that for any choosen "safe distance" parameter $d \in D$ in the Eq. (3), the following bounds hold*

$$I \geq (d - D^*) > 0, \quad S \leq d + D^*.$$

The proof of both theorems is based on the analysis of the chain of equations for $x_k = r_k - d, k > 0$

$$x_k''(t) + \alpha x_k'(t) + \omega^2 x_k(t) = \omega^2 x_{k-1}(t), \ k = 1, 2, ...$$

where

$$x_0(t) = \frac{1}{\omega^2}(z_0''(t) + \alpha z_0'(t))$$

*Density and Currents.* Let $n(t, I)$ be the number of units on the interval $I \subset R$ at time $t$ and $n(T, x)$ be the number of units passing the point $x$ in the time interval $(0, T)$. Then the density and the current through some point $x$ are defined as follows ($|I|$ is the length of $I$)

$$\mu(t) = \lim_{|I| \to \infty} \frac{n(t, I)}{|I|}$$

$$J(x) = \lim_{T \to \infty} \frac{n(T, x)}{T}$$

if these limits exist. For fixed $N$ define also mean length of the chain of cars $0, 1, ..., N$

$$L_N(t) = \frac{z_0(t) - z_N(t)}{N}.$$

**Theorem 3.** *Under the conditions of Theorem 2 assume also that the initial conditions are such that the following finite limits exist*

$$\lim_{N\to\infty} L_N(0) = L(0), \quad \lim_{N\to\infty} \dot{L}_N(0) = \dot{L}(0),$$

*Then for any t there exist*

$$\lim_{N\to\infty} L_N(t) = L(t),$$

*and moreover*

$$L(t) = L(0) + \frac{1}{\alpha}(1 - e^{-\alpha t})\frac{dL}{dt}(0)$$

Note that if moreover $\frac{dz_k(0)}{dt}$ are uniformly bounded then $\frac{dL}{dt}(0) = 0$, that is the mean length does not change with time.

*Instability.* First reason for the instability is the absence of dissipation.

**Theorem 4.** *Let $\alpha = 0$, then for the initial conditions (4) for $k > 0$ and for $z_0(t) = tv + \sin \omega' t$, $v > 0$, $\omega' \neq 0$ and $k \geqslant 2$ we have*

$$\inf_{t\geqslant 0} r_k(t) = -\infty$$

It is sufficiently easy to explain by the resonance effect: $x_1(t)$ is the harmonic movement with frequencies $\omega, \omega'$, and the proper frequency of $x_2$ is $\omega$.

Assume again (4). We will prove instability (even for rather simple behaviour of the leading unit $z_0(t)$) when $k$ and $t$ tend simultaneously to $\infty$ so that $t = \mu k$, $k \to \infty$ for some constant $\mu > 0$. The following theorem consists of two parts: the first one exhibits the possible zero density, the second one exhibits the possibility of collisions.

**Theorem 5.** *Let $\frac{dz_0}{dt}(t) = v$ for all $t \geqslant 0$. Then*
  *1. for any $\alpha > 0$ there exists $\omega > 0$ and constants $q_+ > 1$, $\mu_+ > 0, c_+ > 0$, so that for any $d > 0$*

$$r_k(t) \sim \frac{c_+}{\sqrt{k}} q_+^k$$

*as $t = \mu_+ k$, $k \to \infty$;*
  *2. for any $\alpha > 0$ there exists $\omega > 0$ and constants $q_- > 1$, $\mu_- > 0, c_- < 0$, so that for any $d > 0$*

$$r_k(t) \sim \frac{c_-}{\sqrt{k}} q_-^k$$

*as $t = \mu_- k$, $k \to \infty$.*

# 3    Transport in Networks Without Control

Transportation without control means that any unit moves with its own velocity and does not know anything about the movement of other units. In other words the functions $z_k(t)$ are fixed. We will consider only the case when

$$z_k(t) = c_k + v_k t$$

for some constants $v_k > 0$ and

$$... < c_{k-1} < c_k = z_k(0) < ...$$

The transportation network is defined to be safe if at any time moment $t$ the distance between any two units is greater or equal to some fixed number $d > 0$ called safety distance. It is clear that for the safe network all $v_k$ should be equal to some $v > 0$, their joint constant velocity. It is also clear that $\mu(t) \geq d^{-1}$ and this density is attained in the case when

$$z_i(t) = z_i(0) + vt, \ \ z_{i-1}(0) - z_i(0) = d \implies z_{i-1}(t) - z_i(t) = d$$

It is easy to see that in our case at time $t$ any density $\mu(t)$ is admissible iff

$$\mu(t) \leq d^{-1}$$

The network (of roads) is defined here as the one-dimensional topological space, which is the disjoint union of some number of real lines (roads) $k = 1, ..., M \leq \infty$ with coordinates $z_k$. It is assumed that any pair of real lines has finite number of identified points (crosses). Consider the graph $G$, which vertices are these identified points (intersections of the real lines) and the edges are all segments of the roads in-between the vertices. That is we delete all infinite intervals of the roads. The metrics is defined as usual: the distance $\rho(x_k, x_l)$ between two points $x_k$ and $x_l$ on the network is the minimal length of paths between these two points.

On any road $k$ the transportation units are labeled as $(k, i)$, where $k$ is the road along which it moves and $i$ is its order on this road. Here the sequences are assumed to be infinite to both sides

$$... < z_{k,i}(t) < z_{k,i-1}(t) < ...$$

and we again consider networks where the velocities and safe distance are the same for all roads

**Proposition 1.**

1.  *Assume there are $M < \infty$ roads with only one common cross. In this case the graph $G$ consists of one vertex only. Then at time $t$ any density $\mu(t)$ is admissible iff*

$$\mu(t) \leq (Md)^{-1} \tag{6}$$

2. *Assume there are $M$ roads and let $L$ be maximal multiplicity of their inter- sections. Assume also that the graph $G$ has no cycles. Then at time $t$ any density $\mu(t)$ is admissible iff*

$$\mu(t) \leq (Ld)^{-1}$$

3. *Consider 4 roads $A, B, C, D$ with 4 different crosses $1 = A \cap B, 2 = B \cap C, 3 = C \cap D, 4 = D \cap A$, Then the graph $G$ is a quadruple and has 4 edges $12, 23, 34, 41$. Then the admissible densities are*

$$\mu(t) \leq \frac{3}{8}(d)^{-1}$$

Note that if the number of cycles grows then the coefficient in front of $d^{-1}$ decreases very quickly. It would be interesting to calculate it explicitely.

There is another reason showing that such transportation network is not only of small density but also is strongly unstable. Namely, consider the case when each road $k = 1, ..., M$ has its own safe distance $d_k$ and its own velocity $v_k$, Thus

$$z_{k,i}(t) = z_{k,i}(0) + v_k t, \quad z_{k,i-1}(0) - z_{k,i}(0) = d_k \implies z_{k,i-1}(t) - z_{k,i}(t) = d_k$$

Then the corresponding currents are $J_k = v_k d_k^{-1}$. The following proposition shows extreme instability of the simplest network.

**Proposition 2.** *Consider two roads with one cross. Call the transportation sta- ble if distances between any two units are greater or equal to some $D_0 > 0$ uniformly in $t$. Then:*

*If $\frac{J_1}{J_2}$ is not rational, then such transportation cannot be stable.*

*If $\frac{J_1}{J_2} = \frac{n_1}{n_2}$ for some integers $n_1, n_2$ such that $(n_1, n_2) = 1$. Then stable transportation exists iff*

$$\frac{n_1}{d_1} + \frac{n_2}{d_2} < \frac{1}{D_0}$$

*Remark 1* (Control types). To avoid instability of Proposition 2 some control is necessary. There are two possibilities for the control. The first one we considered in Sect. 2 - local internal control, that is depending only on distances between cars. Second type is the control which forces velocities of cars to change in certain points of the network. The mostly used such control is the organization of traffic lights where the cars should stand still for some time. Other control types are also known, see [1, 4–8], and it could be interesting to find general classification of control types.

# References

1. Blank, M.: Ergodic properties of a simple deterministic traffic flow model. J. Stat. Phys. **111**, 903–930 (2003)
2. Prigogine, I., Herman, R.: Kinetic Theory of Vehicular Traffic. Elsevier, N.Y. (1971)

3. Helbing, D.: Traffic and related self-driven many particle systems. Rev. Mod. Phys. **73**, 1067–1141 (2001)
4. Feintuch, A., Francis, B.: Infinite chains of kinematic points. Automatica **48**, 901–908 (2012)
5. Hui, Q., Berg, J.M.: Semistability theory for spatially distributed systems. In: Proceedings of the IEEE Conference on Decision and Control, January 2009
6. Jovanovic, M.R., Bamieh, B.: On the Ill-Posedness of certain vehicular platoon control problems. IEEE Trans. Autom. Control **50**(9), 1307–1321 (2005)
7. Melzer, S.M., Kuo, B.C.: Optimal regulation of systems described by a countably infinite number of objects. Automatica **7**, 359–366 (1971). Pergamon Press
8. Swaroop, D., Hedrick, J.K.: String stability of interconnected systems. IEEE Trans. Autom. Control **41**(3), 349–357 (1996)
9. Malyshev, V.A., Musychka, S.A.: Dynamical phase transition in the simplest molecular chain model. Theoret. Math. Phys. **179**(1), 123–133 (2014)

# Econometric Models of Controllable Multiple Queuing Systems

Alexander Mandel[(✉)]

V.A.Trapeznikov ICS RAS, Profsoyuznaya 65, Moscow, Russia
`almandel@yandex.ru`

**Abstract.** The paper is concerned with a controllable multiple queuing system (QS) where control with a constant time step reduces to connecting spare service devices or to disconnection of main service devices. The maximal average profit from serving the demands in a multi-step planning period is taken to be the criterion of the QS operation optimality. Two models are considered. In first one the QS is assumed to have a simple demands arrival flow at a random rate that varies with homogeneous Markov chain. It is assumed that it takes one step for a stationary mode to set in QS. The second model does not assume a stationary mode but the simple input flow rate is assumed to vary with the specified time function. For both models algorithms are obtained that determine the optimal switching strategies.

**Keywords:** Multiple QS · Controllable QS · Maximizing the average profit · Simple flow · Homogeneous Markov chain · Deterministic rate variations · Switching strategies

## 1  Introduction

We will concentrate on two models of controllable multiple queuing systems (QS) that are widely used in analyzing the functioning of socio-economic systems such as selling various tickets, booking hotels, or commercial and industrial, transport and numerous other systems. The number of service devices in multiple QS that is selected as a function of the time-dependent input flow rate is taken to be the parameter which can be dynamically controlled and in this particular case optimized.

Such problems are resolved by using a fairly accurate model of similar control systems that is obtained with the use of conventional queuing system theory techniques [1,2] and adequate tools for solution of the associated optimization problems (in this case dynamic).

Controllable queuing systems attracted much research interest in 1970s; in particular a fundamental paper [3] presented a fairly general statement of a QS control problem and suggested approaches to its solution by introducing a discipline of servicing order, choosing its structure and varying the service rate or numerous other variables. Paper [4] has made a significant contribution to the

© Springer International Publishing Switzerland 2016
V. Vishnevsky and D. Kozyrev (Eds.): DCCN 2015, CCIS 601, pp. 296–304, 2016.
DOI: 10.1007/978-3-319-30843-2_31

theory by looking into the possibility of stating and resolving control problems for so-called conflict queuing systems.

The end of the 1980s saw an increasing interest in controllable queuing systems with the advent of an essentially new range of applied systems that could be regarded and analyzed as queuing systems. These were computing and data processing computer networks. Authorities in the queuing system theory concentrated their efforts on developing specific (in particular, network) applied service systems. In Russia followers of Professors V.V. Rykov, V.M. Vishnevskii and V.A. Kashtanov [5–7] made the largest contribution to this field of research.

As a consequence, over approximately two decades from 1985 to 2005 numerous models of controllable queuing systems that could provide an adequate description, analysis and optimization of socio-economic systems remained neglected. What is important is that purely econometric analysis of such systems could provide a thorough and proper recognition of all economic indices that are essential in evaluating the performance of such probabilistic and stationary systems.

Papers [8,9] pioneered in this research and this paper builds on their contribution. We will take up two models. The one, referred to as *stationary*, assumes that the input flow is Poisson with a variable corresponding to a discrete time homogeneous Markov chain rate. In the other, referred to as *nonstationary*, the input rate varies with the specified time function. Control of such QS reduces to developing a spare service device switching strategy or a strategy of disconnecting main service devices (turning them to spare).

## 2   Stationary Model of Controllable QS

The queuing system is assumed to be a multiple QS with a set of switchable spare service devices.

All service devices are assumed to have exponential mutually independent service times with a service rate $\mu$.

Let the cost components be:

$c_1$ —   operating cost of one main service device per time unit;

$c_2$ —   maintenance cost of one spare service device per time unit (it is natural that $c_1 > c_2$, on numerous occasions $c_2 = 0$);

$A_1$ —   price of switching one spare service device into the set of main service devices ("switching on");

$A_2$ —   price of switching one main service device into the set of spare service devices ("switching off");

$d$ —   cost of time unit of one demand awaiting service;

$h$ —   income obtained when one demand is served;

$m_1$ —   number of main service devices;

$m_2$ —   number of spare service devices

The goal is to maximize the total average income of the system while it operates in the planning period $[0, T]$. Here $T = N_\tau$, where $N$ is a sufficiently large natural number and $\tau$ is the time interval between sequential times when decisions are made to switch on new service devices of to switch off some service devices. The interval $\tau$ will be referred to as a "step".

Let the input flow rate $\lambda$ at fixed times $t_n = n\tau, n = 1, 2, \ldots, N$, jumps from $\lambda_i$ to $\lambda_j$ with probabilities $p_{ij}(m_1)$, $i, j = 1, 2, \ldots, L^1$. It is essentially possible that the probabilities also depend on $n$. Then $p_{ij}(m_1)$ should be replaced with $p_{ij}(m_1, n)$). Consequently, at each interval of duration $\tau$ the demand input flow rate is constant and described by a model of random quantity that takes on values from a finite set $\Lambda = \lambda_1, \lambda_2, \ldots, \lambda_L$ assuming that $\lambda_1 < \lambda_2 < \cdots < \lambda_L$. We will further assume that $M\mu > \lambda_L$, where $\mu$ is the service rate in one main service device while $M = m_1 + m_2$.

At times $t_n$ spare service devices can be switched into the set of main ones and the inverse is true; in other words, spare service devices can be made main ones, or devices can be switched on or off. This queuing model is ideologically very much akin to the reliability model proposed in the 1960s and referred to as dynamic reservation [10,11], even though, as will be shown below, it is a much closer relation to two-level inventory control strategies of [12].

We will take up the case of setting stationary modes in a queuing system, i.e. the case wherein in every interval of duration $\tau$ the number of demands that can arrive and be served is sufficient for the queuing system to get into a stationary mode in the probabilistic sense at each of its steps. For this purpose it is sufficient that:

$$\mu\tau >> 1. \tag{1}$$

In this case it is also evident that $\lambda\tau >> 1$.

The assumption of the stationary mode imposes a fairly strong constraint on the proposed model.

Firstly, next inequalities: $\rho_i = \lambda_i/\mu M < 1$, $\lambda_i \in \Lambda$, must hold. In effect, the system can be stable only with a flow for which the condition holds:

$$\lambda_i < \mu M, \quad \forall \lambda_i \in \Lambda. \tag{2}$$

Let $\lambda_m ax$ be the maximal element of the set $\Lambda$ (in point of fact this is the value of $\lambda_L$). Then the inequality (2) can be rearranged into

$$\lambda_{max} < \mu M. \tag{3}$$

Secondly, at every step there is a lower boundary of the number of main service devices $\underline{m}_{1m}^{(i)}$, to which the system must be switched so as to keep its stable functioning (assuming a stationary mode) and this boundary depends on the ordinal number $i$ of the system's state (at the respective step).

---

[1] The dependence of the probability of the jump-wise change in the input flow rate on the number of the service devices m1 is linked to the need of accounting for the effect of the servicing quality (in this case it depends on the time that the customer has to wait for the service to start) on the customer demand, which is exactly what is described by the input flow rate.

Indeed, in state $i$ (in line with the value of the input rate) the stationarity condition $\rho_i = \lambda_i/\mu m_{1m}^{(i)} < 1$, must hold, thence we have:

$$\underline{m}_{1m}^{(i)} > \frac{\lambda_i}{\mu} \Rightarrow, \underline{m}_{1m}^{(i)} = \frac{\lambda_i}{\mu} + 1 \quad \forall i = \overline{1, N}. \tag{4}$$

Consequently, the number of devices operating in the state $i$ may vary from $\underline{m}_{1m}^{(i)}$ to M.

# 3  Problem Statement and Solution for a Stationary Model

Let $C^{(1)}(\lambda_i, m_1, m_{1m})$ denote the average value of the income at one step of duration $\tau$ if at the start of the step with the input flow rate $\lambda_i$; the number of main service devices with which the QS arrives at this step is equal to $m_1$, and a decision is made to switch on at this time $m_{1m}$ main service devices. This is equivalent to: maintaining the existing number of service devices if $m_1 = m_{1m}$; the need to switch off $(m_1 - m_{1m})$ service devices if $m_1 - m_{1m} > 0$; and the requirement to switch on $(m_{1m} - m_1)$ service devices if $m_1 - m_{1m} < 0$.

It is evident that $C^{(1)}(\lambda_i, m_1, m_{1m}) =$ averaged one-step profit averaged one-step costs.

The stationarity assumption suggests that all initial requirements are met and so the averaged one-step profit amounts to $h\lambda_i$.

The averaged costs include switching costs incurred in switching on spare and switching off main service devices, the cost of demands awaiting their turn in the queue and the cost of operating main and spare service devices.

The switching costs are:

$$C_{switch}(m_1, m_{1m}) = \begin{cases} A_1(m_{1m} - m_1) & \text{if } m_{1m} - m_1 > 0, \\ 0, & \text{if } m_{1m} = m_1, \\ A_2(m_1 - m_{1m}) & \text{if } m_{1m} - m_1 < 0. \end{cases} \tag{5}$$

The cost of demands staying in a queue at a step of duration $\tau$ is equal to

$$C_{queue}(\lambda_i, m_{1m}) = \mathbf{E} \int_0^\tau dk(t, m_{1m}, \lambda_i)dt$$

$$= \mathbf{E}\, \mathrm{d} \int_0^\tau k(t, m_{1m}, \lambda_i)dt = d\overline{k}(m_{1m}, \lambda_i)\tau, \tag{6}$$

where $\mathbf{E}$ is the mean value, $k\lfloor t, m_{1m}, \lambda_i \rfloor$ is the number of demands in the queue to the QS with parameters $m_{1m}$ and $\lambda_i$ at time $t$, while $\overline{k}(m_{1m}, \lambda_i)$ is the average length of the queue to the QS with parameters $m_{1m}$ and $\lambda_i$. By virtue of the stationarity assumption, the integration is made over the interval $[0, \tau]$ rather than over the current interval $[t_n, t_{n+1}]$. According to [1] we have for the stationary case:

$$\bar{k}(m_{1m}, \lambda_i) = \pi_0 \frac{(m_{1m}\rho_i)^{m_{1m}} \rho_i}{m_{1m}!(1 - \rho_i)^2}, \tag{7}$$

where $\rho_i = \frac{\lambda_i}{\mu m_{1m}}$, while $\pi_0 = \left[ \sum_{i=0}^{m_{1m}} \frac{(m_{1m}\rho_i)^i}{i!} + \frac{(m_{1m}\rho_i)^{m_{1m}}}{m_{1m}!(1-\rho_i)} \right]^{-1}$ is the stationary probability that there are no demands in the system.

The parameter $d$ characterizes the indirect costs of the QS due to the longer demands stay in the system which reduces the preference of the system by the clients. A mathematical model may be devised that would relate the level of preference loss to the average time of demand staying in the queue. That model could lead to a realistic estimate of the variable $d$. This paper will not address the problem. A remark is, however, in order that in most cases the value of the variable $d$ is 1,5-2 orders of magnitude lower than that of the above variable $h$.

The per step cost of operating main and spare service devices totals:

$$C_{running}(m_{1m}) = c_1 m_{1m} + c_2(M - m_{1m}). \tag{8}$$

In effect the average per step pure income is equal to:

$$C^{(1)}(\lambda_i, m_1, m_{1m}) = h\lambda_i - C_{switch} - C_{queue} - C_{running}. \tag{9}$$

Let $\Pi_s^*(\lambda_i, m_1)$ is the maximal value of the average income over an interval that starts $s$ steps before the end of the planning period $[0, T]$, $s + n = N$ with the input flow rate value $\lambda_i$ and $m_1$ main service devices that are switched on before the control decision is made to switch on $m_{1m}$ devices. What follows now is derivation of a dynamic programming equation for the functional $\Pi_s^*(\lambda_i, m_1)$ in light of the condition (4).

One step before the end of the planning period and with a random value of the input flow rate $\lambda_i$ the value of the above functional $\Pi_s^*(\lambda_i, m_1)$ will be

$$\Pi_1^*(\lambda_i, m_1) = \max_{m_{1m}^{(i)} \leq m_{1m} \leq M} C^{(1)}(\lambda_i, m_1, m_{1m}), \tag{10}$$

where $\lambda_i \in \Lambda$.

Just $s$ steps before the end of the planning period we have

$$\Pi_s^*(\lambda_i, m_1) = \max_{m_{1m}^{(i)} \leq m_{1m} \leq M} \left\{ C^{(1)}(\lambda_i, m_1, m_{1m}) \right.$$
$$\left. + \sum_{j=1}^{L} p_{ij}(m_{1m}) \Pi_{s-1}^*(\lambda_j, m_{1m}), \tag{11} \right.$$

where $\forall s \in \overline{2, N-1}$, $\lambda_i, \lambda_j \in \Lambda$ are input flow rates at times $s$ and $s-1$ steps, respectively, before the end of the planning period.

Equations (10) and (11) are used in computing an optimal control program $m_{1m}^*(i, s)$, where $1 \leq s \leq N - 1$, $i = 1, 2, \ldots, L$.

# 4   A Non-stationary Model of Controllable QS

Let us now take up the case of a non-stationary queuing system mode, or the case where every interval of duration $\tau$ is too short for a stationary mode in the probabilistic sense to set in within the QS.

For simplicity assume that the input flow rate can vary only at times when the QS[2] state is monitored. In order to avoid making the model unduly cumbersome assume also that whatever control decisions are made they should not make the QS leave the states in which QS load factor $\rho \geq 1$.

Let us also assume that at every step $n = 1, 2, \ldots, N$, the QS receives a simple input flow with the rate of $\lambda_n$ where the set $\Lambda'$ of possible values of the rate $\lambda_n$ is a finite set of the form $\Lambda' = \{\lambda^{(1)}, \lambda^{(2)}, \ldots, \lambda^{(M)}\}$ and the plot of the input flow rates over discrete time steps $n$ is specified in the form:

$$\lambda(t) = \lambda^{(j)} \text{ if } t \in [t_{N-n+1}, t_{N-n}], \quad n = 1, 2, \ldots, N, \tag{12}$$

where the indices of step time boundaries use a structure of "inverse" time in steps to the end of the planning period $[0, T]$. In essence, the plot of input flow rates is described as a totality of indices $J = \{j_n\}_{(n=1)}^{N}$.

Now the system's state at every time $t_n$ is described as the number $i$, of demands currently inside the QS. We will assume that the inequality (3) still holds where this time $\lambda_{max} = \lambda^{(M)}$. In the state $j_n$ (associated with the value of the input flow rate), as noted above, the stationarity condition $\rho^{(j_m)} = \lambda^{(j_m)}/\mu M < 1$, $\lambda^{(j_m)} \in \Lambda'$, must hold, whence:

$$\underline{m}_{1m}^{(j_m)} > \frac{\lambda^{(j_m)}}{\mu} \Rightarrow \underline{m}_{1m}^{(j_m)} = \frac{\lambda^{(j_m)}}{\mu} + 1 \qquad \forall \lambda^{(j_m)} \in \Lambda'. \tag{13}$$

In effect, in the state $j_n$ the number of operating main devices may take on values from $\underline{m}_{1m}^{(j_m)}$ to $M$. Because the case is non-stationary, in further discussion we will discard the constraint (13).

# 5   Statement and Solution of the Problem for the Non-stationary Case

Introduce the quantity $C^{(1)}(\lambda^{(j)}, i, m_1, m_{1m})$, which is equal to the averaged income value at one step of duration $\tau$ if at the start of the step the input flow rate value $\lambda^{(j)}$ sets in and there are $i$ demands in the QS and the number of main devices with which the QS arrives at this step is equal to $m_1$ and a control decision is made to switch on $m_{1m}$ main devices at the time when the step starts. While doing so, formulae (5) and (8) remain in force while formulae (6), (9), (10) and (11) need adjustment.

Indeed, no stationary mode arises at every step now and so the formulae for computing the responses of the stationary mode do not hold. This applies to the

---

[2] For that purpose the step length $\tau$ must be made small enough.

formulae for the number of demands processed at the step and for the average queue length. The key elements of the calculation become now the probabilities of the QS state at the latest step, or the probabilities $P_i(t)$ that at time $t$ (recall that $t \in [t_{N-n+1}, t_{N-n}]$) there are $i$ demands in the QS.

A set of differential equations for the probabilities $P_i(t)$ [1]:

$$
\begin{aligned}
&\text{for } i = 0: && \frac{dP_0(t)}{dt} = -\lambda P_0(t) + \mu P_1(t). \\
&\text{for } 1 \le i < m_{1m}: && \frac{dP_i(t)}{dt} = \lambda P_{(i-1)}(t) - (\lambda + i\mu)P_i(t) + (i+1)\mu P_{(i+1)}(t). \quad (14) \\
&\text{for } i \ge m_{1m}: && \frac{dP_i(t)}{dt} = \lambda P_{(i-1)}(t) - (\lambda + m_{1m}\mu)P_i(t) + m_{1m}\mu P_{(i+1)}(t),
\end{aligned}
$$

with an initial condition

$$
P_j(t) = \begin{cases} 1 \ if \ j = i, \\ 0 \ if \ j \ne i. \end{cases} \tag{15}
$$

To avoid resolving an infinite-dimensional set of differential equation (14)–(15), we use an approximating set of finite dimension $I$:

$$
\begin{aligned}
&\text{for } i = 0: && \frac{dP_0^{(I)}(t)}{dt} = -\lambda P_0^{(I)}(t) + \mu P_1^{(I)}(t). \\
&\text{for } 1 \le i < m_{1m}: && \frac{dP_i(t)}{dt} = \lambda P_{i-1}(t) - (\lambda + i)P_i(t) + (i+1)P_{i+1}(t). \\
&\text{for } m_{1m} \le i \le I: && \frac{dP_i^{(I)}(t)}{dt} = \lambda P_{i-1}^{(I)}(t) - (\lambda + m_{1m}\mu)P_{i+1}^{(I)}(t) + m_{1m}\mu P_{i+1}^{(I)}(t),
\end{aligned}
$$
$$(16)$$

with an initial condition

$$
P_j^{(I)}(t) = \begin{cases} 1 \ if \ j = i, \\ 0 \ if \ j \ne i. \end{cases} \tag{17}
$$

Assume that we have a "solver" of the equation set (16)–(17). Denote a solution to the set (16) with an initial condition (17) as $\mathbf{P}^{(I)}(i, \tau)$. The second argument $\tau$ "reminds" one that the set (16)–(17) is resolved in the interval $[0, \tau]$. Now, while assuming that we have an approximate solution to the set (14)–(15) in the form of probabilities $P_i^{(I)}(t)$, $i = 1, 2, \ldots, I$, $t \in [0, \tau]$, proceed to computation of the other characteristics.

In further discussion the computation will be performed in inverse time $s$ and so the plot of the input flow rate will be represented by the values of $\lambda^{(N-s)}$, $s = 1, 2, \ldots, N$. Re-denote these rates as $\lambda^{(N-s)} = \lambda_{mod}^{(s)}$. Let $\Pi_s^*(\lambda, i, m_1)$ be the maximal value of the averaged income over an interval that starts $s$ steps before the end of the planning period $[0, T]$, $s + n = N$ with the value $\lambda$ (naturally $\lambda = \lambda_{mod}^{(s)}$) of the input flow rate, the number of demands inside the system $i$ and $m_1$ of the switched on (before a control decision is made to switch on the $m_{1m}$ main devices. Dynamic programming equations are derived below for the functional $\Pi_s^*(\lambda, i, m_1)$ but this time in disregard of the condition (13).

Evidently that one step before the end of the planning period and with a random value of the input flow rate $\lambda$ the value of the functional $\Pi_1^*(\lambda, i, m_1)$ introduced above will look like

$$
\Pi_1^*(\lambda, i, m_1) = \max_{1 \le m_{1m} \le M} C^{(1)}(\lambda, i, m_1, m_{1m}), \quad \text{where} \quad \lambda \in \Lambda', \tag{18}
$$

Proceeding now to adjustment of formula (6) for queuing cost. The average queue length $k(t, m_{1m}, \lambda)$ at time $t \in [t_s, t_{s-1}]$, where $s$ is inverse time may be represented in the form:

$$k(t, m_{1m}, \lambda) = \sum_{i=m_{1m}+1}^{I} (i - m_{1m}) P_i^{(I)}(t). \tag{19}$$

Then for the cost of demands staying in the queue we have

$$C_{queue}(\lambda, m_{1m}) = d \int_{t_s}^{t_{s-1}} k(t, m_{1m}, \lambda) dt. \tag{20}$$

The income from serving the demands over the interval $[t_s, t_{s-1}]$ will be estimated by using the Wald identity [2]. The average number of demands served over the interval $[t_s, t_{s-1}]$, will be estimated by an integral over the instantaneous service rate at time $t$ which is equal to

$$\mu(t) = \mu \sum_{i=0}^{m_{1m}-1} P_i^{(I)}(t)i + m_{1m}\mu \sum_{i=m_{1m}}^{I} P_i^{(I)}(t) \tag{21}$$

Consequently, the income from serving demands over the interval $[t_s, t_{s-1}]$ is evaluated as $h \int_{t_s}^{t_{s-1}} \mu(t) dt$ . This expression makes it possible to rearrange formula (9) for the average net income within one step as

$$C^{(1)}(\lambda, i, m_1, m_{1m}) = h \int_{t_s}^{t_{s-1}} \mu(t) dt - C_{switch} - C_{queue} - C_{running}, \tag{22}$$

where three cost components in the right-hand part of Eq. (22) are computed by formulae (5), (20) and (8), respectively.

Now, $s$ steps before the end of the planning period we have:

$$\Pi_s^*(\lambda_{mod}^{(s)}, i, m_1) = \max_{1 \leq m_{1m} \leq M}\{C^{(1)}(\lambda_{mod}^{(s)}, i, m_1, m_{1m})$$
$$+ \sum_{j=o}^{I} P_j^{(I)}(t) \Pi_{s-1}^*(\lambda_{mod}^{(s-1)}, j, m_{1m})\} \tag{23}$$

where $\forall s \in \overline{2, N-1}$.

## 6   Conclusions

Models are proposed for control of controllable multiple queuing systems. Those models are helpful in analyzing and optimizing the functioning of various socio-economic systems such as booking air tickets and hotels, trading and purchasing, public transport, billing and numerous others. The simple input flow rate varies with a homogeneous Markov chain or as a specified deterministic time function. With these assumptions algorithms are obtained of device switching strategies from spare to main and vice versa. Case studies suggest that optimal switching strategies have features of threshold strategies the control science refers to as two-level inventory control strategies or $(S, s)$-strategies.

# References

1. Gnedenko, B.V., Kovalenko, I.N.: Introduction to Queuing Theory. Nauka Publisher, Moscow (1966). (in Russian)
2. Saaty, T.L.: Elements of Queuing Theory with Applications. McGraw-Hill, New York (1961). no ISBN (translated into Russian, Spanish and German)
3. Rykov, V.V.: Controllable queuing systems. probability theory. mathematical statistic. theoretical cybernetic, vol. 12, pp. 43-153 (1975) (in Russian)
4. Neymark, Y.I.: Dynamic Systems and Controllable Processes. Nauka Publisher, Moscow (1978). (in Russian)
5. Golysheva, N.M., Fedotkin, M.A.: Conflict flows cyclic control under critical length queues birth-death process. Automation and Remote Control, No.4 (1990)
6. Vishnevskii, V.M.: Theoretical Foundations of Computer Networks Design. Technosphera Publisher, Moscow (2003). (in Russian)
7. Zaharov, P.P.: Computer-aided software complex development for quality and efficiency research of technical and controllable queuing systems models. Engineering-Science C and Dissertation. MGIEM, Moscow (2006) (in Russian)
8. Barladyan, I.I., Kuznetsov, A.V., Mandel, A.S.: Critical parameters analysis and simulation of controllable queuing system. Problemy upravleniya, vol. 6, pp. 21–25 (2007) (in Russian)
9. Kuznetsov, A.V., Mandel, A.S., Tokmakova, A.B.: One model of controllable queuing system. Problemy upravleniya, vol. 6, pp. 39–43 (2007) (in Russian)
10. Raikin, A.L.: Reliability Theory Topics for Technical Systems Design, p. 256. Sov. Radio Publisher, Moscow (1967) (in Russian)
11. Mandel, A.S., Raikin, A.L.: Spare units switching optimal plan construction. Automation·and Remote Control, vol. 5 (1967)
12. Headly, J., Whiting, T.: Inventory Control Systems Analysis. Wiley, New York (1967)

# A Cluster Caching Rule in Next Generation Networks

N. Markovich[(✉)]

V.A. Trapeznikov Institute of Control Sciences of Russian Academy of Sciences,
Moscow, Russia
nat.markovich@gmail.com, markovic@ipu.rssi.ru

**Abstract.** Probabilistic aspects of caching are considered. The caching serves to keep popular contents inside a memory unit called 'cache' to be able to access them quickly. Using extreme value theory we propose a caching strategy called Cluster Caching Rule driven by content popularity that may change in time. A non-Poisson request arrival process is used when requests are statistically correlated. The idea of the new approach is to locate in cache only contents whose popularity exceeds a sufficiently large threshold. Due to dependence and possible heavy-tailed distribution of inter-requests and inter-arrival times of documents, the popularity process builds clusters of exceedances. The cluster and inter-cluster sizes are geometrically distributed as derived in Markovich (2014). We use it to calculate means of the cache utilization and occupancy. We escape assumptions like a constant size of content and a Poisson request process that are typical in the literature.

**Keywords:** Caching · Content · Popularity process · Cluster of exceedances · Cache utilization · Cache occupancy

## 1 Introduction

The paper is devoted to probabilistic aspects of caching. A caching policy serves to keep popular contents inside a short memory unit called 'cache' to be able to access them as soon as possible. We consider it from new perspectives using achievements of extreme value theory.

Let $C$ be the size of a cache and $d_1, ..., d_M$ be a catalog of documents. Since the popularity of contents may change over time and place, the hitting of documents inside the cache has to be changed, too. Cache filling depends on the input (the arrival) process of document requests, the cache size and the replacement policy. The latter may be controled by the time that is allowed for the document to be in the cache. This called TTL (the Time-to-Live) may be individual for each document depending on its popularity [1] or cache dependent [2], or the TTL may be fixed, [3]. If a requested document is found in the cache then the cache content remains unchanged. Then we say that the document request hits the cache. If the document is new, i.e. it cannot be found in the cache,

© Springer International Publishing Switzerland 2016
V. Vishnevsky and D. Kozyrev (Eds.): DCCN 2015, CCIS 601, pp. 305–313, 2016.
DOI: 10.1007/978-3-319-30843-2_32

then the missing content is brought in from the outside world (a long memory). In this case, we say that the document request misses the cache. The previous cache content may remain unchanged in this case too if the cache was not full. Otherwise, the new document evicts some content from the cache according to the replacement rule.[1]

There are several identification problems in the caching such as size and location of caches, statistical estimation of the content popularity, a convenient modeling of the inter-request time (IRT) process, the optimality evaluation of the replacement rule, e.g., by hit/miss probabilities as well as the occupancy and the utilization of the cache. The most popular replacement rule is the Least-Recently-Used (LRU) caching, [3,4], where the requested document hits the first position of the cache. The object at the last position may leave the cache if the requested document is new or it stays longer otherwise.

Despite of an easy idea, the analytical analysis of the LRU is difficult. Usually, it is assumed for simplicity that the content size is deterministic despite it is naturally random. However, one can combine several content units into chunks of a constant size. Such chunks are assumed to be claimed.

The requesting point process of the object is often assumed to be a renewal Poisson process and the IRTs are then independent exponentially distributed, see [1,4] among others. A superposition of renewal Poisson processes is also renewal Poisson. This simplifies the use of cache trees where superposed request processes from leaf-caches arrive to the root-cache, [1]. Generally, the superposition of two renewal processes is not a renewal process. Markov modulated processes are closed under superposition. Hence, Markov and semi-Markov modulated processes are typically used as alternative models of arrival request processes, [1,5,6]. In contrast to renewal Poisson processes such superposed request processes are correlated.

Usually, the probability of the content popularity is modeled by Zipf's law that has a Pareto tail, [7,8]. In [9] the mixture of Zipf and heavy-tailed Weibull distributions is found as an appropriate model of the web content popularity. In [4] real web traces from the National Laboratory for Applied Network Research (NLANR) were analyzed. It was found that about 70 % of all documents are one-time requested. The caching of such documents is not reasonable. However, the LRU and TTL rules allow to cache such documents.

Assuming that the document popularity can be modeled by a generalized Zipf's law, in [5] it was derived that the cache missing probability is asymptotically, for sufficiently large cache sizes, the same for the LRU replacement both with dependent and independent identically distributed IRTs. In [6] similar result was proved for the Least-Frequently-Used (LFU) rule which only caches the most popular objects. The IRTs of each object may be dependent and heavy-tailed distributed which reflects the heavy-tailed popularity of the documents.

According to our new caching rule called *Cluster Caching Rule* (CCR), only clusters of frequently requested contents whose popularity exceeds a sufficiently

---

[1] If the document size is fixed then a new document evicts one document from the full cache.

large threshold may hit the cache. This is similar to TTL and LFU rules, [10] but there are novelties based on the extreme value theory. The latter allows us to explain the impact of the dependence and heavy-tailed quantities on the caching. Due to dependence and a possibly heavy-tailed distribution of IRTs of the object, the popularity and IRT processes build clusters of exceedances, see Fig. 1. Such assumptions like (1) the content size is constant; (2) the IRT process is renewal Poisson; (3) IRTs are independent are avoided.

We consider clusters of exceedances of a popularity process of random sized content whose IRTs may be statistically correlated. Inter-arrival times of documents can be heavy-tailed. The clusters provide objects to hit/miss the cache.

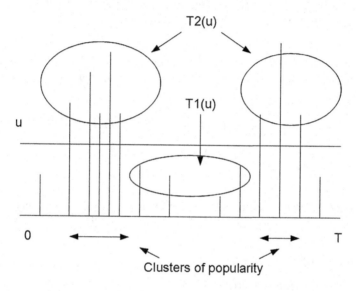

**Fig. 1.** Clusters of exceedances where the popularity of requested documents exceeds the threshold $u$ within the time frame $[0, T]$; $T_1(u)$ and $T_2(u)$ denote inter-cluster and cluster sizes for the threshold $u$, respectively.

We obtain the cache utilization and occupancy by means of clusters and explain the impact of correlated requests on the fault probability from perspectives of clustering. The CCR allows to recommend the cache size.

We focus on clusters of exceedances of the popularity process. The popularity may change over time. When the next cluster of the most popular documents is requested, it is sent to the cache. The former cache content is evicted. If some popular content is requested repeatedly then it remains in the cache without replacement. In case, the cache size is smaller than the cluster size, the threshold $u$ may be increased until the cluster size becomes smaller or equal to the cache size. The cache content may be changed completely by a new cluster. In our approach, the document size can be variable, i.e. a random variable (r.v.). We concern only probabilistic aspects of fitting the cache avoiding statistical and combinatorial methods.

The paper is organizing as follows. In Sect. 2 necessary results from the extreme value theory are mentioned. Our achievements are given in Sect. 3. We finalize our exposition with conclusions.

## 2    Facts from Extreme Value Theory

We focus on clusters of exceedances of an underlying process over a sufficiently high threshold as a source for caching. The cluster contains a set of consecutive exceedances of the process between two consecutive non-exceedances, [11]. The mean cluster size (i.e. the mean number of exceedances per cluster) is approximated by the reciprocal of the extremal index $\theta$. $\theta$ is the dependence measure of extremes since it allows to represent the distribution of maxima of $n$ dependent r.v.s.

**Definition 1.** *The stationary sequence $\{X_n\}_{n\geq 1}$ is said to have extremal index $\theta \in [0,1]$ if for each $0 < \tau < \infty$ there is a sequence of real numbers $u_n = u_n(\tau)$ such that*

$$\lim_{n\to\infty} n(1 - F(u_n)) = \tau \qquad and \qquad (1)$$

$$\lim_{n\to\infty} P\{M_n \leq u_n\} = e^{-\tau\theta}$$

*hold ([12], p.53).*

For an iid sequence $\theta = 1$ holds. $\theta = 0$ indicates a total dependence. Clusters form a compound Poisson process with the rate $\theta\tau$, [13].

### 2.1    Clusters of Exceedances

Let us consider the inter-cluster size as

$$T_1(u) = \min\{j \geq 1 : M_{1,j} \leq u, X_{j+1} > u | X_1 > u\}$$

and the cluster size as

$$T_2(u) = \min\{j \geq 1 : L_{1,j} > u, X_{j+1} \leq u | X_1 \leq u\},$$

where $M_{1,j} = \max\{X_2, ..., X_j\}$, $M_{1,1} = -\infty$, $L_{1,j} = \min\{X_2, ..., X_j\}$, $L_{1,1} = +\infty$, Fig. 1.

In [11] asymptotically equivalent distributions of $T_1(u_n)$ and $T_2(u_n)$ are derived for sequence of increasing thresholds $\{u_n\}$. It was proposed to take sufficiently high quantiles $x_{\rho_n}$ of the level $(1 - \rho_n)$ ($\rho_n \to 0$ as $n \to \infty^2$) of the process $X_t$ as $u_n$. If specific mixing conditions are fulfilled uniformly in $j \in [a, n]$, $a = a(n) \to \infty$, $n \to \infty$ then geometric-like models.

$$P\{T_1(x_{\rho_n}) = j\} \sim \rho_n(1 - \rho_n)^{(j-1)\theta}, \qquad P\{T_2(x_{\rho_n}) = j\} \sim q_n (1 - q_n)^{(j-1)\theta} \tag{2}$$

---

[2] This follows from (1) since $\rho_n = 1 - F(x_{\rho_n}) \sim \tau/n$.

hold as $n \to \infty$, where $q_n = 1 - \rho_n$ and $n$ is a sample size.[3] The required mixing conditions are equivalent to the condition

$$|P\{X_{j+1} \le x_{\rho_n}|X_1 \le x_{\rho_n}\} - P\{X_{j+1} \le x_{\rho_n}\}| = o(1), \qquad n \to \infty. \quad (3)$$

In case $j = 1$ the random sequence $\{X_i\}$ must be independent with $\theta = 1$ to satisfy (3). Equation (3) is valid, for example, for the wide class of $m$-dependent processes for $j > m$ and Markov chains for $j \ge 2$.

The asymptotic expectation of $T_2(x_{\rho_n})$ was obtained in [11]. It was derived that if for some $\varepsilon > 0$

$$\sup_n E(T_2^{1+\varepsilon}(x_{\rho_n}))/\Lambda_{n,2} < \infty, \qquad \Lambda_{n,2} = q_n/(1 - (1 - q_n)^\theta)^2$$

holds, then it follows

$$\lim_{n \to \infty} E(T_2(x_{\rho_n}))/\Lambda_{n,2} = 1. \quad (4)$$

In [14] it was derived that if for some $\varepsilon > 0$

$$\sup_n E(T_1^{1+\varepsilon}(x_{\rho_n}))/\Lambda_{n,1} < \infty, \qquad \Lambda_{n,1} = 1/(1 - (1 - \rho_n)^\theta)^2$$

holds, then it follows

$$\lim_{n \to \infty} E(T_1(x_{\rho_n}))/(\Lambda_{n,1}\rho_n) = 1. \quad (5)$$

It is remarkable that models (2) are valid for any distribution of the underlying process $\{X_t\}$ if the dependence structure (3) holds.

## 3    Main Results

Let $\{X_i\}_{1 \le i \le M}$ be a stationary process of the content popularity with marginal cumulative distribution function $F(x)$ and $M_n = \max\{X_1, ..., X_n\}$. We assume $P\{X_1 = x_F\} = 0$, where $x_F = \sup\{x : F(x) < 1\}$ is the right end-point of $F(x)$. This indicates that $x_F$ is infinite and $F(x)$ is heavy-tailed. Let $\{Y_i\}$ be iid sizes of documents from the catalog with finite mean $EY_i = \alpha_1$.

Let $\{\tau_n\}_{n \ge 1}$ and $\{\tau_{i,n}\}_{n \ge 1}$ be stationary processes of inter-arrival times (IATs) of documents and IRTs of the $i$th document, $i = 1, .., M$, respectively. The frequency of requests changes over time. We determine the popularity of the $i$th document at its $j$th request time $T_{i,j} = \sum_{n=1}^{j} \tau_{i,n}$ by

$$X_i = j/N_{T_{i,j}}, \qquad i = 1, ..., M, \qquad j \ge 1,$$

where $N_{T_{i,j}}$ is the total number of requests in time interval $[0, T_{i,j}]$.

---

[3] $f_n \sim g_n$ implies $\lim_{n \to \infty} f_n/g_n = 1$.

## 3.1  Cluster Caching Rule

Let us introduce the CCR. Only highly requested documents corresponding to clusters of exceedances of the popularity process may hit the cache. The new cluster content renews the cache apart of objects that could be found both in previous and (or) in present clusters and thus, they could hit the cache earlier. Frequently requested documents may appear in the popularity cluster several times. We denote the number of exceedances in the cluster corresponding to different objects as $T_2^*(u)$ such as $T_2^*(u) \leq T_2(u)$. $T_1^*(u)$ denotes the number of non-exceedances in the inter-cluster corresponding to different objects ($T_1^*(u) \leq T_1(u)$). The cluster content or more exactly $T_2^*(u)$ may be smaller or larger than the cache size $C$. This may lead to a not complete cache utilization or to loss of documents exceeding the cache in size, respectively. One can decrease the cluster content by increasing of $u$ or vice versa. In case $u$ is fixed, one can send the excess documents to the next cache if there is a line or a tree of caches. We focus here on the single-cache control.

The $CCR$ is similar to the TTL and the LFU rules, respectively. Really, LFU caches only the most popular documents and TTL caches only frequent documents with sufficiently small IRTs. In contrast to the latter rules, the CCR may lead to a long random cache occupancy by popular objects if they are found in several consecutive clusters. By the LRU rule and the TTL rules any requested documents may hit the cache including one time requested ones. It leads to a un-effective use of the cache. The CCR does not cache rarely requested objects. The mean cluster size $ET_2(u) \approx 1/\theta$ may be proposed as a cache size.

## 3.2  Cache Utilization

Let us consider the cache utilization for fixed and random content sizes. Let $Y_i$ be fixed. The utilization may be determined by the ratio $T_2^*(x_{\rho_n})/C$ of the cluster and the cache sizes. Clusters may contain changing popularity of the same documents. Since $ET_2(u) \approx 1/\theta$ holds we get

$$E(T_2^*(x_{\rho_n})/C) \leq E(T_2(x_{\rho_n})/C) \approx 1/(\theta C). \tag{6}$$

Hence, the mean utilization may reach $100\%$ if $C = 1/\theta$.

Let $Y_i$ be random. The cluster content and the inter-cluster duration are determined respectively by

$$S_{T_2(u)} = \sum_{i=1}^{T_2(u)} Y_i, \qquad S_{T_1(u)} = \sum_{i=1}^{T_1(u)} \tau_i. \tag{7}$$

From independence of $Y_i$ and $T_2(u)$ and from Wald's equation we get

$$E(S_{T_2^*(u)}/C) \leq E(S_{T_2(u)}/C) = E(T_2(x_{\rho_n}))E(Y_i)/C \approx \alpha_1/(\theta C) \tag{8}$$

Thus, the mean utilization may reach $100\%$ if $C = \alpha_1/\theta$. The cache size requires a statistical estimation of the extremal index $\theta$. Since $\theta$ is a function of the threshold $u$, one can use (4) to determine $C$. Equation (6) can be rewritten as

$$E(T_2(x_{\rho_n})/C) \sim q_n/((1-(1-q_n)^\theta)^2 C)$$

for $q_n \sim 1 - \tau/n.$[4] Then $C = q_n/(1-(1-q_n)^\theta)^2$ can be selected. If the content popularity $\{X_i\}$ are iid and $\theta = 1$ holds then $C = 1/q_n$ may be selected.

## 3.3  Cache Occupancy

The cache occupancy shows how long an object stays in cache within some time interval. Let $C(t)$ denote a binary process that shows the presence of the $i$th document in the cache. Denoting independent renewal processes of IRTs of individual objects and their TTLs as $\{\tau_{i,k}\}_{k\geq 1}$ and $\{T_{i,k}\}_{k\geq 1}$, respectively, in [1] the following formula of the cache occupancy

$$\lim_{n\to\infty} \int_0^{t_n} C(s)ds/t_n = \lim_{n\to\infty} \sum_{k=1}^n \min\{\tau_{i,k}, T_{i,k}\}/\sum_{k=1}^n \tau_{i,k} \qquad (9)$$

was proposed. Here, $t_n = \sum_{k=1}^n \tau_{i,k}$ is the time of the $n$th request of the object and $C(s)$ is a binary process indicating the presence of the object in cache. Equation (9) corresponds to the TTL rule with $R$-policy when the stopping time to be in cache is equal to $\min\{t : \tau_t > T_t\}$.

The $CCR$-approach is different from (9). Once the object hits the cache, it will stay there at least the time equal to the inter-cluster duration $S_{T_1(u)}$. Thus, assuming that the IATs $\{\tau_i\}$ of the documents are iid r.v.s with $\beta_1 = E(\tau_i) < \infty$ and the mutual independence of $\tau_i$ and $T_1(u)$, we obtain from Wald's equation and (5), (7)

$$E(S_{T_1(x_{\rho_n})}) = E(T_1(x_{\rho_n}))E(\tau_i) \sim \beta_1/(1-(1-\rho_n)^\theta)^2$$

with $\rho_n \sim \tau/n$. More precisely, the cache occupancy of the $i$th object may be longer or equal to

$$O_i \geq \sum_{j=1}^{L_{i,T}} \left( \sum_{k=1}^{T_{1,j}(u)} \tau_k + \sum_{k=k^*}^{T_{2,j}(u)} \tau_k \right),$$

where $k^* = \min\{t : X_t \geq u\}$ is the first hitting time of the cache by the $i$th object located in the cluster, $L_{i,T}$ is the number of consecutive clusters containing the $i$th document and, $T_{1,j}(u)$ and $T_{2,j}(u)$ are inter- and cluster sizes, respectively, at time interval $[0, T]$.

---

[4] This follows from (1) since $1 - F(x_{\rho_n}) = \rho_n = 1 - q_n \sim \tau/n$ as $n \to \infty$.

### 3.4  Dependent and Independent Requests

If one uses clusters of the popularity exceedances over the threshold to fit the cache, then for any stationary popularity distribution (2) is valid. A consistent statistical estimation of $\theta$ is also required.

Using (2) one can approximate the probability to hit the whole cache by one cluster as

$$P\{T_2(x_{\rho_n}) = C\} = q_n\left(1 - q_n\right)^{(C-1)\theta} + o(1), \qquad n \to \infty. \tag{10}$$

Looking on the right-hand side probability of (10) one may conclude that $(1 - q_n)^{(C-1)\theta}$ tends to 0 as $q_n \to 1$ and $q_n$ does not influence much if $C \to \infty$. Since the dependence is accumulated in the extremal index $\theta$, this explains why correlations of IRTs of some document do not impact on the fault probability asymptotically as $C \to \infty$. The latter conclusion was done in [5,6] regarding the LRU and the LFU rules. It was concluded that 1) the LRU and the LFU caching work the same both for iid and correlated requests, hence one should not take the dependence into account; 2) the fault probability is generated by the distribution tail of the requests. We can explain these features by the clustering and (2). One cannot exclude $\theta$ as the dependence measure from the consideration for a fixed $C$. The iid requests imply $\theta = 1$. Furthermore, clusters of exceedances are generated by rare events, namely, by exceedances over a sufficiently high threshold. The latter relate to the tail of the popularity or IRT distributions. The higher the threshold the more chance to have statistically independent clusters. Such clusters contain likely new observations that cannot be found in the cache. Cluster contents cause missing objects in the cache and the fault probability.

## 4  Conclusions and Future Work

A new caching rule $CCR$ based on clusters of exceedances by an underlying process over a sufficiently large threshold is proposed. Considering the content popularity as underlying process, its clusters of exceedances indicate popular frequently requested documents . The latter are to be located in the cache. The CCR generalizes the TTL and LFU rules. The CCR is formulated for a single cache. The content size may be random. The IRTs may be correlated. Future work will concern further investigation of the CCR and its extension to lines and trees of caches.

**Acknowledgments.** The author was partly supported by the Russian Foundation for Basic Research, grant 13-08-00744 A.

## References

1. Berger, D.S., Gland, P., Singla, S., Ciucu, F.: Exact analysis of TTL cache networks: the case of caching policies driven by stopping times. In: The ACM International Conference on Measurement and Modeling of Computer Systems, SIGMETRICS 2014, pp. 595–596 (2014)

2. Foback, N.C., Nain, P., Neglia, G., Towsley, D.: Analysis of ttl-based cache networks. In: IEEE VALUETOOLS, pp. 1–10 (2012)
3. Friecker, C., Robert, P., Roberts, J.: A versatile and accurate approximation for LRU cache performance. In: Proceedings of ITC, pp. 1–8 (2012)
4. Che, H., Tung, Y., Wang, Z.: Hierarchical web caching systems: modeling, design and experimental results. IEEE JSAC **20**(7), 1305–1314 (2002)
5. Jelenković, P.R., Radovanović, A.: Least-recently-used caching with dependent requests. Theoret. Comput. Sci. **326**(1–3), 293–327 (2004)
6. Jelenković, P.R., Radovanović, A.: Asymptotic optimality of the static frequency caching in the presence of correlated requests. Oper. Res. Lett. **37**(5), 307–311 (2009)
7. Clauset, A., Shalizi, C.R., Newman, M.E.J.: Power-law distributions in empirical data. SIAM Rev. **51**(4), 661–703 (2009)
8. Newman, M.E.J.: Power laws, Pareto distributions and Zipfs law [cond-mat.stat-mech] (2006). arxiv:cond-mat/0412004v3
9. Imbrenda, C., Muscariello, L., Rossi, D.: Analyzing cacheable traffic in ISP access networks for micro CDN applications via content-centric networking. In: ACM SIGCOMM Information Centric Networks (ICN), September 2014
10. Lee, D., Choi, J., Kim, J.-H., Noh, S.H., Min, S.L., Cho, Y., Kim, C.S.: LRFU: a spectrum of policies that subsumes the least recently used and least frequently used policies. IEEE Trans. Comput. **50**(12), 1352–1362 (2001)
11. Markovich, N.M.: Modeling clusters of extreme values. Extremes **17**(1), 97–125 (2014)
12. Leadbetter, M.R., Lingren, G., Rootzén, H.: Extremes and Related Properties of Random Sequence and Processes. Springer, New York (1983)
13. Beirlant, J., Goegebeur, Y., Teugels, J., Segers, J.: Statistics of Extremes: Theory and Applications. Wiley, Chichester, West Sussex (2004)
14. Markovich, N.M.: Clusters of extremes: models, estimation and applications (2015, to submitted)

# On Initial Width of Contention Window Influence on Wireless Network Station IEEE 802.11 Characteristics

P. Mikheev[✉] and S. Suschenko

Tomsk State University, Tomsk, Russia
doka.patrick@gmail.com, ssp.inf.tsu@gmail.com

**Abstract.** A mathematical model of access method "carrier sense multiple access with collision avoidance" for two active stations was proposed. The effect of carrier capture and unimodal dependence of the operating characteristics from the initial width of the contention window was detected. Measures of preventing the effect of carrier capture, based on the modifications of the standard protocol were proposed.

## 1  Random Multiple Access Method in 802.11 Wireless Networks

Let us analyze the wireless local area network (LAN) based on the IEEE 802.11 standard. The fundamental access method of such LANs is called DCF (Distributed Coordination Function) [1,2] known as *carrier sense multiple access with collision avoidance* (CSMA/CA) [2–4].

This mechanism is based upon the fact that the transmitting station checks whether the carrier signal is present in the medium, and, before starting transmission of a data frame, expects release of the communication medium. IEEE 802.11 stations, in contrast to wired Ethernet, are not capable of detecting collisions in a communication medium [1,5]. Due to this fact, detection of collisions and non-conflict transmissions of protocol-based data units is based on the time-outs mechanism and on the algorithm of positive decision feedback.

Let us analyze the cycle of a data frame transmission from the sending station to the recipient station. First and foremost, the sending station senses the medium to determine if another station is transmitting. Thereafter, at the end of the inter-frame interval, the random delaying algorithm is initiated to select a random backoff interval (the number of a slot in which the data transmission may be started). The slot number is selected with equal probability from the interval $[0, S_n - 1]$, where $S_n$ is the size of the contention window measured in slot intervals $t_c$ and determined by the relation $S_n = 2^{N_0+m}$, $m = n$ if $n \leq 10 - N_0$ and $m = 10 - N_0$ if $n \geq 10 - N_0$. Here $N_0 = \overline{1,10}$ is the initial value predetermining the width of the contention window during the first attempt of a sender to transfer data, and $n \geq 0$ is the number of retransmission. The width

© Springer International Publishing Switzerland 2016
V. Vishnevsky and D. Kozyrev (Eds.): DCCN 2015, CCIS 601, pp. 314–322, 2016.
DOI: 10.1007/978-3-319-30843-2_33

of the contention window may not exceed the maximum value established by the standard. For all physical layers and methods of modulation, the IEEE 802.11 standard has established the maximum width of the contention window equal to $S_{max} = 1024$ [2] The number of a selected slot shall be assigned to the backoff interval counter $t_o$, after which the countdown of slot intervals begins. At the end of each slot interval, the backoff interval counter shall decrement as long as medium is idle. If the medium is determined to be busy at any time during a backoff slot, then the backoff procedure is suspended. Decrementing is resumed when the medium is idle again. Transmission shall commence when the backoff interval counter reaches zero ($t_o = 0$). When the transmission is completed, the sender waits for a acknowledgement during the time $t_{out}$, after which it is considered that a conflict has occurred, and stations having got into such conflict increase the $n$ value by one, and the actions targeted at data transmission are repeated. The width of the contention window is doubled with each attempt of data frame transmission, until the maximum value is achieved; and the width of the contention window remains equal to $S_{max}$ with each subsequent attempt of data frame transmission. After successful transmission, the window width obtains the initial value $S_0$.

Thus, the wireless access technology, due to lack of possibility to detect collisions in a communication medium, has three significant differences from the random access method implemented in the wired medium. Firstly, the wireless transmission method employs the mechanism of positive feedback (positive acknowledgements). Secondly, in contrast to the random access method, in wired networks the WiFi technology employs the random delay mechanism as early as during the first transmission. And at last, the wireless access protocol employs the mechanism of "suspension" of the delaying timer from the time of detection of the medium occupation until expiration of the random delay timer.

## 2    Mathematic Modelling of 802.11 Wireless LAN

Let us analyze the operation of a wireless local area network until the first error-free data frame transmission with obtained acknowledgement on successful delivery of data. Let us suppose that the wireless LAN contains $K$ stations which are data sources. Consider that all the sources are independent and equal, and always have data frames for sending, and all interval spaces are expressed in slot intervals $t_c$. Let all the stations exchange frames of equal sizes. Then, according to the sequence of protocol actions, the elementary cycle of data frame transfer to the recipient will be determined by the size of the interframe space $t_m$, random delay period $t_o$, duration of "suspension" of the random delay timer $t_z$, time of data frame transmission $t_k$, and the value of time-out for expecting a positive acknowledgement $t_{out}$, which consists of a short interframe space plus the time of transmission of a positive acknowledgement [2,4]. The average time of data frame transmission $T(K, N_0)$ consists of the weighted sum of average periods of waiting for failed transmissions and the time of successful transmission [6]:

$$T(K, N_0) = d + \sum_{N=0}^{\infty} \left[ Nd + \sum_{n=0}^{N-1} t(n, K, N_0) + \tau(N, K, N_0) \right] f(N, K, N_0). \quad (1)$$

Here $d = t_m + t_k + t_{out}$, $t(n, K, N_0)$ and $\tau(N, K, N_0)$ are the average conditional times until failed and successful $N$-th repeated attempts to send a data frame by a subscriber, and $f(N, K, N_0)$ is the function of probability [7] of the duration of competition between subscribers for the medium, which is determined by the probability of successful data frame transmission on the $N$-th repeated step after $N - 1$ failures [6]:

$$f(N, K, N_0) = P(N, K, N_0) \prod_{n=0}^{N-1} \pi(n, K, N_0).$$

Along with the average time of data frame transmission, one of the main indicators showing the efficiency of functioning the data transfer network is the throughput performance. In the case under analysis, we will look for an individual throughput performance, the standardized value of which shall be determined as a ratio between the time necessary for data frame transmission $t_k$ and the average time of data frame transmission $T(K, N_0)$:

$$C(K, N_0) = \frac{t_k}{T(K, N_0)}. \quad (2)$$

## 3   The Competition of Two Wireless Stations

Let us analyze the competition of two wireless stations ($K = 2$) of a local area network. We denote the competing (conflicting) stations through $A$ and $B$. Let us find the probability timing characteristics of the data transmission process executed by the $A$ station. Let us denote via $p_n(i)$ the probability of selection of random backoff interval with a duration equal to $i$ slot intervals on the $n$-th repeated transmission by the $A$ station, and via $f_n(j)$ the probability of selection of random backoff interval with a duration equal to $j$ slot intervals on the $n$-th repeated transmission by the $B$ station. Then the conditional probability of a conflict on the $n$-th repeated transmission for the $A$ station is determined by the relation

$$\pi(n, 2, N_0)$$

$$= \begin{cases} \sum_{i=0}^{S_0-1} p_0(i) \sum_{j=0}^{i} f_0(j) L_{i-j}, & n = 0; \\ \sum_{k=1}^{n} E_k(n) \left[ \sum_{i=0}^{S_k-1} p_n(i) \sum_{j=0}^{i} f_k(j) L_{i-j} + \sum_{i=S_k}^{S_n-1} p_n(i) \sum_{j=0}^{S_k-1} f_k(j) L_{i-j} \right], & n \geq 1. \end{cases}$$

$$(3)$$

Here $L_k$ represents recurrent probabilities of movement of the $B$ station "bottom-up" from originally selected slot interval $j$ to a conflict slot interval $i$ selected by the $A$ station ($k$ is a difference between $j$-th and $i$-th slots), for many steps with successful transmissions:

$$L_k = \begin{cases} \sum\limits_{i=0}^{\infty} f_0^i(0) \sum\limits_{i=1}^{k} f_0(i) L_{k-i}, & k = \overline{1, S_0 - 1}, \ L_0 = 1; \\ \sum\limits_{i=0}^{\infty} f_0^i(0) \sum\limits_{i=1}^{S_0-1} f_0(i) L_{k-i}, & k = \overline{S_0, S_n - 1}. \end{cases}$$

In other words elements $L_k$ include probabilities of all possible actions of the $B$ station before collision with the $A$ station, if the $B$ station originally selected slot interval $j$ and the $A$ station selected slot interval $i$. From this point, it is not difficult to see that, before the conflict with the $A$ rival, the competing $B$ station may carry out an unlimited number of successful transmissions in case of "fallout" of random delay having zero duration. Using the relations for the arithmetic-geometrical progression [8] for $L_k$ with $k = \overline{1, S_0 - 1}$, we obtain the final relation:

$$L_k = \frac{S_0^{k-1}}{(S_0 - 1)^k}, \quad k = \overline{1, S_0 - 1}. \tag{4}$$

Inserting (4) into (3), we find the probability of a conflict on the first attempt of data frame transmission:

$$\pi(0, 2, N_0) = \frac{S_0 - 1}{S_0^2} \left[ \left( \frac{S_0}{S_0 - 1} \right)^{S_0} - 1 \right].$$

The coefficients $E_k(n)$ in the relation (3) are the probabilities that on the $n$-th repeated transmission by the $A$ station, the $B$ station will be in the condition of the $k$-th repeated transmission:

$$E_1(1) = 1; \quad E_1(n) = \sum_{k=1}^{n-1} \frac{E_k(n-1)}{\pi(n-1, 2, N_0)} \left[ \sum_{i=1}^{S_k-1} p_{n-1}(i) \sum_{j=0}^{i-1} f_k(j) L_{i-j} \right.$$

$$\left. + \sum_{i=S_k}^{S_{n-1}-1} p_{n-1}(i) \sum_{j=0}^{S_k-1} f_k(j) L_{i-j} \right], \quad n \geq 2;$$

$$E_k(n) = \frac{E_{k-1}(n-1) \sum\limits_{i=0}^{S_{k-1}-1} p_{n-1}(i) f_{k-1}(i)}{\pi(n-1, 2, N_0)}, \quad n \geq 2, \ k = \overline{2, n}.$$

The average conditional times until failed and successful $n$-th attempt of data transmission $t(N, K, N_0)$ and $\tau(N, K, N_0)$ consist of the average duration of random delay $N_s(n)$ (average number of slots until the start of transmission) and the average number of suspensions caused by medium capture by the $B$ station, $Z_t(n, N_0)$ in case of failure and $Z_\tau(n, N_0)$ in case of success, respectively:

$$t(n, 2, N_0) = N_s(n) + Z_t(n, N_0)d, \quad \tau(n, 2, N_0) = N_s(n) + Z_\tau(n, N_0)d.$$

Here $N_s(n) = \sum_{i=0}^{S_n-1} i p_n(i) = (S_n - 1)/2$, and the average numbers of suspensions $Z_t(n, N_0)$ and $Z_\tau(n, N_0)$ look similar:

$$Z_t(n, N_0)$$
$$= \begin{cases} \sum_{i=1}^{S_0-1} p_0(i) \sum_{j=0}^{i-1} f_0(j) M_{i-j}, & n = 0; \\ \sum_{k=1}^{n} E_k(n) \left[ \sum_{i=1}^{S_k-1} p_n(i) \sum_{j=0}^{i-1} f_k(j) M_{i-j} + \sum_{i=S_k}^{S_n-1} p_n(i) \sum_{j=0}^{S_k-1} f_k(j) M_{i-j} \right], & n \geq 1; \end{cases}$$

(5)

$$Z_\tau(n, N_0)$$
$$= \begin{cases} \sum_{i=1}^{S_0-1} p_0(i) \sum_{j=0}^{i-1} f_0(j) V_{i-j}, & n = 0; \\ \sum_{k=1}^{n} E_k(n) \left[ \sum_{i=1}^{S_k-1} p_n(i) \sum_{j=0}^{i-1} f_k(j) V_{i-j} + \sum_{i=S_k}^{S_n-1} p_n(i) \sum_{j=0}^{S_k-1} f_k(j) V_{i-j} \right], & n \geq 1. \end{cases}$$

(6)

The elements $M_k$ and $V_k$ are indicators of the average number of suspensions of the delaying timer for the $A$ station after selection of random delay with the duration $i$ on the $n$-th repeated transmission upon selection of the $j$-th slot preceding to the $i$-th one by the competing $B$ station ($k$ is a difference between $j$-th and $i$-th slots):

$$M_k = \begin{cases} \sum_{m=1}^{k} f_0(m) \sum_{i=0}^{\infty} (i + 1 + M_{k-m}) f_0^i(0), & k = \overline{1, S_0 - 1}, \ M_0 = 0; \\ \sum_{m=1}^{S_0-1} f_0(m) \sum_{i=0}^{\infty} (i + 1 + M_{k-m}) f_0^i(0), & k = \overline{S_0, S_n - 1}; \end{cases}$$

$$V_k$$
$$= \begin{cases} \sum_{i=0}^{\infty} (i + 1) f_0^i(0) \sum_{m=k+1}^{S_0-1} f_0(m) + \sum_{m=1}^{k-1} f_0(m) \sum_{i=0}^{\infty} (i + 1 + V_{k-m}) f_0^i(0), & k = \overline{1, S_0 - 1}; \\ \sum_{m=1}^{S_0-1} f_0(m) \sum_{i=0}^{\infty} (i + 1 + V_{k-m}) f_0^i(0), & k = \overline{S_0, S_n - 1}. \end{cases}$$

After inserting here the probabilities of fallout of delay duration $f_0(m)$, we obtain the following relations:

$$M_k = \begin{cases} \dfrac{S_0}{S_0 - 1} \left[ \left( \dfrac{S_0}{S_0 - 1} \right)^k - 1 \right], & k = \overline{1, S_0 - 1}; \\ \dfrac{S_0}{S_0 - 1} + \dfrac{\sum_{m=1}^{S_0-1} M_{k-m}}{S_0 - 1}, & k = \overline{S_0, S_n - 1}. \end{cases}$$

$$V_k = \begin{cases} \dfrac{S_0 - 2}{S_0 - 1} \left( \dfrac{S_0}{S_0 - 1} \right)^k, & k = \overline{1, S_0 - 1}; \\ \dfrac{S_0}{S_0 - 1} + \dfrac{\sum_{m=1}^{S_0-1} V_{k-m}}{S_0 - 1}, & k = \overline{S_0, S_n - 1}. \end{cases}$$

The indicator of the general throughput performance can be found by analogy with individual operational speed (2), therewith the numerator of such relation should be adjusted not only for the package successfully transferred by the $A$ station, but also for the average number of packages transferred by the $B$ station for the concerned period:

$$C_g(2, N_0) = \frac{(G(N_0) + 1)t_k}{T(2, N_0)},$$

where $G(N_0)$ will be determined by the weighted amount of the average number of suspensions of the delaying timer of the $A$ station in expectation of failed and successful transmissions, which are determined by the relations (5) and (6):

$$G(N_0) = \sum_{N=0}^{\infty} \left[ \sum_{n=0}^{N-1} Z_t(n, N_0) + Z_\tau(N, N_0) \right] f(N, 2, N_0).$$

## 4   Numerical Results

The numeric research into the average time of data frame transmission by the $A$ station shows that the function (1) has a strongly manifested minimum at the coordinate $N_0$ (see Fig. 1) determining the initial size of the competition window and, subsequently, the degree of scattering of stations by durations of delays before the start of the competition procedure. For two competing stations, the minimum is reached at $N_0 = 4$. It is obvious that the value $N_0$ minimizing the average time of data frame transmission maximizes the individual throughput (see Fig. 1). Moreover, as early as at the stage of formalization of the task, the probability of capture of the communication medium by one of the subscribers mentioned in [9, 10] has become obvious. This effect manifests itself especially strongly with small values $N_0$. The effect of capturing the communication medium causes discrimination-related individual indicators against a good level of the general throughput performance of the network (see Fig. 1).

As early as at the first attempt of competition between two stations, capture of the communication medium becomes possible (e.g. by the $B$ station), and its probability will be determined by the probabilities that for one of the stations ($B$) the delay duration will turn out to be shorter than the duration of delay of the other station ($A$); then the "succeeded" station ($B$) will have fallout of zero duration, which will alternate with shorter delays than the residual value of the station's $A$ delaying timer:

$$P_z(0, 2, N_0) = \sum_{i=1}^{S_0-1} p_0(i) L_{i-1} \sum_{k=1}^{\infty} f_0^k(0) = \frac{1}{S_0^2} \left( \frac{S_0}{S_0 - 1} \right)^{S_0-1}.$$

From this point, it is not difficult to see that the probability of capture is considerably determined by the initial width of the contention window $S_0$ (see Fig. 2). After several conflicts, the possibility of capture for the "succeeded" station becomes yet more probable.

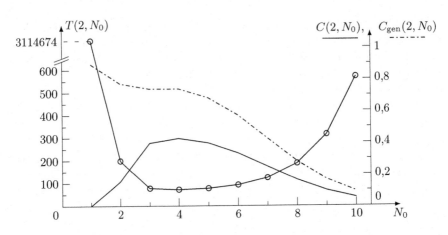

**Fig. 1.** Average time of data frame transmission, and individual and general throughput performances

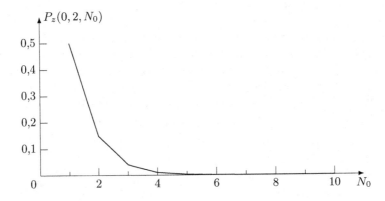

**Fig. 2.** Probability of the medium capture by one of the stations

The main reason for the effect of capturing the communication medium is the protocol action — "suspension of delay", because this results in a fact that after a non-conflict transmission the station may capture the communication medium for an infinitely long time, getting into the delay interval from 0 to the residual value of the delay of other stations.

Another reason for an increase in the probability of capturing the communication medium by one of the subscribers after several conflicts, consists in various sizes of the contention window for stations withdrawn from the conflict and stations continuing resolution of the conflict in the condition of waiting for expiration of delay time and suspension periods. After a positive resolution of the conflict by one of the stations (or by several stations), the size of its contention window is reduced in multiples down to the initial value $S_0 < S_n$, which gives this station a priority right in subsequent competition for the medium with "conflicting" stations, because the shorter duration of an occasional delay for such station has a significantly higher probability as compared with the similar operational indicator of the "conflicting" station.

It is obvious that to reduce the probability of the effect of medium capturing for an infinitely long time, it is possible to offer, on one hand, to fix the size of the contention window for the first and all subsequent transmissions, and on the other hand – the duration of random delay $t_o$ should be selected within the interval from 1 to $2^{N_0} - 1$ of slot periods $t_c$, thus excluding the delay of the zero size. Therewith, medium capturing by one station will never exceed $2^{N_0} - 2$ successful transmissions until the subsequent conflict or its resolution.

## 5   Conclusion

The performed analysis is targeted at studying the method of carrier sense multiple access with collision avoidance. Analytic correlations have been obtained for probability timing characteristics of the competition process between two stations. The "medium capture effect" and the extreme dependence of operational parameters on the initial contention window size have been revealed.

It has been suggested to change the parameters of the protocol procedure of competition, ensuring prevention of the capture effect by saving high values of individual and integral indices of operational speed.

It has been shown that the optimal initial width of the contention window $(S_0)$ is determined by the active size of the network (the number of competing stations), and it ensures almost uniform distribution of a jointly used time resource of the medium between competing subscribers.

## References

1. Tanenbaum, A.S., Wetherall, D.J.: Computer Networks, 5th edn, p. 960. Prentice Hall, Boston (2010)
2. IEEE Std 802.11 — 2007, Revision of IEEE Std 802.11 — 1999. IEEE Std 802.11 — 2007, IEEE Standard for Information Technology, Telecommunications, information exchange between systems, Local, metropolitan area networks, Specific requirements. Part 11: Wireless LAN Medium Access Control (MAC) and Physical Layer (PHY) Specifications, p. 1184. IEEE Computer Society (2007)
3. Geier, J.: Wireless Networks First-step. Cisco Press, Indianapolis (2005). p. 192
4. Roshan P., Leary J.: 802.11 Wireless LAN Fundamentals, p. 312. Cisco Press, Indianapolis (2004)
5. Bianchi, G.: Performance analysis of the IEEE 802.11 distributed coordination function. IEEE J. Sel. Areas Commun. 18(3), 535–547 (2000)
6. Kustov, N.T., Suschenko, S.P.: Capacity of the random multiple access method. Autom. Remote Control. 62(1), 76–85 (2001)
7. Hastings, N.A.J., Peacock, J.B.: Statistical Distributions in Scientific Work Series: A Handbook for Students and Practitioners, p. 130. Butterworth, London (1975)
8. Prudnikov, A.P., Brychkov, Y.A., Marichev, O.I.: Integrals and Series: Special functions, p. 751. Overseas Publishers Association, Amsterdam (1986)

9. Bononi, L., Conti, M., Donatiello, L.: Design and performance evaluation of distributed contention control (DCC) mechanism for IEEE 802.11 wireless local area network. J. Parallel Distrib. Comput. **60**(4), 407–430 (2000)
10. Vishnevsky, V.M., Lyakhov, A.I.: IEEE 802.11 wireless LAN: saturation throughput analysis with seizing effect consideration. Cluster Comput. **5**, 133–144 (2002)

# Tandem of Infinite-Server Queues with Markovian Arrival Process

Alexander Moiseev$^{(\boxtimes)}$ and Anatoly Nazarov

Tomsk State University, Lenin ave. 36, 634050 Tomsk, Russia
moiseev.tsu@gmail.com, nazarov.tsu@gmail.com

**Abstract.** We consider a tandem queueing system with infinite number of servers and Markovian arrival process. Service times at the system stages are i.i.d. and given by distribution functions individually for each stage. The study is performed under the asymptotic condition of the arrivals rate growth. It is shown that multi-dimensional probability distribution of customers number at the system stages can be approximated by multi-dimensional Gaussian distribution which parameters are obtained in the paper.

**Keywords:** Tandem of queues · Infinite-server · Markovian arrival process · Asymptotic analysis

## 1 Introduction

Tandem queues are used for modelling of modern telecommunications and data processing systems [1] in such cases when it is need to perform a processing of calls or data packages in the several stages. Most researchers consider such models for case of Poisson arrivals [2] or for some specific configuration of the tandem [3–5], usually only with two stages. However, there are results [6] which prove that the Poisson model can be adequate only in a few cases of modern telecommunication streams. Therefore, many researchers use more complex models of the streams such as Markovian arrival process [7] or semi-Markov process [8].

We consider the tandem of queues with infinite number of servers at each stage and total number of stages $K \geq 2$. Infinite-server models are popular in the queueing theory because they allow to obtain some results which can be applied to real finite-server systems after some assumptions. More information about infinite-server models can be found in [9]. In [10], we performed some studies about how the results for infinite-server model can be applied to the models with limited number of servers.

A model and a problem under study are described in the Sect. 2. In the Sect. 3, we present a method for constructing Kolmogorov equations (they a derived in the Sect. 4) which allow to solve the problem. In the Sect. 5, we make the asymptotic solution of derived equations under condition of growth of arrivals rate. Also in that section, we obtain a Gaussian approximation for the multidimensional probability distribution of customers number at the tandem stages.

© Springer International Publishing Switzerland 2016
V. Vishnevsky and D. Kozyrev (Eds.): DCCN 2015, CCIS 601, pp. 323–333, 2016.
DOI: 10.1007/978-3-319-30843-2_34

Numerical results are presented in the Sect. 6 and they allow to determine the boundaries of the approximation applicability.

Similar results was obtained for the case of renewal arrival process [11].

## 2    Mathematical Model

The object of the study is a tandem of $K$ infinite-server queues which we call 'stages'. A customer that comes into the system enters the first stage for service with the cumulative distribution function $B_1(x)$. After the service is complete at the first stage the customer moves to the next stage for the service, and so on while the service at the last $K$-th stage will be complete. After that the customer leaves the system. Denote by $B_k(t)$ for $k = 1, \ldots, K$ cumulative distribution functions of service time at the respective stages of tandem.

Process of customers' arriving is Markovian arrival process (MAP) [12,13] with representation $(D_0, D_1)$ of order $M$. The matrix $D = D_0 + D_1$ is a generator of the underlying Markov chain of MAP. Denote the stationary distribution of the chain states by row vector $\theta$. This vector satisfies the following equations:

$$\begin{cases} \theta D = 0, \\ \theta e = 1. \end{cases} \tag{1}$$

Here $0$ is a row vector with zeros and $e$ is column vector with entries all equal to 1. The value

$$\lambda = \theta D_1 e \tag{2}$$

is called as fundamental rate of the Markovian arrival process [7]. It characterizes the average rate of the arrivals.

Denote by $i_k(t)$ the number of customers at the $k$-th stage of the tandem at the instant $t$. The goal of study is to find the multi-dimensional joint probability distribution of the tandem state $i(t) = \{i_1(t), \ldots, i_K(t)\}$.

## 3    Method of Multi-dimensional Dynamic Screening

Direct analysis of the multi-dimensional non-Markovian process $i(t)$ doesn't seem possible. Therefore, we will use a special technique [14] known as the method of multi-dimensional dynamic screening. We will introduce basic principles of this approach here.

Consider $K + 1$ time axes (Fig. 1) which are numbered from 0 to $K$. We will mark on the axis number 0 the epochs of customers' arrivals. The axes with numbers from 1 to $K$ correspond to respective stages of the tandem. On these axes, we will mark the events of components of multi-dimensional screened point process. A generation of these events is defined as follows.

Let the system be empty at some initial time moment $t_0$. Let's fix some epoch $T > t_0$. Consider a customer that arrives at the system at the time moment $t_0 < t < T$. Denote by $S_0(t)$ a probability that the customer departs from the system before the moment $T$, and denote by $S_k(t)$ the probability that

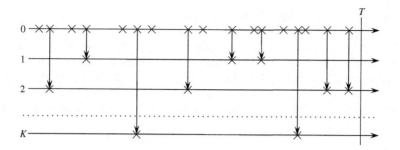

**Fig. 1.** Generating events in the screened streams on base of the customers arrivals

the customer is served at the $k$-th stage at the instant $T$ $(k = 1, \ldots, K)$. It is obvious that

$$S_0(t) = 1 - \sum_{k=0}^{K} S_k(t). \tag{3}$$

We will say that a customer, which comes to the system at the moment $t < T$, generates an event in the $k$-th component of the screened process with the probability $S_k(t)$. And it doesn't generate any event in the screened process with probability $S_0(t)$.

Denote by $\boldsymbol{n}(t) = \{n_1(t), \ldots, n_K(t)\}$ a counting process of the described screened point process. Here $n_k(t)$ is a number of events generated in the $k$-th component of the screened process before the instant $t$. The basic idea of the multi-dimensional dynamic screening method is that the probability distribution of customers numbers in the system $\boldsymbol{i}(t)$ and the probability distribution of value of multi-dimensional counting process $\boldsymbol{n}(t)$ are coincide at the instant $t = T$:

$$\mathrm{P}\{\boldsymbol{i}(T) = \boldsymbol{i}\} = \mathrm{P}\{\boldsymbol{n}(T) = \boldsymbol{i}\} \tag{4}$$

for all non-negative vectors $\boldsymbol{i}$.

So, if we obtain the probability distribution for process $\boldsymbol{n}(t)$, then applying $t = T$, we can simple derive a probability distribution of the process $\boldsymbol{i}(t)$ at the arbitrary fixed moment $T$ (non-stationary distribution). For the purpose of finding the stationary distribution, we will find the final distribution by a performing the asymptotic transition $T \to \infty$.

Expressions for the screening probabilities $S_k(t)$ were obtained in the paper [11] where the analysis of the tandem $GI/(GI/\infty)^K$ was performed. It is obvious that these probabilities only depend on the service times at the stages of the tandem but don't depend on the parameters of the arrival process. So, we have the following expression [11] for the screening probabilities

$$S_k(t) = B_{k-1}^*(T - t) - B_k^*(T - t). \tag{5}$$

Here $j = \sqrt{-1}$ is imaginary unit, $B_k^*(\tau) = (B_1 * \cdots * B_k)(\tau)$ is a convolution of functions $B_1(\tau), \ldots, B_k(\tau)$, and we assume that $B_0^*(\tau) = 1$, $B_1^*(\tau) = B_1(\tau)$.

## 4    Kolmogorov Equations

Denote by $m(t)$ the state of the underlying Markov chain of the Markovian arrival process. Let's consider the $(K+1)$-dimensional Markovian process $\{n(t), m(t)\}$. Using the notation $p(n, m, t) = \mathrm{P}\{n(t) = n, m(t) = m\}$ and applying the formula of total probability, after some derivations we can write the following system of Kolmogorov differential equations for the probability distribution of the process $\{n(t), m(t)\}$:

$$\frac{\partial p(n, m, t)}{\partial t} = \sum_{\eta=1}^{M} p(n, \eta, t)(D_0)_{\eta m}$$
$$+ \sum_{\eta=1}^{M} \left[ p(n, \eta, t)(D_1)_{\eta m} S_0(t) + \sum_{k=1}^{K} p(n - e_k, \eta, t)(D_1)_{\eta m} S_k(t) \right]$$

for $m = 1, \ldots, M$ and $n \geq 0$. Here $e_k$ is a vector that consists of zeros except the $k$-th entry which is equal to 1.

Using vector notation $p(n, t) = \{p(n, 1, t), \ldots, p(n, M, t)\}$ we can rewrite this system in the form

$$\frac{\partial p(n, t)}{\partial t} = p(n, t)D_0 + p(n, t)D_1 S_0(t) + \sum_{k=1}^{K} p(n - e_k, t)D_1 S_k(t) \qquad (6)$$

where we assume $p(n, t) = 0$ if any entry of vector $n$ is negative.

Let's consider vector characteristic function

$$h(u, t) = \sum_{n_1=0}^{\infty} \cdots \sum_{n_K=0}^{\infty} e^{j u_1 n_1 + \cdots + j u_K n_K} p(n, t).$$

Using the system (6), we can write the following matrix differential equation for the function $h(u, t)$:

$$\frac{\partial h(u, t)}{\partial t} = h(u, t) \left[ D_0 + D_1 S_0(t) + D_1 \sum_{k=1}^{K} e^{j u_k} S_k(t) \right].$$

Taking into account formula (3), we reduce this equation to the form

$$\frac{\partial h(u, t)}{\partial t} = h(u, t) \left[ D + D_1 \sum_{k=1}^{K} \left( e^{j u_k} - 1 \right) S_k(t) \right] \qquad (7)$$

where $D = D_0 + D_1$. The initial condition is determined by the expression

$$h(u, t_0) = 0. \qquad (8)$$

# 5    Asymptotic Analysis

We can not directly solve the problem (7), (8) using a matrix exponent because the matrices $D$ and $D_1$ are not commutative: $D \cdot D_1 \neq D_1 \cdot D$. So, here we obtain an asymptotic solution of this problem under a condition of growth of the fundamental rate of arrivals using the technique [15]. To do this, we make in the Eq. (7) the substitutions $ND_0$ instead the matrix $D_0$ and $ND_1$ instead the matrix $D_1$. Here $N > 0$ is some parameter. The fundamental rate of such arrival process is equal to $N\lambda$. If value of the parameter $N$ grows infinitely, then value of the fundamental rate of the arrival process grows infinitely too. Such arrival processes we call as high-rate or high-intensive processes [11,14].

As a result of described transforms we obtain the following matrix differential equation

$$\frac{1}{N} \frac{\partial h(u,t)}{\partial t} = h(u,t) \left[ D + D_1 \sum_{k=1}^{K} \left( e^{ju_k} - 1 \right) S_k(t) \right]. \tag{9}$$

The initial condition for this asymptotic problem is the same as expression (8).

## 5.1    The First-Order Asymptotic Analysis

Consider an asymptotic condition of high-rate arrivals: $N \to \infty$. Let's make the following changes of variables in the problem (9), (8):

$$\frac{1}{N} = \varepsilon, \quad u = \varepsilon w, \quad h(u,t) = f_1(w,t,\varepsilon). \tag{10}$$

Using these new variables, the problem can be written in the form

$$\varepsilon \frac{\partial f_1(w,t,\varepsilon)}{\partial t} = f_1(w,t,\varepsilon) \left[ D + D_1 \sum_{k=1}^{K} \left( e^{j\varepsilon w_k} - 1 \right) S_k(t) \right], \tag{11}$$

$$f_1(w,t_0,\varepsilon) = \theta. \tag{12}$$

Denote an asymptotic solution of this problem under a condition $\varepsilon \to 0$ by $f_1(w,t)$:

$$f_1(w,t) = \lim_{\varepsilon \to 0} f_1(w,t,\varepsilon).$$

Let's prove the following statement.

**Theorem 1.** *The asymptotic solution $f_1(w,t)$ of the problem (11)–(12) under a condition $\varepsilon \to 0$ is the following*

$$f_1(w,t) = \theta \exp \left\{ \lambda \sum_{k=1}^{K} jw_k \int_{t_0}^{t} S_k(\tau) \, d\tau \right\} \tag{13}$$

*where $\theta$ and $\lambda$ are defined by expressions (1) and (2) respectively.*

*Proof.* We perform the proof by two stages.

*Stage 1.* Let $\varepsilon \to 0$ in the Eq. (11). We obtain the equation

$$f_1(w,t)D = 0.$$

Comparing this equation with the first one in the system (1), we can draw a conclusion that the function $f_1(w,t)$ can be represented in the form

$$f_1(w,t) = \theta\Phi_1(w,t), \tag{14}$$

where $\Phi_1(w,t)$ is some scalar function which satisfies the condition

$$\Phi_1(w,t_0) = 1. \tag{15}$$

*Stage 2.* Let's perform the following operations:

1. Sum all rows of the matrix Eq. (11) by multiplication on vector $e$ and take into account that
$$De = 0. \tag{16}$$

2. Divide the result by $\varepsilon$.
3. Use the expression (14).
4. Perform an asymptotic transition $\varepsilon \to 0$.

We obtain the following differential equation

$$\theta\frac{\partial\Phi_1(w,t)}{\partial t}e = \theta\Phi_1(w,t)D_1e\sum_{k=1}^{K}jw_kS_k(t).$$

Taking into account that $\theta e = 1$ and using the expression (2), we reduce this equation to the form

$$\frac{\partial\Phi_1(w,t)}{\partial t} = \Phi_1(w,t)\lambda\sum_{k=1}^{K}jw_kS_k(t).$$

As a result we obtain the following solution of this differential equation under an initial condition (15):

$$\Phi_1(w,t) = \exp\left\{\lambda\sum_{k=1}^{K}jw_k\int_{t_0}^{t}S(\tau)\,d\tau\right\}.$$

Substituting this expression into the formula (14), we obtain the expression for the function $f_1(w,t)$ in the form (13). The theorem is proved. $\quad\square$

## 5.2    The Second-Order Asymptotic Analysis

Denote by $h_2(u, t)$ a function which satisfies the following expression

$$h(u, t) = h_2(u, t) \exp \left\{ N\lambda \sum_{k=1}^{K} ju_k \int_{t_0}^{t} S_k(\tau) \, d\tau \right\}. \tag{17}$$

Substituting this expression into the formulas (9), (8), we obtain the following Cauchy problem for the function $h_2(u, t)$.

$$\begin{cases} \frac{1}{N} \frac{\partial h_2(u,t)}{\partial t} + h_2(u, t)\lambda \sum\limits_{k=1}^{K} ju_k S_k(t) \\ \quad = h_2(u, t) \left[ D + D_1 \sum\limits_{k=1}^{K} \left( e^{ju_k} - 1 \right) S_k(t) \right], \\ h_2(u, t_0) = \boldsymbol{0}. \end{cases} \tag{18}$$

Let's make here the following changes of variables

$$\frac{1}{N} = \varepsilon^2, \quad u = \varepsilon w, \quad h_2(u, t) = f_2(w, t, \varepsilon). \tag{19}$$

Using new variables, the problem (18) can be rewritten in the form

$$\begin{cases} \varepsilon^2 \frac{\partial f_2(w,t,\varepsilon)}{\partial t} + f_2(w, t, \varepsilon)\lambda \sum\limits_{k=1}^{K} j\varepsilon w_k S_k(t) \\ \quad = f_2(w, t, \varepsilon) \left[ D + D_1 \sum\limits_{k=1}^{K} \left( e^{j\varepsilon w_k} - 1 \right) S_k(t) \right], \\ f_2(w, t_0, \varepsilon) = \boldsymbol{0}. \end{cases} \tag{20}$$

Denote an asymptotic solution of this problem under a condition $\varepsilon \to 0$ by $f_2(w, t)$:

$$f_2(w, t) = \lim_{\varepsilon \to 0} f_2(w, t, \varepsilon). \tag{21}$$

Let's prove the following statement.

**Theorem 2.** *The asymptotic solution $f_2(w, t)$ of the problem* (20) *under a condition $\varepsilon \to 0$ is the following*

$$f_2(w, t) = \boldsymbol{\theta} \exp \left\{ \lambda \sum_{k=1}^{K} \frac{(jw_k)^2}{2} \int_{t_0}^{t} S_k(\tau) \, d\tau + \kappa \sum_{k=1}^{K} \sum_{\nu=1}^{K} \frac{j^2 w_k w_\nu}{2} \int_{t_0}^{t} S_k(\tau) S_\nu(\tau) \, d\tau \right\} \tag{22}$$

*where*

$$\kappa = 2g(D_1 - \lambda I)e. \tag{23}$$

*Here the row vector $g$ satisfies the following matrix linear equation*

$$gD = \boldsymbol{\theta}(\lambda I - D_1). \tag{24}$$

*Proof.* We perform the proof by three stages.

*Stage 1.* Substituting $\varepsilon \to 0$ in the problem (20), we obtain the system

$$\begin{cases} \boldsymbol{f}_2(\boldsymbol{w}, t) \boldsymbol{D} = \boldsymbol{0}, \\ \boldsymbol{f}_2(\boldsymbol{w}, t_0) = \boldsymbol{\theta}. \end{cases}$$

This system has a form similar to the Eq. (1). So, we can make a conclusion that the function $\boldsymbol{f}_2(\boldsymbol{w}, t)$ can be written in the form

$$\boldsymbol{f}_2(\boldsymbol{w}, t) = \boldsymbol{\theta} \Phi_2(\boldsymbol{w}, t), \tag{25}$$

where $\Phi_2(\boldsymbol{w}, t)$ is some scalar function that satisfies the initial condition

$$\Phi_2(\boldsymbol{w}, t_0) = 1. \tag{26}$$

*Stage 2.* Taking into account formula (25) and relation (21), we can represent the function $\boldsymbol{f}_2(\boldsymbol{w}, t, \varepsilon)$ in the following expansion form

$$\boldsymbol{f}_2(\boldsymbol{w}, t, \varepsilon) = \Phi_2(\boldsymbol{w}, t) \left[ \boldsymbol{\theta} + \boldsymbol{g} \sum_{k=1}^{K} j\varepsilon w_k S_k(t) \right] + O(\varepsilon^2) \tag{27}$$

where $\boldsymbol{g}$ is some row vector and $O(\varepsilon^2)$ is a row vector that consists of the infinitesimals of the order $\varepsilon^2$. Let's substitute the expansion (27) and the expansion $e^{j\varepsilon w_k} = 1 + j\varepsilon w_k + O\left(\varepsilon^2\right)$ into the first equation of the problem (20). We obtain the equality

$$\boldsymbol{\theta} \lambda \sum_{k=1}^{K} j\varepsilon w_k S_k(t) = \boldsymbol{\theta} \boldsymbol{D} + \boldsymbol{\theta} \boldsymbol{D}_1 \sum_{k=1}^{K} j\varepsilon w_k S_k(t) + \boldsymbol{g} \boldsymbol{D} \sum_{k=1}^{K} j\varepsilon w_k S_k(t) + O(\varepsilon^2).$$

Performing here the asymptotic transition $\varepsilon \to 0$ and taking into account the Eq. (1), we obtain the following matrix equation for the row vector $\boldsymbol{g}$:

$$\boldsymbol{g} \boldsymbol{D} = \boldsymbol{\theta} (\lambda \boldsymbol{I} - \boldsymbol{D}_1)$$

which coincides with expression (24).

*Stage 3.* Multiplying both parts of the first equation of the problem (20) by vector $\boldsymbol{e}$, using the expansion $e^{j\varepsilon w_k} = 1 + j\varepsilon w_k + \frac{(j\varepsilon w_k)^2}{2} + O(\varepsilon^3)$, taking into account the properties (16), (1) and the expressions (2), (23), under an asymptotic condition $\varepsilon \to 0$ we obtain the following linear differential homogeneous (on variable $t$) equation for the function $\Phi_2(\boldsymbol{w}, t)$

$$\frac{\partial \Phi_2(\boldsymbol{w}, t)}{\partial t} = \Phi_2(\boldsymbol{w}, t) \left[ \lambda \sum_{k=1}^{K} \frac{(jw_k)^2}{2} S_k(t) + \kappa \sum_{k=1}^{K} \sum_{\nu=1}^{K} \frac{j^2 w_k w_\nu}{2} S_k(t) S_\nu(t) \right]$$

where the value of $\kappa$ is defined by expression (23). Under the initial condition (26), the solution of this equation is the following

$$\Phi_2(\boldsymbol{w}, t) = \exp \left\{ \lambda \sum_{k=1}^{K} \frac{(jw_k)^2}{2} \int_{t_0}^{t} S_k(\tau) \, d\tau + \kappa \sum_{k=1}^{K} \sum_{\nu=1}^{K} \frac{j^2 w_k w_\nu}{2} \int_{t_0}^{t} S_k(\tau) S_\nu(\tau) \, d\tau \right\}.$$

Substituting this expression into (25), we obtain the final form of the function $f_2(w, t)$ as the expression (22). The theorem is proved.    □

## 5.3    Probability Distribution of Customers Number at the Stages of the Tandem

Let's make in the formula (22) substitutions that are inverse to the changes (19). Using expression (17), we obtain the following expression for the vector characteristic function $h(u, t)$:

$$h(u, t) = \theta \exp \left\{ N\lambda \sum_{k=1}^{K} \left[ ju_k + \frac{(ju_k)^2}{2} \right] \int_{t_0}^{t} S_k(\tau) d\tau \right.$$

$$\left. + N\kappa \sum_{k=1}^{K} \sum_{\nu=1}^{K} \frac{ju_k ju_\nu}{2} \int_{t_0}^{t} S_k(\tau) S_\nu(\tau) d\tau \right\}.$$

Let's perform a transition to the random vector $i(T)$ which is the main object of the study. Denote by $h(u)$ the characteristic function of the probability distribution of the random vector $i(T)$. Due to the main formula of the multi-dimensional dynamic screening method (4), we have the following approximation $h^{(2)}(u)$ for the characteristic function $h(u)$ under a condition that $N$ is large enough:

$$h(u) \approx h^{(2)}(u) = \exp \left\{ N\lambda \sum_{k=1}^{K} \left[ ju_k + \frac{(ju_k)^2}{2} \right] \int_{t_0}^{T} S_k(\tau) d\tau \right.$$

$$\left. + N\kappa \sum_{k=1}^{K} \sum_{\nu=1}^{K} \frac{ju_k ju_\nu}{2} \int_{t_0}^{T} S_k(\tau) S_\nu(\tau) d\tau \right\}.$$

Considering the initial time moment $t_0 = 0$ and performing the asymptotic transition $T \to \infty$, we obtain the following approximation for the characteristic function $h(u)$ for stationary regime of the system evolution:

$$h(u) \approx h^{(2)}(u) = \exp \left\{ juN\lambda Se + \frac{1}{2} ju \left[ N\lambda S + N\kappa V \right] ju^T \right\} \qquad (28)$$

where the matrix $S$ is diagonal with entries $\int_{0}^{\infty} S_k(\tau) d\tau$ for $k = 1, \ldots, K$ and the matrix $V$ has entries $\int_{0}^{\infty} S_k(\tau) S_\nu(\tau) d\tau$ for $k, \nu = 1, \ldots, K$.

So, the multi-dimensional stationary probability distribution of the number of customers at the stages of the tandem system $MAP/(GI/\infty)^K$ under a condition of high-rate arrivals can be approximated by multi-dimensional Gaussian distribution with vector of means $N\lambda Se$ and covariance matrix $N[\lambda S + \kappa V]$.

# 6    Numerical Results

To determine the accuracy of the approximation (28) we have performed simulations of the system evolution for numerous concrete examples and then compare resulting empiric frequencies with probabilities of a distribution constructed on base of the approximation (28). The accuracy of the approximation we characterize by the Kolmogorov distance [16] which has the following form for discrete distributions:

$$d = \max_{x \geq 0} \left| \sum_{l=0}^{x} [\tilde{p}(l) - p(l)] \right|.$$

Here $p(l)$ ($l \geq 0$) is the distribution of Gaussian approximation in discrete points $0, 1, 2, \ldots$, and $\tilde{p}(l)$ is empiric distribution of customers number at the system stage constructed on base of simulation results. Parameters of the model are chosen in a such way to make the average number of customers at the considered stage of the system be equal to $N$. So, we can see the common trend of the approximation applicability bounds in relation to value of parameter $N$ as the average number of customers at the stage.

Obtained values of the Kolmogorov distance $d$ for various values of the parameter $N$ are presented in the Table 1. It is obvious that growth of the arrivals rate (parameter $N$) causes the Gaussian approximation (28) be more accurate. If we choose the acceptable value of the Kolmogorov distance $d \leq 0.05$, we can draw a conclusion that this approximation is applicable when value of the parameter $N$ (or the average number of customers at the node) is about 10 or more.

**Table 1.** Kolmogorov distance $d$ for marginal distributions of the customers number at one stage of tandem for various values of parameter $N$.

| $N$ | 1 | 2.5 | 5 | 10 | 25 | 50 | 100 |
|---|---|---|---|---|---|---|---|
| $d$ | 0.183 | 0.110 | 0.069 | 0.043 | 0.0026 | 0.016 | 0.011 |

# 7    Conclusions

A study of tandem of infinite-server queues with Markovian arrival process is presented in the paper. The study is performed under an asymptotic condition of high rate of arrival process. It is shown that the multi-dimensional joint stationary probability distribution of the customers number at the stages of tandem can be approximated by a multi-dimensional Gaussian distribution. Parameters of this approximation are derived in the paper. Numerical examples show that the approximation is enough accurate when values of the arrivals rate parameter $N$ (or, in other words, the average number of customers at the stage) is about 10 or more.

**Acknowledgments.** The work is performed under the state order of the Ministry of Education and Science of the Russian Federation (No. 1.511.2014/K).

# References

1. Grachev, V.V., Moiseev, A.N., Nazarov, A.A., Yampolsky, V.Z.: Tandem queue as a model of the system of distributed data processing. Proceedings of TUSUR 2 (26), Part 2, 248–251 (2012). (in Russian)
2. Gnedenko, B.W., König, D. (eds.): Handbuch der Bedienungstheorie: II. Formeln und andere Ergebnisse. Akademie-Verlag, Berlin (1984)
3. Kim, C., Dudin, A., Klimenok, V., Taramin, O.: A Tandem $BMAP/G/1 \to$ $\bullet/M/N/0$ queue with group occupation of servers at the second station. Math. Prob. Eng. **2012**, Article ID 324604 (2012)
4. Gopalan, M.N., Anantharaman, N.: Stochastic modelling of a two stage transfer line production system with end buffer and random demand. Microelectron. Reliab. **32**(1–2), 11–15 (1992)
5. Genadis, T.: The distribution of the passage time in a twostation reliable production line: an exact analytic solution. Int. J. Qual. Reliab. Manag. **14**(9), 12–25 (1997)
6. Heyman, D.P., Lucantoni, D.: Modelling multiple IP traffic streams with rate limits. IEEE/ACM Trans. Netw. **11**, 948–958 (2003)
7. Chakravarthy, S.R.: Markovian arrival processes. Wiley Encyclopedia of Operations Research and Management Science (2010)
8. Moiseev, A.: Asymptotic Analysis of the Queueing Network $SM - (GI/\infty)^K$. In: Dudin, A., et al. (eds.) ITMM 2015. CCIS, vol. 564, pp. 73–84. Springer, Heidelberg (2015)
9. Massey, W.A., Whitt, W.: Networks of infinite-server queues with nonstationary poisson input. Queueing Syst. **13**, 183–250 (1993)
10. Nazarov, A., Moiseev, A.: Calculation of the probability that a gaussian vector falls in the hyperellipsoid with the uniform density. In: International Conference on Application of Information and Communication Technology and Statistics in Economy and Education (ICAICTSEE-2013), pp. 519–526. University of National and World Economy, Sofia, Bulgaria (2014)
11. Moiseev, A.N., Nazarov, A.A.: Asymptotic analysis of a multistage queuing system with a high-rate renewal arrival process. Optoelectron. Instrum. Data Process. **50**(2), 163–171 (2014)
12. Neuts, M.F.: A versatile markovian arrival process. J. Appl. Prob. **16**, 764–779 (1979)
13. Lucantoni, D.M.: New results on the single server queue with a batch markovian arrival process. Stochast. Models **7**, 1–46 (1991)
14. Nazarov, A., Moiseev, A.: Analysis of an open non-markovian $GI - (GI|\infty)^K$ queueing network with high-rate renewal arrival process. Prob. Inf. Transm. **49**(2), 167–178 (2013)
15. Nazarov, A.A., Moiseeva, S.P.: Asymptotic Analysis Method in the Queueing Theory. NTL, Tomsk (2006). (in Russian)
16. Kolmogorov, A.: Sulla determinazione empirica di una Legge di distribuzione. Giornale dell' Intituto Italiano degli Attuari **4**, 83–91 (1933)

# A Number of Customers in the System with Server Vacations

Anatoly Nazarov and Svetlana Paul[✉]

Tomsk State University, Lenina pr.36, Tomsk 634050, Russia
nazarov.tsu@gmail.com, paulsv82@mail.ru

**Abstract.** We review the queueing system with one service device and a queue with an unlimited number of waiting seats with server vacations. By using method of asymptotic analysis under large load, we found asymptotic distribution of probabilities of a number of customers in the system. It is shown that this probability distribution can be approximated by exponential distribution.

**Keywords:** Queueing system with server vacations · Asymptotic analysis · Exponential distribution

## 1  Introduction

Single-line queueing systems with server vacations are mathematical models of telecommunication systems, which are pretty common in practice [1]. In real systems "vacations" are considered as a temporal suspension of service either for device other applications or for its breakdown or repair [2].

The mathematical model under study is presented in the Sect. 2. Kolmogorov equations for the investigated processes can be found in the Sect. 3. Asymptotic analysis of the obtained equations is performed in the Sect. 4.

## 2  Mathematical Model

Let's review the queueing system with one service device and a queue with an unlimited number of waiting seats [3]. The system receives Poisson process of requests with intensity $\lambda$. Device operation mode consists of two consecutive intervals. During first intervals customers are handled at the device for random time, distributed by exponential law with parameter $\mu$. If there are no requests in the queue at the beginning of this interval or if the device has handled all customers that were in the queue at that interval of time, then device is still operating in the same mode, waiting for requests. When this interval ends, the server goes on a vacation. During vacations, all requests that came into system are gathered in the queue and are waiting for device to return to operating mode.

Durations of these intervals are random and determined by distribution functions $T_1(x)$ and $T_2(x)$ respectively. We will review systems with customers priority service [4].

© Springer International Publishing Switzerland 2016
V. Vishnevsky and D. Kozyrev (Eds.): DCCN 2015, CCIS 601, pp. 334–343, 2016.
DOI: 10.1007/978-3-319-30843-2_35

Let's denote:

$i(t)$ is number of customers in the system at the time $t$.

$k(t)$ is device mode: 1 device is on the service, 2 device is on vacations.

$z(t)$ is remaining time of device staying in corresponding mode.

## 3   Kolmogorov Equations

Let's review three-dimensional Markov process $\{i(t), k(t), z(t)\}$ and for distribution of probabilities

$$P_k(i, z, t) = P\{i(t) = i, k(t) = k, z(t) < z\}$$

Let's set up the direct system of Kolmogorovs differential equations.

$$
\begin{aligned}
P_1(i, z - \Delta t, t + \Delta t) &= (P_1(i, z, t) - P_1(i, \Delta t, t))\,(1 - \lambda \Delta t)\,(1 - \mu \Delta t) \\
&\quad + P_1(i - 1, z, t)\lambda \Delta t + P_1(i + 1, z, t)\mu \Delta t \\
&\quad + P_2(i, \Delta t, t)T_1(z) + o(\Delta t), \\
P_2(i, z - \Delta t, t + \Delta t) &= (P_2(i, z, t) - P_2(i, \Delta t, t))\,(1 - \lambda \Delta t) + P_2(i - 1, z, t)\lambda \Delta t \\
&\quad + P_1(i, \Delta t, t)T_2(z) + o(\Delta t).
\end{aligned}
$$

$$
\begin{aligned}
\frac{\partial P_1(i, z, t)}{\partial t} &= \frac{\partial P_1(i, z, t)}{\partial z} - \frac{\partial P_1(i, 0, t)}{\partial z} - P_1(i, z, t)\,(\lambda + \mu) \\
&\quad - P_2(i, z, t)\lambda + P_2(i - 1, z, t)\lambda + \frac{\partial P_1(i, 0, t)}{\partial z}T_2(z).
\end{aligned}
$$

We will assume that system operates in stationary mode:

$$
\begin{cases}
\frac{\partial P_1(i, z)}{\partial z} - \frac{\partial P_1(i, 0)}{\partial z} \\
+ (P_1(i - 1, z) - P_1(i, z))\,\lambda + (P_1(i + 1, z) - P_1(i, z))\,\mu + \frac{\partial P_2(i, 0)}{\partial z}T_1(z) = 0, \\
\frac{\partial P_2(i, z)}{\partial z} - \frac{\partial P_2(i, 0)}{\partial z} \\
+ \lambda\,(P_2(i - 1, z) - P_2(i, z)) + \frac{\partial P_1(i, 0)}{\partial z}T_2(z) = 0.
\end{cases}
$$

Let's introduce partial characteristic functions

$$H_k(u, z) = \sum_{i=0}^{\infty} e^{jui} P_k(i, z),$$

for which we will rewrite the direct system of Kolmogorovs differential equations in this form:

$$
\begin{cases}
\frac{\partial H_1(u, z)}{\partial z} - \frac{\partial H_1(u, 0)}{\partial z} + \left(\lambda\left(e^{ju} - 1\right) + \mu\left(e^{-ju} - 1\right)\right) H_1(u, z) \\
+ \frac{\partial H_2(u, 0)}{\partial z}T_1(z) + \left(1 - e^{-ju}\right)\mu P_1(0, z) = 0, \\
\frac{\partial H_2(u, z)}{\partial z} - \frac{\partial H_2(u, 0)}{\partial z} + \lambda H_2(u, z)\left(e^{ju} - 1\right) + \frac{\partial H_1(u, 0)}{\partial z}T_2(z) = 0.
\end{cases}
\tag{1}
$$

$P_1(0, z)$ is probability of situation where device stays in the service mode, and there are no customers in the system.

We will solve system (1) by using method of asymptotic analysis in conditions of large system load [5–7].

# 4   Method of Asymptotic Analysis

Let's denote:

$$\lambda = (1 - \varepsilon)S\mu, u = \varepsilon w,$$

$$H_k(u, z) \quad = \quad F_k(w, z, \varepsilon), P_1(0, z) \quad = \quad \varepsilon \pi_1(z, \varepsilon). \tag{2}$$

S is the systems load. It's value will be found below.

Then we will rewrite system (1) in this form:

$$\begin{cases} \frac{\partial F_1(w,z,\varepsilon)}{\partial z} - \frac{\partial F_1(w,0,\varepsilon)}{\partial z} \\ + \left((1 - \varepsilon) S\mu \left(e^{j\varepsilon w} - 1\right) + \mu \left(e^{-j\varepsilon w} - 1\right)\right) F_1(w, z, \varepsilon) \\ + \left(1 - e^{-j\varepsilon w}\right) \mu \varepsilon \pi_1(z, \varepsilon) + \frac{\partial F_2(w,0,\varepsilon)}{\partial z} T_1(z) = 0, \\ \frac{\partial F_2(w,z,\varepsilon)}{\partial z} - \frac{\partial F_2(w,0,\varepsilon)}{\partial z} \\ + \left(1 - \varepsilon\right) S\mu F_2(w, z, \varepsilon) \left(e^{j\varepsilon w} - 1\right) + \frac{\partial F_1(w,0,\varepsilon)}{\partial z} T_2(z) = 0. \end{cases} \tag{3}$$

**Theorem 1.** *The limit value for $\varepsilon \to 0$* $\quad F_k(w) \quad = \quad F_k(w, \infty)$ *solutions $F_k(w, z, \varepsilon)$ of the system (3) has the form*

$$F_k(w, z) = R_k(z)\Phi(w),$$

*where*

$$R_1 = S, \quad R_2 = 1 - S, \quad S = \frac{T_1}{(T_1 + T_2)},$$

*and asymptotic characteristic function $\Phi(w)$ is defined by*

$$\Phi(w) = \frac{S}{S - jw\{S + \Delta\}},$$

*the value of $\Delta$ is*

$$\Delta = -R_1 R_2 \mu \frac{T_1 T_2}{T_1 + T_2} + R_1^2 R_2 \mu \frac{T_2^{(2)}}{2T_2} + R_2^2 R_1 \mu \frac{T_1^{(2)}}{2T_1}, \tag{4}$$

*where*

$$\int_0^\infty (1 - T_k(x))dx = T_k$$

*is the average time of staying in the corresponding mode. $T_k^{(2)}$ is the second initial moment of the time of devices staying in mode k.*

*Prelimit characteristic function $H(u)$ the number of customers $i(t)$ can be written as*

$$H(u) = \frac{d}{d - ju},$$

*where*

$$d = \frac{S}{S + \Delta}. \tag{5}$$

*I.e in the form of the characteristic function of exponential distribution the parameter d.*

*Proof.* We perform the proof by three stages.

*Stage 1.* By tending $\varepsilon$ to zero $\varepsilon \to 0$ in (3) we will obtain:

$$\begin{cases} \frac{\partial F_1(w,z)}{\partial z} - \frac{\partial F_1(w,0)}{\partial z} + \frac{\partial F_2(w,0)}{\partial z} T_1(z) = 0, \\ \frac{\partial F_2(w,z)}{\partial z} - \frac{\partial F_2(w,0)}{\partial z} + \frac{\partial F_1(w,0)}{\partial z} T_2(z) = 0. \end{cases}$$

We will seek solution for this system in this form

$$F_k(w, z) \quad = \quad R_k(z)\Phi(w).$$

We will obtain following system

$$\begin{cases} R_1'(z) - R_1'(0) - R_2'(0)T_1(z) = 0, \\ R_2'(z) - R_2'(0) - R_1'(0)T_2(z) = 0. \end{cases}$$

solution for which we will write down in this form:

$$\begin{cases} R_1(z) = \int\limits_0^z (R_1'(0) - R_2'(0)T_1(x))dx, \\ R_2(z) = \int\limits_0^z (R_2'(0) - R_1'(0)T_2(x))dx. \end{cases}$$

$\{R_1(z), R_2(z)\}$ is a two-dimensional distribution of devices mode and value of remaining time of devices staying in that mode.

By tending z to infinity $z \to \infty$, considering that $T(\infty) \quad = 1$, we will obtain:

$$\begin{cases} R_1(\infty) = \int\limits_0^\infty (R_1'(0) - R_2'(0)T_1(x))dx, \\ R_2(\infty) = \int\limits_0^\infty (R_2'(0) - R_1'(0)T_2(x))dx. \end{cases}$$

For improper integral to be convergent it is necessary that the following condition is met

$$R_1'(0) - R_2'(0)T_1(\infty) = 0,$$

then we will obtain

$$R_1'(0) = R_2'(0) = R'(0).$$

We have

$$R_k(z) = R'(0) \int\limits_0^z (1 - T_k(x))dx.$$

Let's denote

$$\int\limits_0^\infty (1 - T_k(x))dx = T_k$$

as the average time of staying in the corresponding mode.

Then, from the one side

$$R_1(\infty) + R_2(\infty) = 1,$$

and from the other

$$R_1(\infty) + R_2(\infty) = R'(0) \int\limits_0^\infty (1 - T_1(x))dx + \int\limits_0^\infty (1 - T_2(x)) = dx R'(0) (T_1 + T_2).$$

We will obtain

$$R'(0) = \frac{1}{(T_1 + T_2)}.$$

Then

$$\begin{cases} R_1(\infty) = \frac{1}{T_1+T_2} \int\limits_0^\infty (1 - T_1(x))dx = \frac{T_1}{T_1+T_2}, \\ R_2(\infty) = \int\limits_0^\infty (R'_2(0) - R'_1(0)T_2(x))dx = \frac{T_2}{T_1+T_2}. \end{cases}$$

*Stage 2.* We will substitute the following expansion in the system (3).

$$F_k(w, z, \varepsilon) = \Phi(w) \{R_k(z) + j\varepsilon w f_k(z)\} + O(\varepsilon^2). \tag{6}$$

We will expand the exponent in a row to obtain this system

$$\frac{\partial R_1(z)}{\partial z} + j\varepsilon w \frac{\partial f_1(z)}{\partial z} - \frac{\partial R_1(0)}{\partial z} - j\varepsilon w \frac{\partial f_1(0)}{\partial z} + (S - 1)\, \mu j\varepsilon w \{R_1(z) + j\varepsilon w f_1(z)\}$$

$$+ \frac{\partial R_2(0)}{\partial z} T_1(z) + j\varepsilon w \frac{\partial f_2(0)}{\partial z} T_1(z) = O\left(\varepsilon^2\right),$$

$$\frac{\partial R_2(z)}{\partial z} + j\varepsilon w \frac{\partial f_2(z)}{\partial z} - \frac{\partial R_2(0)}{\partial z} - j\varepsilon w \frac{\partial f_2(0)}{\partial z} + S\mu j\varepsilon w \{R_2(z) + j\varepsilon w f_2(z)\}$$

$$+ \frac{\partial R_1(0)}{\partial z} T_2(z) + j\varepsilon w \frac{\partial f_1(0)}{\partial z} T_2(z) = O\left(\varepsilon^2\right).$$

Then

$$\frac{\partial R_1(z)}{\partial z} + j\varepsilon w \frac{\partial f_1(z)}{\partial z} - \frac{\partial R_1(0)}{\partial z} - j\varepsilon w \frac{\partial f_1(0)}{\partial z} + j\varepsilon w (S - 1)\, \mu R_1(z)$$

$$+ \frac{\partial R_2(0)}{\partial z} T_1(z) + j\varepsilon w \frac{\partial f_2(0)}{\partial z} T_1(z) = O\left(\varepsilon^2\right),$$

$$\frac{\partial R_2(z)}{\partial z} + j\varepsilon w \frac{\partial f_2(z)}{\partial z} - \frac{\partial R_2(0)}{\partial z} - j\varepsilon w \frac{\partial f_2(0)}{\partial z} + j\varepsilon w S\mu R_2(z) + \frac{\partial R_1(0)}{\partial z} T_2(z)$$

$$+ j\varepsilon w \frac{\partial f_1(0)}{\partial z} T_2(z) = O\left(\varepsilon^2\right).$$

The last equations can be rewritten in the form:

$$R_1'(z) - R'(0)(1 - T_1(z)) + j\varepsilon w \left(f_1'(z) - f_1'(0) + f_2'(0)T_1(z)\right)$$
$$+ jw\varepsilon (S - 1)\mu R_1(z) + O(\varepsilon^2) = 0,$$
$$R_2'(z) - R'(0)(1 - T_2(z)) + j\varepsilon w \left(f_2'(z) - f_2'(0) + f_1'(0)T_2(z)\right)$$
$$+ j\varepsilon w S\mu R_2(z) + O(\varepsilon^2) = 0.$$

After simple transformations we will obtain

$$\begin{cases} j\varepsilon w \left(f_1'(z) - f_1'(0) + f_2'(0)T_1(z)\right) + jw\varepsilon (S - 1)\mu R_1(z) + O(\varepsilon^2) = 0, \\ j\varepsilon w \left(f_2'(z) - f_2'(0) + f_1'(0)T_2(z)\right) + j\varepsilon w S\mu R_2(z) + O(\varepsilon^2) = 0. \end{cases}$$

By dividing both equations by $\varepsilon$ and be tending e to zero $\varepsilon \to 0$, we will obtain:

$$\begin{cases} f_1'(z) - f_1'(0) + f_2'(0)T_1(z) + (S - 1)\mu R_1(z) = 0, \\ f_2'(z) - f_2'(0) + f_1'(0)T_2(z) + S\mu R_2(z) = 0. \end{cases}$$

Let's write down

$$\begin{cases} f_1(z) = \int_0^z \{f_1'(0) - f_2'(0)T_1(x) + R_1(x)\mu(1 - S)\} \, dx, \\ f_2(z) = \int_0^z \{f_2'(0) - f_1'(0)T_2(x) - S\mu R_2(x)\} dx. \end{cases}$$

Let's find $f_1(\infty)$, $f_2(\infty)$

$$\begin{cases} f_1(\infty) = \int_0^z \{f_1'(0) - f_2'(0)T_1(x) + R_1(x)\mu(1 - S)\} \, dx, \\ f_2(\infty) = \int_0^z \{f_2'(0) - f_1'(0)T_2(x) - S\mu R_2(x)\} dx. \end{cases}$$

It is necessary that the following is true

$$\begin{cases} f_1'(0) - f_2'(0)T_1(\infty) + R_1(\infty)\mu(1 - S) = 0, \\ f_2'(0) - f_1'(0)T_2(\infty) - S\mu R_2(\infty) = 0. \end{cases}$$

Then

$$R_1(\infty)\mu(1 - S) - S\mu R_2(\infty) = 0.$$

Let's denote

$$R_1 = R_1(\infty), \quad R_2 = R_2(\infty),$$

and we will obtain:

$$R_1 = S, \quad R_2 = 1 - S.$$

From the other side

$$f_1'(0) - f_2'(0) = S\mu(S-1) = \mu R_1(R_1 - 1) \tag{7}$$

$$\begin{cases} f_1(\infty) = \int\limits_0^\infty \{f_2'(0) - f_2'(0)T_1(x) + \mu R_1(R_1 - 1) + R_1(x)\mu(1 - R_1)\}\, dx, \\ f_2(\infty) = \int\limits_0^\infty \{f_1'(0) - f_1'(0)T_2(x) - \mu R_1(R_1 - 1) - \mu R_1 R_2(x)\}dx. \end{cases}$$

Let's write down:

$$\begin{cases} f_1(\infty) = f_2'(0) \int\limits_0^\infty (1 - T_1(x))\, dx - \mu R_2 \int\limits_0^\infty (R_1 - R_1(x))\, dx, \\ f_2(\infty) = f_1'(0) \int\limits_0^\infty (1 - T_2(x))\, dx + \mu R_1 \int\limits_0^\infty (R_2 - R_2(x))dx. \end{cases}$$

Let's denote

$$\Delta_k = \int\limits_0^\infty (R_k - R_k(x))\, dx.$$

$$\begin{cases} f_1(\infty) = f_2'(0)T_1 - \mu R_2 \Delta_1, \\ f_2(\infty) = f_1'(0)T_2 + \mu R_1 \Delta_2. \end{cases} \tag{8}$$

*Stage 3.* In the Eq. (3) let's tend z to infinity $z \to \infty$.

$$\begin{cases} -\frac{\partial F_1(w,0,\varepsilon)}{\partial z} + ((1-\varepsilon)\, S\mu\left(e^{j\varepsilon w} - 1\right) + \mu\left(e^{-j\varepsilon w} - 1\right))F_1(w,\infty,\varepsilon) \\ + \left(1 - e^{-j\varepsilon w}\right)\mu\varepsilon\pi_1(\infty,\varepsilon) + \frac{\partial F_2(w,0,\varepsilon)}{\partial z}T_1(\infty) = 0, \\ -\frac{\partial F_2(w,0,\varepsilon)}{\partial z} \\ + (1-\varepsilon)\, S\mu F_2(w,\infty,\varepsilon)\left(e^{j\varepsilon w} - 1\right) + \frac{\partial F_1(w,0,\varepsilon)}{\partial z}T_2(\infty) = 0. \end{cases}$$

Let's sum the equations of the last system

$$\begin{aligned} ((1-\varepsilon)\, S\mu\left(e^{j\varepsilon w} - 1\right) + \mu\left(e^{-j\varepsilon w} - 1\right))F_1(w,\infty,\varepsilon) \\ + (1-\varepsilon)\, S\mu\left(e^{j\varepsilon w} - 1\right)F_2(w,\infty,\varepsilon) + \left(1 - e^{-j\varepsilon w}\right)\mu\varepsilon\pi_1(\infty,\varepsilon) = 0. \end{aligned}$$

We will substitute expansions (6) in the equality which we got from above

$$\left((1-\varepsilon)\, S\mu\left(e^{j\varepsilon w} - 1\right) + \mu\left(e^{-j\varepsilon w} - 1\right)\right)\Phi(w)\{R_1(\infty) + j\varepsilon w f_1(\infty)\}$$

$$+ (1-\varepsilon)\, S\mu\left(e^{j\varepsilon w} - 1\right)\Phi(w)\{R_2(\infty) + j\varepsilon w f_2(\infty)\}$$

$$+ \left(1 - e^{-j\varepsilon w}\right)\mu\varepsilon\pi_1(\infty,\varepsilon) = 0.$$

$$\left((1-\varepsilon)\, S\mu\left(j\varepsilon w + \frac{(j\varepsilon w)^2}{2}\right) + \mu\left(-j\varepsilon w + \frac{(j\varepsilon w)^2}{2}\right)\right)\Phi(w)\{R_1 + j\varepsilon w f_1(\infty)\}$$

$$+ (1-\varepsilon)\, S\mu\left(j\varepsilon w + \frac{(j\varepsilon w)^2}{2}\right)\Phi(w)\{R_2 + j\varepsilon w f_2(\infty)\}$$

$$+ j\varepsilon^2 w\mu\pi_1(\infty,\varepsilon) = o\left(\varepsilon^2\right).$$

$$\left[\frac{(j\varepsilon w)^2}{2}S\mu - j\varepsilon^2 wS\mu + \frac{(j\varepsilon w)^2}{2}S\mu + (j\varepsilon w)^2(S-1)\mu f_1(\infty) + (j\varepsilon w)^2 S\mu f_2(\infty)\right]\Phi(w)$$
$$+ j\varepsilon^2 w\mu\pi_1(\infty,\varepsilon) = o\left(\varepsilon^2\right).$$
$$\left[(j\varepsilon w)^2 S\mu - j\varepsilon^2 wS\mu + (j\varepsilon w)^2(S-1)\mu f_1(\infty) + (j\varepsilon w)^2 S\mu f_2(\infty)\right]\Phi(w)$$
$$+ j\varepsilon^2 w\mu\pi_1(\infty,\varepsilon) = o\left(\varepsilon^2\right).$$

By dividing both parts of the equations by $\varepsilon^2$ and by tending $\varepsilon^2$ to zero $\varepsilon \to 0$ we will obtain:

$$\left[jwS\mu - S\mu + jw(S-1)\mu f_1(\infty) + jwS\mu f_2(\infty)\right]\Phi(w)$$

$$+ \mu\pi_1(\infty) = 0.$$

Let's assume that $w = 0$ and considering that

$$\Phi(0) = 1,$$

we will obtain:

$$\pi_1 = S, \qquad \pi_1 = \lim_{\varepsilon \to 0}\pi_1(\varepsilon^2).$$

Then

$$\Phi(w) = \frac{S}{S - jw\{S + (S-1)f_1(\infty) + Sf_2(\infty)\}}.$$

Let's review the expression in the denominator separately

$$(S-1)f_1(\infty) + Sf_2(\infty).$$

Considering (7)–(8) we will obtain and denote the following

$$(S-1)f_1(\infty) + Sf_2(\infty) = R_1 f_2(\infty) - R_2 f_1(\infty)$$
$$= R_1 f_1'(0)T_2 + R_1^2\mu\Delta_2 - R_2 f_2'(0)T_1 + R_2^2\mu\Delta_1$$
$$= \frac{T_1 T_2}{T_1 + T_2}(f_1'(0) - f_2'(0)) + R_1^2 R_2\mu\frac{T_2^{(2)}}{2T_2} + R_2^2 R_1\mu\frac{T_1^{(2)}}{2T_1}$$
$$= -R_1 R_2\mu\frac{T_1 T_2}{T_1 + T_2} + R_1^2 R_2\mu\frac{T_2^{(2)}}{2T_2} + R_2^2 R_1\mu\frac{T_1^{(2)}}{2T_1} = \Delta.$$

Here

$$\Delta_k = \int\limits_0^\infty (R_k - R_k(x))\,dx$$

$$= R_k \int\limits_0^\infty\left\{1 - \frac{1}{T_k}\int\limits_0^z (1 - T_k(x))\,dx\right\}dz$$

$$= R_k\left\{1 - \frac{1}{T_k}\int\limits_0^z (1 - T_k(x))\,dx\right\}z\Bigg|_{z=0}^\infty - R_k\int\limits_0^\infty zd\left\{1 - \frac{1}{T_k}\int\limits_0^z (1 - T_k(x))\,dx\right\}$$

$$= R_k \int_0^\infty z \frac{1}{T_k} \left(1 - T_k(z)\right) dz$$

$$= R_k \int_0^\infty \frac{1}{T_k} \left(1 - T_k(z)\right) d\frac{z^2}{2} = R_k \frac{1}{T_k} \left(1 - T_k(z)\right) \left. \frac{z^2}{2} \right|_{z=0}^\infty$$

$$- R_k \int_0^\infty \frac{1}{T_k} \frac{z^2}{2} d\left(1 - T_k(z)\right) = R_k \frac{1}{T_k} \int_0^\infty \frac{z^2}{2} d\left(T_k(z)\right) = R_k \frac{T_k^{(2)}}{2T_k}.$$

where

$$\int_0^\infty \left\{ 1 - \frac{1}{T_k} \int_0^z \left(1 - T_k(x)\right) dx \right\} dz$$

is the average value of the remaining time of staying in mode $k$. $T_k^{(2)}$ is the second initial moment of the time of devices staying in mode $k$. $R_k(z)$ is written down in this form:

$$R_k(z) = \frac{T_k}{T_1 + T_2} \left\{ \frac{1}{T_k} \int_0^z \left(1 - T_k(x)\right) dx \right\}.$$

Then

$$\Phi(w) = \frac{S}{S - jw\left\{S + \Delta\right\}},$$

where $\Delta$ is determined by equation (4). I.e. the function $\Phi(w)$ is a characteristic function of the exponentially distributed random variable. Let's denote

$$d = \frac{S}{S + \Delta}.$$

The theorem was proved.

By making backward substitutions, we will tend $z$ to infinity and $\varepsilon$ to zero, $z \to \infty$, $\varepsilon \to 0$, and well get the characteristic function of a number of customers in the system

$$H(u) = \sum_k F(w, \varepsilon) = \sum_k \Phi(w) R_k(\infty) + o(\varepsilon) = \frac{d}{d - ju}.$$

## 5   Conclusions

In this work we have researched mathematical model of the system with server vacations. By using method of asymptotic analysis under large load, we have found asymptotic distribution of probabilities of values of a number of customers in the system. It is shown that this distribution is exponential with the parameter determined by the equality (5).

**Acknowledgments.** The work is performed under the state order of the Ministry of Education and Science of the Russian Federation (No. 1.511.2014/K).

# References

1. Pechinkin, A.V., Sokolov, I.A.: Queueing system with an unreliable device in discrete time. J. Inform. Appl. **5**(4), 6–17 (2005). (in Russian)
2. Saksonov, E.A.: Method of the calculation of probabilities of modes for one-line queueing systems with the server vacation. J. Automat. Tele-Mech. **1**, 101–106 (1995). (in Russian)
3. Nazarov, A.A., Terpugov, A.F.: Queueing theory: educational material. NTL, Tomsk (2004). (in Russian)
4. Nazarov A.A., Paul S.V.: Research of queueing system with the server vacation that is controlled by T-strategy. In: Proceedings of the International Science Conference Theory of Probabilies, Random Processes, Mathematical Statistics and Applications, pp. 202–207. Minsk (2015). ( in Russian)
5. Nazarov, A.A., Moiseeva, S.P.: Method of Asymptotic Analysis in Queueing Theory. NTL, Tomsk (2006). (in Russian)
6. Moiseeva, E.A., Nazarov, A.A.: Research of RQ-system MMP—GI—1 by using method of asymptotic analysis under large load. TSUs herald/messenger. Adm. Calculating Tech. Inform. **4**(25), 83–94 (2013)
7. Nazarov, A.A., Moiseev, A.N.: Analysis of an open non-Markovian GI(GI-)K queueing network with high-rate renewal arrival process. Prob. Inform. Transm. **49**(2), pp. 167–178. doi:10.1134/S0032946013020063

# Stationary Blocking Probability in Multi-server Finite Queuing System with Ordered Entry and Poisson Arrivals

Rostislav Razumchik[1,2]([⊠]) and Ivan Zaryadov[2]

[1] Institute of Informatics Problems of the Federal Research Center
"Computer Science and Control" of the Russian Academy of Sciences,
Vavilova street, 44-2, 119333 Moscow, Russia
rrazumchik@ipiran.ru
[2] Peoples' Friendship University of Russia,
Ordzhonikidze street, 3, 117198 Moscow, Russia
izariadov@gmail.com

**Abstract.** Analysis of performance characteristics of industry conveyor systems is a well-known research topic, which can be treated from queueing systems perspective. This paper is devoted to the analysis of the conveyor model, which is represented by markovian heterogeneous $N$-server queueing system with ordered entry ($N \geq 2$). Servers are labelled (from 1 to $N$), each has a dedicated finite-capacity queue, and serve customers according to exponential distribution. Customer arrive at the system according to a Poisson flow and each customer, which finds more than one server idle, selects the queue with the lowest (server's) number. After service completion customer leaves the system. If all the queues are full upon customer's arrival, it is lost (no recirculation). It is shown that the stationary blocking probability can be found firstly by analyzing corresponding 2-server system, then finding inter-overflow time distributions and applying well-known results for finite-capacity $GI/M/1$ queue. Exact (not matrix-geometric) method for recursive computation of joint stationary probability in the 2-server system is given.

**Keywords:** Queueing system · Ordered entry · Finite capacity · Blocking probability

## 1 Introduction

Queuing systems with ordered entry service discipline is a well-known type of multi-server queueing systems in which customers are dispatched to servers according to a selection rule known as the "ordered entry rule". This rule implies, that if, for example, a queueing system has $N$ servers, numbered from 1 to $N$, no queues (i.e. no waiting room) then any arriving customer who finds more than

This work was supported by the Russian Foundation for Basic Research (grants 15-07-03007, 15-07-02354, 15-07-02341).

V. Vishnevsky and D. Kozyrev (Eds.): DCCN 2015, CCIS 601, pp. 344–357, 2016.
DOI: 10.1007/978-3-319-30843-2_36

one server idle, selects the one with the lowest number. An arriving customer which finds all servers busy is lost. Since the seminal papers [1, 2], such systems have received significant attention from the research community. They are considered to be useful in the manufacturing processes for performance modelling of different variations of conveyor systems (open and closed-loop) with multiple unloading stations. Some of the early theoretical results and ideas of practical implementations one can find, for example, in [1, 3–6].

Up to nowadays queuing systems with ordered entry have already been analysed in different setting (see [7–21]). Some of them prove to be relatively easy to analyse and others are notoriously hard and thus approximations are used. The main assumption in which most of queuing systems with ordered entry considered in the literature differ from one another is the assumption about the presence of waiting rooms (queues) for the arriving customers. The analysis is usually carried out under one of the three main assumptions: (i) there is no waiting room for customers, (ii) there is a single queue where customers can wait for service, (iii) each server has its own dedicated queue and any arriving customer which finds more than one queue with free waiting room selects the one with the lowest (server's) number. With respect to the blocked customers (i.e. those which find all servers busy and no free waiting room on arrival) in the majority of cases, considered in the literature, it is assumed that the blocked customers are lost and never return to the system. But it is also possible, as mentioned in [20, 21], for the blocked customers to recirculate i.e. each blocked customer after some time arrives again at the system and follows the "ordered entry rule". Consecutive moments of such arrivals constitute the new, additional flow of "recirculating" customers.

Note that assumptions (i)–(iii) follow directly from those problem settings, which one usually comes across in conveyor theory. So far, most of the analytical results for the queueing systems with ordered entry have been obtained for the steady state under assumptions (i) and (ii). For the respective latest, though short, review of these results, one can refer to [19]. Usually the characteristics under study are the stationary blocking probability, the utilization of each server and the joint stationary distribution of the number of customers in the system.

This paper is devoted to the steady-state analysis of the finite capacity Markovian queueing system with ordered entry under assumption (iii). Such systems have not received a great deal of attention (see [9, 13, 15–17, 20]), though are reported to adequately model several types of real-life manufacturing systems (see [3, 4, 9, 10]). From the theoretical point of view, there is no general method for their stationary analysis (except for the celebrated QBD approach) and the presence of finite capacity queues in front of each server makes the analytical methods used in the analysis of systems under assumptions (i) and (ii) hardly applicable.

In [20] authors consider the system with queues at each server (with different capacities) and possible recirculation of blocked customers. Though authors present only the results of the thorough simulation analysis, their conclusions contain many insightful observations concerning the interrelation of queues capacities, recirculation times and (stationary) performance characteristics of the system. The paper [15] is devoted to the numerical transient and steady-state analysis of the Markovian ordered entry queueing system with Poisson flow of customers and arbitrary finite number of servers with finite capacity queue at each. Queues' capacities are different and service times at each server are exponentially distributed with different parameters. Blocked customers leave the system and never come back. The authors use a Runge-Kutta method to solve the difference-differential equations and draw several conclusions concerning optimal server arrangement and general effects that initial parameters have on system's performance. One of the main conclusions is that the servers should be arranged in descending order with respect to the service rates to decrease the blocking probability. In [16] the authors extend the numerical analysis to the case when customers arrive from a finite source and also develop the cost models (for both finite and infinite source) to determine the optimal allocation of queues' capacities. In [13] one considers Markovian ordered entry queueing system with three servers and finite capacity queue at each server. Total system's capacity is fixed and is shared among the three queues, i.e. the capacity of all three queues cannot exceed total system's capacity. Assuming the system is in the steady-state, authors numerically (using exhaustive search) find the queue capacities assignments, which minimize the blocking probability and compare the performance of the ordered entry rule with other rules (for example, "join the shortest queue"). Non-markovian two-server ordered entry queueing system with deterministic arrivals has been considered in [9]. Here servers are heterogeneous and with exponentially distributed service times and queues' capacities at each server are different. Using QBD approach authors obtain joint stationary distribution of the number of customers (just before the arrival) in different queues. Paper [17] is also worth noticing in our context because there authors introduce a sort of generalization of the "ordered entry rule" which they call "semi-ordered entry rule". According to it the arriving customer at first checks the length of the queue at each server and chooses the one with the minimum length and, in case of ties, – the queue with the lowest (server's) number.

To authors knowledge these are the only references on ordered entry queueing system under assumption (iii) available in the literature. The purpose of this paper is two-fold. Firstly it is shown that for the multi-server heterogeneous Markovian ordered entry queueing system under assumption (iii) one can obtain stationary blocking (loss) probability, in algorithmic manner. It means that the $N$-server ordered entry queueing system is analyzed sequentially: at the first step one finds the joint stationary distribution in the 2-server system; at the second step one finds blocking probability for the 3-server system; at the third step – in the 4-server system and so on.

Secondly this paper presents the case study, which shows that the completely algebraic method based on generating and special functions, previously used in a number of papers (see, for example, [27–29]), allows one to find the recursive solution for the joint stationary distribution in the two-dimensional Markov chain with finite state-space and certain type of transitions. We note that recursive solution for the considered Markov chain was considered to be intractable (see [26, Chap. 3]).

The paper is structured as follows. In the next section we give the detailed description of the system and after that, in Sect. 3, the stationary blocking probability is being obtained at first for the 2-server system and then it is shown how the procedure can be generalized for the $N$-server system ($N \geq 3$). In the conclusion we briefly discuss the pros and cons of the proposed method and give directions of further research.

## 2 System Description

Consider the $N$-server ordered entry queuing system ($N \geq 2$), where servers are labelled by numbers $1, 2, \ldots, N$ without repetitions. The $i^{th}$ server has a buffer of finite capacity ($R_i - 1$) in front of it. Customers arrive according to a Poisson process of rate $\lambda$. The ordered entry discipline implies the following admission rule for incoming customers. A new customer upon arrival goes to the $1^{st}$ server or, if it is busy, occupies one place in the queue in front of it. If a new customer upon arrival sees the queue in front of the $1^{st}$ server full, it goes to the $2^{nd}$ server or, if it is busy, enters its queue. Next, if upon arrival of a new customer queues in front of the $1^{st}$ and $2^{nd}$ servers are full, it goes to the $3^d$ server or, if it is busy, to its queue and so on. Whenever newly arriving customer, sees queues in front of each server full, it leaves the system and never comes back (i.e. recirculation is not permitted).

In order to complete the description we specify that customers from each queue are served according to FIFO (or LIFO, Random) service discipline and $i^{th}$ server, $1 \leq i \leq N$, serves customers for exponentially distributed times at rate $\mu_i$.

Denote by $Y_i(t)$ the number of customers in $i^{th}$ server and its queue at the time $t$. The random process $\{Y(t) = (Y_1(t), Y_2(t), \ldots, Y_N(t)), \ t \geq 0\}$ describing the stochastic behaviour of the considered system is Markovian with continuous time and discrete finite state space. The set of states of the process $\{Y(t), \ t \geq 0\}$ is $\mathcal{Y} = \{(i_1, i_2, \ldots, i_N), \ 0 \leq i_j \leq R_j, \ 1 \leq j \leq N\}$. The state $(0, 0, \ldots, 0)$ corresponds to the empty system. The state $(i_1, i_2, \ldots, i_N)$ of the process means that at the instant $t$ there are $i_1$ customers in the $1^{st}$ server and its queue, $i_2$ customers in the $2^{nd}$ server and its queue etc. The process is $\{Y(t), \ t \geq 0\}$ is clearly ergodic and its stationary distribution exists. Denote by $\{p_{i_1, i_2, \ldots, i_N}, \ 0 \leq i_n \leq R_n, 1 \leq n \leq N\}$, the stationary probability of the fact that there are $i_1$ in the $1^{st}$ server and its queue, $i_2$ customers in the $2^{nd}$ server and its queue etc. i.e.

$$p_{i_1, i_2, \ldots, i_N} = \lim_{t \to \infty} \mathbf{P}\{Y_1(t) = i_1, Y_2(t) = i_2, \ldots, Y_N(t) = i_n\}.$$

In order not to introduce additional notation henceforth we use the agreement that the joint stationary distribution of the number of customers in the first $k$ servers (and their queues) is denoted by $p_{i_1, i_2, \ldots, i_k}, \ 2 \leq k \leq N$.

# 3  Computation of the Blocking Probability $p_{R_1,R_2,...,R_N}$

The computation of $p_{R_1,R_2,...,R_N}$ consists of $N-1$ steps. At the first step one obtains the joint stationary distribution $p_{i,j}$ of the fact that there are $i$ customers in $1^{st}$ server and its queue and $j$ customers in the $2^{nd}$ server and its queue. At the second step one computes Laplace-Stieltjes transform (LST) of the inter-overflow time distribution (i.e. distribution of time between successive arrivals of new customers which see the $1^{st}$ and $2^{nd}$ servers busy and their queues full) and then determines $p_{R_1,R_2,R_3}$. At the fourth step one again computes Laplace-Stieltjes transform of the inter-overflow time distribution but this time it is the distribution of time between successive arrivals of new customers which see the $1^{st}$, $2^{nd}$ and $3^d$ servers busy and their queues full, and then one computes $p_{R_1,R_2,R_3,R_4}$, etc. The last step is the computation of $p_{R_1,R_2,...,R_N}$.

## 3.1  Computation of the Joint Stationary Distribution in the First Two Queues

Due to the lack of space we will present the results in concise form and omit detailed derivations. Some of the details can be recovered from [22]. It can be verified by direct calculation from the global balance equations that probabilities $p_{i,j}$ in terms of the generating function

$$P(u,v) = \sum_{i=0}^{R_1}\sum_{j=0}^{R_2} p_{ij}u^i v^j, \ 0 \le u,v \le 1,$$

have the form

$$-[\lambda u^2 v-[(\lambda + \mu_1 + \mu_2)v - \mu_2]u + \mu_1 v]P(u,v)$$

$$= \lambda u^{R_1+1}v^{R_2+1}(1-v)p_{R_1,R_2} + \mu_1 v(u-1)\sum_{j=0}^{R_2} p_{0,j}v^j$$

$$+ \mu_2 u(v-1)\sum_{i=0}^{R_1} p_{i,0}u^i + \lambda u^{R_1+1}v(v-u)\sum_{j=0}^{R_2} p_{R_1,j}v^j. \quad (1)$$

The expression in the square brackets on the left hand-side of (1) is a polynomial in $u$ with roots

$$u_{1,2}(v) = u_{1,2} = \frac{(\lambda + \mu_1 + \mu_2)v - \mu_2 \mp \sqrt{((\lambda + \mu_1 + \mu_2)v - \mu_2)^2 - 4\lambda\mu_1 v^2}}{2\lambda v}.$$

Taking into consideration that we have defined the generating function $P(u,v)$ for the values $u$ and $v$ both in $[0,1]$, it can be seen that roots $u_1$ and $u_2$ are real whenever

$$v \in V = \left(0, \frac{\mu_2}{(\sqrt{\lambda} + \sqrt{\mu_1})^2 + \mu_2}\right) \cup \left(\frac{\mu_2}{(\sqrt{\lambda} - \sqrt{\mu_1})^2 + \mu_2}, 1\right]. \quad (2)$$

For each $v \in \mathcal{V}$ the generating function $P(u,v)$ is the continuous function of $u$ on the whole set of real numbers. As the left part of (1) vanishes at points $(u_1(v), v)$ and $(u_2(v), v)$, then the right part of (1) must vanish at these points too. This leads to the following two equations:

$$\lambda u_1^{R_1+1} v^{R_2+1}(1-v)p_{R_1,R_2} + \mu_1 v(u_1 - 1)\sum_{j=0}^{R_2} p_{0,j}v^j$$

$$+ \mu_2 u_1(v-1)\sum_{i=0}^{R_1} p_{i,0}u_1^i + \lambda u_1^{R_1+1}v(v-u_1)\sum_{j=0}^{R_2} p_{R_1,j}v^j = 0, \qquad (3)$$

$$\lambda u_2^{R_1+1} v^{R_2+1}(1-v)p_{R_1,R_2} + \mu_1 v(u_2 - 1)\sum_{j=0}^{R_2} p_{0,j}v^j$$

$$+ \mu_2 u_2(v-1)\sum_{i=0}^{R_1} p_{i,0}u_2^i + \lambda u_2^{R_1+1}v(v-u_2)\sum_{j=0}^{R_2} p_{R_1,j}v^j = 0. \qquad (4)$$

Now we cancel out from these equations the term with $p_{R_1,R_2}$ and then the term with $\sum_{j=0}^{R_2} p_{R_1,j}v^j$. By expressing $p_{R_1,R_2}$ from (3) and substituting it into (4) we obtain the following relation (after collecting the common terms):

$$\left(\frac{u_2^{R_1+1} - u_1^{R_1+1}}{u_2 - u_1} - \frac{\mu_1}{\lambda} \cdot \frac{u_2^{R_1} - u_1^{R_1}}{u_2 - u_1}\right)\sum_{j=0}^{R_2} p_{0,j}v^{j+1} - \frac{\mu_2(v-1)}{\lambda}$$

$$\times \sum_{i=0}^{R_1} p_{i,0}\left(\frac{\mu_1}{\lambda}\right)^i \left(\frac{u_2^{R_1-i} - u_1^{R_1-i}}{u_2 - u_1}\right) - \left(\frac{\mu_1}{\lambda}\right)^{R_1}\sum_{j=0}^{R_2} p_{R_1,j}v^{j+1} = 0. \qquad (5)$$

By analogy, expressing $\sum_{j=0}^{R_2} p_{R_1,j}v^j$ from (3) and substituting it into (4), yields:

$$(1-v)\lambda^2 p_{R_1,R_2} v^{R_2+1}\left(\frac{\mu_1}{\lambda}\right)^{R_1+1} + \lambda\mu_1\left(v\left[\frac{u_2^{R_1+1} - u_1^{R_1+1}}{u_2 - u_1} + \frac{\mu_1}{\lambda}\cdot\frac{u_1^{R_1} - u_2^{R_1}}{u_2 - u_1}\right]\right.$$

$$+ \frac{\mu_1}{\lambda}\cdot\frac{u_2^{R_1+1} - u_1^{R_1+1}}{u_2 - u_1} + \frac{u_1^{R_1+2} - u_2^{R_1+2}}{u_2 - u_1}\right)\sum_{j=0}^{R_2} p_{0,j}v^{j+1}$$

$$+ \mu_1\mu_2(v-1)v\sum_{i=0}^{R_1} p_{i,0}\left(\frac{\mu_1}{\lambda}\right)^i\left(\frac{u_1^{R_1-i} - u_2^{R_1-i}}{u_2 - u_1}\right)$$

$$+ \mu_1\mu_2(v-1)\sum_{i=0}^{R_1} p_{i,0}\left(\frac{\mu_1}{\lambda}\right)^i\left(\frac{u_2^{R_1+1-i} - u_1^{R_1+1-i}}{u_2 - u_1}\right) = 0. \qquad (6)$$

It is well-known that such fractions like $(u_2^i - u_1^i)/(u_2 - u_1)$, $i \geq 1$, are polynomial functions of a single variable (of integer degree). Following the derivation lines from [23] one can show that the terms $(u_2^i - u_1^i)/(u_2 - u_1)$, $i \geq 1$, which are present in both Eqs. (5) and (6), can be written in the form:

$$\frac{u_2^i - u_1^i}{u_2 - u_1} = \left(\sqrt{\frac{\mu_1}{\lambda}}\right)^{i-1} \sum_{j=0}^{i-1} a_{i,j} v^{-j}, i \geq 1, \tag{7}$$

where

$$a_{i,j} = (-1)^j \left(\frac{\mu_2}{\lambda + \mu_1 + \mu_2}\right)^j \left[\sum_{m=j}^{i-1} C_{i-m-1}^{m+1}(0) \left(\frac{\lambda + \mu_1 + \mu_2}{\sqrt{\lambda \mu_1}}\right)^m \binom{m}{j}\right],$$

and $\binom{m}{j}$ is the binomial coefficient, and $C_n^m(x)$ are the Gegenbauer polynomials. The values of $C_n^m(0)$ are equal to (see, for example, [24, p. 175]):

$$C_n^m(0) = \begin{cases} 0, & n \neq 2k, \\ \frac{(-1)^{\frac{n}{2}} \Gamma(m+\frac{n}{2})}{(\frac{n}{2})! \Gamma(m)}, & n = 2k. \end{cases} \tag{8}$$

Here $\Gamma(\cdot)$ denotes gamma function. Now we can simplify (5) and (6) by substituting in each term, where necessary, representation (7) instead of $(u_2^i - u_1^i)/(u_2 - u_1)$. Eventually by doing this one can show that (5) and (6) are indeed polynomial functions of single variable $v$ of integer degree. This is purely a technical issue and thus we omit the intermediate calculations and just state the final result. Note that after the above-mentioned substitution one has to perform the multiplication of the sums with different upper bounds of summation. If we assume, without the loss of generalization, that $R_1$ and $R_2$ are chosen such that $R_2 \geq R_1 + 2$, then Eqs. (5) and (6) can be rewritten in the form

$$\frac{1}{\sqrt{\rho_1}} \sum_{j=1}^{R_1} v^{-j+1} \sum_{k=0}^{R_1-j} p_{0,k} r_1(k+j) + \frac{1}{\sqrt{\rho_1}} \sum_{j=0}^{R_2} v^{j+1} \sum_{k=j}^{\min(R-1+j,R_2)} p_{0,k} r_1(k-j)$$

$$- \frac{1}{\rho_2}(v-1) \sum_{j=0}^{R_1-1} r_2(j)v^{-j} - \frac{1}{\sqrt{\rho_1}^{R_1+1}} \sum_{j=0}^{R_2} p_{R_1,j} v^{j+1} = 0, \tag{9}$$

$$\frac{p_{R_1,R_2}}{\sqrt{\rho_1}^{R_1}}(1-v)v^{R_2+1}$$

$$+ r_1(0) \sum_{j=0}^{R_2} p_{0,j} v^{j+2} + \sum_{j=0}^{R_2} v^{j+1} \sum_{k=j}^{\min(R_1+j+1,R_2)} p_{0,k} q_1(k-j)$$

$$+ \sum_{j=0}^{R_1} v^{-j} \sum_{k=0}^{R_1-j} p_{0,k} q_1(k+j+1) - \frac{1}{\rho_2} \sum_{j=0}^{R_1} q_2(j)v^{-j} - \frac{\sqrt{\rho_1}}{\rho_2} \sum_{j=0}^{R_1-1} r_2(j)v^{-(j-2)}$$

$$+ \frac{\sqrt{\rho_1}}{\rho_2} \sum_{j=0}^{R_1-1} r_2(j)v^{-(j-1)} + \frac{1}{\rho_2} \sum_{j=0}^{R_1} q_2(j)v^{-(j-1)} = 0, \tag{10}$$

where $\rho_i = \lambda/\mu_i$, $i = 1, 2$, and

$$r_1(R_1) = a_{R_1+1,R_1}, \quad r_1(j) = a_{R_1+1,j} - \frac{a_{R_1,j}}{\sqrt{\rho_1}}, \quad j = \overline{0, R_1 - 1},$$

$$q_1(j) = a_{R_1+1,j+1} - \frac{a_{R_1,j+1}}{\sqrt{\rho_1}} + \frac{a_{R_1+1,j}}{\sqrt{\rho_1}^2} - \frac{a_{R_1+2,j}}{\sqrt{\rho_1}}, \quad j = \overline{0, R_1 - 2},$$

$$q_1(R_1 - 1) = a_{R_1+1,R_1} + \frac{a_{R_1+1,R_1-1}}{\sqrt{\rho_1}^2} - \frac{a_{R_1+2,R_1-1}}{\sqrt{\rho_1}},$$

$$q_1(R_1) = \frac{a_{R_1+1,R_1}}{\sqrt{\rho_1}^2} - \frac{a_{R_1+2,R_1}}{\sqrt{\rho_1}}, \quad q_1(R_1 + 1) = -\frac{a_{R_1+2,R_1+1}}{\sqrt{\rho_1}},$$

$$r_2(j) = \sum_{i=0}^{R_1-1-j} p_{i,0} \frac{a_{R_1-i,j}}{\sqrt{\rho_1}^i}, \quad j = \overline{0, R_1 - 1}, \quad q_2(j) = \sum_{i=0}^{R_1-j} p_{i,0} \frac{a_{R_1+1-i,j}}{\sqrt{\rho_1}^i}, \quad j = \overline{0, R_1}.$$

The left parts of (9) and (10) are the single-variable polynomial functions in $v$ of integer degree. From the fact that they are equal to zero for all $v \in \mathcal{V}$ (what guaranties that roots $u_1$ and $u_2$ are real), all their coefficients are equal to zero. Hence one obtains two systems of algebraic equations with constant coefficients. The first system follows from (9) by equating to zero each coefficient at $v^i$. It holds that

$$\mu_2 r_2(R_1 - 1) + \sqrt{\mu_1 \lambda} p_{0,0} r_1(R_1) = 0, \tag{11}$$

$$\mu_2 r_2(j) + \sqrt{\mu_1 \lambda} \sum_{k=0}^{R_1-1-j} p_{0,k} r_1(k+1+j) - \mu_2 r_2(j+1) = 0, \quad j = \overline{0, R_1 - 2}, \tag{12}$$

$$\sqrt{\mu_1 \lambda} \sum_{k=0}^{R_1} p_{0,k} r_1(k) - \mu_2 r_2(0) - \frac{\sqrt{\mu_1 \lambda}}{\sqrt{\rho_1}^{R_1}} p_{R_1,0} = 0, \tag{13}$$

$$\sqrt{\mu_1 \lambda} \sum_{k=j}^{\min(R_1+j,R_2)} p_{0,k} r_1(k-j) - \frac{\sqrt{\mu_1 \lambda}}{\sqrt{\rho_1}^{R_1}} p_{R_1,j} = 0, \quad j = \overline{1, R_2}. \tag{14}$$

By analogy, the second system follows from (10) by equating to zero each coefficient at $v^i$. For $i = \overline{2, R_2 + 2}$ we have

$$r_1(0) p_{0,R_2} - \frac{1}{\sqrt{\rho_1}^{R_1}} p_{R_1,R_2} = 0, \tag{15}$$

$$\frac{1}{\sqrt{\rho_1}^{R_1}} p_{R_1,R_2} + r_1(0) p_{0,R_2-1} + q_1(0) p_{0,R_2} = 0, \tag{16}$$

$$r_1(0) p_{0,j-2} + \sum_{k=j-1}^{\min(R_1+j,R_2)} p_{0,k} q_1(k-j+1) = 0, \quad j = \overline{3, R_2}. \tag{17}$$

$$\sqrt{\lambda\mu_1}r_1(0)p_{0,0} + \sqrt{\lambda\mu_1} \sum_{k=1}^{\min(R_1+2,R_2)} p_{0,k}q_1(k-1) - \mu_2 r_2(0) = 0. \qquad (18)$$

Equations (15)–(18) constitute the system of linear equations in $R_1 + R_2 + 2$ unknowns. Indeed the unknown quantities are $p_{R_1,R_2}$, $p_{0,j}$, $j = \overline{0, R_2}$ and $r_2(0)$, which depends on $p_{i,0}$, $i = \overline{0, R_1 - 1}$. After summation of Eq. (13) with Eq. (14), one obtains

$$\mu_2 r_2(0) = \sqrt{\lambda\mu_1}\left( p_{0,0}r_1(0) + \sum_{k=1}^{R_1} p_{0,k}r_1(k) + \sum_{j=1}^{R_2} \sum_{k=j}^{\min(R_1+j,R_2)} p_{0,k}r_1(k-j) \right.$$
$$\left. - \frac{\sqrt{\rho_1}^{R_1}(1-\rho_1)}{1-\rho_1^{R_1+1}} \right), \qquad (19)$$

where we used the equality

$$\sum_{j=0}^{R_2} p_{R_1,j} = \frac{\rho_1^{R_1}(1-\rho_1)}{1-\rho_1^{R_1+1}},$$

which immediately follows from the local balance principle. Such representation for $r_2(0)$ makes the system (15)–(18) of $R_2 + 1$ equations have only $R_2 + 1$ unknowns and thus it can be solved. Substitution of (19) into (18) yields

$$\sum_{k=1}^{\min(R_1+2,R_2)} p_{0,k}q_1(k-1) - \sum_{k=1}^{R_1} p_{0,k}r_1(k) - \sum_{j=1}^{R_2} \sum_{k=j}^{\min(R_1+j,R_2)} p_{0,k}r_1(k-j)$$
$$+ \frac{\sqrt{\rho_1}^{R_1}(1-\rho_1)}{1-\rho_1^{R_1+1}} = 0.$$

Recalling that all probabilities are usually found up to a constant, then the substitution of $p_{0,j} = x_j p_{R_1,R_2}$ (where $x_j$ are yet unknown constants) into the previous equation yields the expression for the probability $p_{R_1,R_2}$ in the form

$$p_{R_1,R_2} = \frac{\frac{\sqrt{\rho_1}^{R_1}(1-\rho_1)}{1-\rho_1^{R_1+1}}}{\sum_{k=1}^{R_1} x_k r_1(k) - \sum_{k=1}^{\min(R_1+2,R_2)} x_k q_1(k-1) + \sum_{j=1}^{R_2} \sum_{k=j}^{\min(R_1+j,R_2)} x_k r_1(k-j)}.$$

The constants $x_j$ can be found straightforwardly from (15)–(17) by dividing each equation by $p_{R_1,R_2}$ and expressing each $x_j$ through $x_{j+1}, \ldots, x_{R_2}$. Once the probability $p_{R_1,R_2}$ is found one can calculate probabilities $p_{R_1,j}$, $j = \overline{0, R_2 - 1}$ recursively from Eqs. (13)–(14). Probability $p_{0,0}$ is found by summing up all equations of the system (11)–(14), which eventually gives

$$p_{0,0} = \frac{1}{\sum\limits_{k=0}^{R_1} r_1(k)} \left( \frac{\sqrt{\rho_1}^{R_1}(1-\rho_1)}{1-\rho_1^{R_1+1}} - \sum_{j=1}^{R_2} \sum_{k=j}^{\min(R_1+j,R_2)} p_{0,k} r_1(k-j) \right.$$

$$\left. - \sum_{j=0}^{R_1-2} \sum_{k=1}^{R_1-1-j} p_{0,k} r_1(k+1+j) + \sum_{k=1}^{R_1} p_{0,k} r_1(k) \right).$$

Once probabilities $p_{R_1,j}$, $p_{0,j}$, $j = \overline{0,R_2}$ and $p_{0,0}$ are known the whole joint stationary distribution $p_{i,j}$ can be calculated recursively from the system of global balance equations.

## 3.2  Distribution of the Inter-overflow Times

If customer arrives at the $N$-server ordered entry queueing system and finds queues in front of the $1^{st}$ and $2^{nd}$ servers full, it is said to have overflowed and it tries to enter $3^d$ server etc. Denote by $\tau_1, \tau_2, \ldots$ the successive moments at which customer overflows from $2^{nd}$ server and its queue. From the independence of arriving and service process, and memoryless property of the exponential service time distribution the time to the next overflow is independent of the past behaviour of the system and hence the sequence $\{\tau_1, \tau_2, \ldots\}$ forms a renewal process with distribution function which we denote by $G(t)$[1]. This renewal process governs the successive time instants at which new customers arrive at the $3^d$ server and its queue. We will find the distribution of time between overflows $G(t)$ in terms of LST $\tilde{G}(s) = \int_0^\infty e^{-sx} dG(t)$.

Let $\tilde{G}_{ij}(s)$, $0 \le i \le R_1$, $0 \le j \le R_2$, – be the time to the next overflow in terms of LST, given $i$ customers in the $1^{st}$ server and its queue and $j$ customers in the $2^{nd}$ server and its queue. Clearly $\tilde{G}_{R_1,R_2}(s) = \tilde{G}(s)$. By applying first step analysis it can be shown that $\tilde{G}_{ij}(s)$ satisfy the following system of linear algebraic equations:

$$\tilde{G}_{00}(s) = \frac{\lambda}{\lambda + s} \tilde{G}_{10}(s), \tag{20}$$

$$\tilde{G}_{i0}(s) = \frac{1}{\lambda + \mu_1 + s} \left( \mu_1 \tilde{G}_{i-1,0}(s) + \lambda \tilde{G}_{i+1,0}(s) \right), \quad 1 \le i \le R_1 - 1, \tag{21}$$

$$\tilde{G}_{R_1,0}(s) = \frac{1}{\lambda + \mu_1 + s} \left( \mu_1 \tilde{G}_{R_1-1,0}(s) + \lambda \tilde{G}_{R_1,1}(s) \right), \tag{22}$$

$$\tilde{G}_{0,j}(s) = \frac{1}{\lambda + \mu_2 + s} \left( \mu_2 \tilde{G}_{0,j-1}(s) + \lambda \tilde{G}_{1,j}(s) \right), \quad 1 \le j \le R_2, \tag{23}$$

---

[1] Following [25], if one denotes the state space of the ordered entry queueing system, consisting only of $1^{st}$ and $2^{nd}$ servers with their queues, by $S = \{(i,j), 0 \le i \le R_1, 0 \le j \le R_2\}$ and adds virtual (absorbing) state $k^*$, then $G(t)$ equals the distribution of the time until absorption of the process when the initial state is $(R_1, R_2)$. In other words, $G(t)$ can be interpreted as the first passage time distribution from state $(R_1, R_2)$ into state $k^*$.

$$\tilde{G}_{i,j}(s) = \frac{1}{\lambda + \mu_1 + \mu_2 + s} \left( \lambda \tilde{G}_{i+1,j}(s) + \mu_1 \tilde{G}_{i-1,j}(s) + \mu_2 \tilde{G}_{i,j-1}(s) \right),$$

$$1 \le i \le R_1 - 1, \ 1 \le j \le R_2, \quad (24)$$

$$\tilde{G}_{R_1,j}(s) = \frac{1}{\lambda + \mu_1 + \mu_2 + s} \left( \lambda \tilde{G}_{R_1,j+1}(s) + \mu_1 \tilde{G}_{R_1-1,j}(s) + \mu_2 \tilde{G}_{R_1,j-1}(s) \right),$$

$$1 \le j \le R_2 - 1, \quad (25)$$

$$\tilde{G}_{R_1,R_2}(s) = \frac{1}{\lambda + \mu_1 + \mu_2 + s} \left( \lambda + \mu_1 \tilde{G}_{R_1-1,R_2}(s) + \mu_2 \tilde{G}_{R_1,R_2-1}(s) \right). \quad (26)$$

The system (20)–(26) can be solved recursively: firstly express $\tilde{G}_{2,0}(s)$ through $\tilde{G}_{1,0}(s)$ using (20) and (21), then express $\tilde{G}_{i,0}(s)$ through $\tilde{G}_{i-1,0}(s)$ for each $3 \le i \le R_1$ from (21). After that find $\tilde{G}_{R_1,1}(s)$ (22) through $\tilde{G}_{R_1,0}(s)$, and then obtain $\tilde{G}_{1,1}(s)$ and $\tilde{G}_{0,1}(s)$ from (24) and (23) and so on. The solution can also be easily written out in the matrix form but due to the lack of space we will not dwell on this and in the next subsection we show that one needs to be able to solve the system (20)–(26) only up to such extent, that allows accurate estimation of derivatives of $\tilde{G}(s)$ at point $s = 0$.

## 3.3   Relation of $p_{R_1,R_2,R_3}$ to Blocking Probability in $GI/M/1/(R_3 - 1)$ Queueing System

As it was mentioned in the previous subsection the overflow process with distribution function $G(t)$ and LST $\tilde{G}(s)$ governs the successive arrival instants of customers at the $3^d$ server and its queue. Thus one can view the $3^d$ server and the queue in front of it as the queueing system with inter-arrival distribution function $G(t)$, exponential service times with rate $\mu_3$ and capacity $R_3$ i.e. as standard $GI/M/1/(R_3 - 1)$ queue. Denote by $\pi$ the probability that just before an arrival of the customer at $GI/M/1/(R_3 - 1)$ its queue is full. Due to the fact that the arriving at the $N$-server ordered entry queueing system customer goes to the $4^{th}$ server whenever on arrival it sees the queues in front of the $1^{st}$, $2^{nd}$ and $3^d$ servers full (which happens with the probability $\pi$) and also sees both queues in front of the $1^{st}$ and $2^{nd}$ server full (which happens with the probability $p_{R_1,R_2}$) then, once $\pi$ is known, the probability $p_{R_1,R_2,R_3}$ is equal to $p_{R_1,R_2,R_3} = \pi p_{R_1,R_2}$. The stationary probability $\pi$ that the arriving customer is lost in $GI/M/1/(R_3 - 1)$ can be calculated using the duality[2] of queue $GI/M/1/(R_3-1)$ to queue $M/GI/1/R_3$ from [23], where it is shown that

$$\pi = \frac{1}{\sum_{j=0}^{R_3} p_j^*},$$

---

[2] Paper [23] gives even more general results, but here we need only this one.

where $p_j^*$ is the (non-normalized with $p_0^* = 1$) stationary probability of $j$ customers in $M/GI/1$ queueing system with arrival rate $\mu$ and service time distribution function $G(t)$. The values $p_j^*$ can be computed using any of the well-known formulas for $M/G/1$ queue, which require values of the derivatives of $\tilde{G}(s)$ at $s = 0$.

It is worth noticing here that using the results from [23] one can also calculate the probabilities $p_{R_1,R_2,k}$, $0 \le k \le R_3 - 1$, from relation

$$p_{R_1,R_2,k} = \frac{p_{R_3-k}^* p_{R_1,R_2}}{\sum\limits_{j=0}^{R_3} p_j^*}, \ 0 \le k \le R_3 - 1.$$

Knowing how to calculate all the probabilities $p_{R_1,R_2,k}$, one can *recursively* solve the global balance equations for the joint stationary distribution $p_{i,j,k}$.

### 3.4   Generalization to the $N$-server Case

In order to compute loss probability $p_{R_1,R_2,R_3,...,R_N}$ one obtains at first the probability $p_{R_1,R_2,R_3}$ (following Subsects. 3.1, 3.2 and 3.3). Then one needs to compute inter-overflow time distribution from $1^{st}$, $2^{nd}$ and $3^d$ server (and their queues) following Subsect. 3.2. Then, following Subsect. 3.3, one finds $p_{R_1,R_2,R_3,R_4}$. This procedure continues until one finds $p_{R_1,R_2,R_3,...,R_N}$.

## 4   Conclusion

Remarkably that starting from calculation of probability $p_{R_1,R_2,R_3}$ and on, one does not need to able to calculate the whole joint stationary distribution in order to find the blocking probability $p_{R_1,R_2,R_3,...,R_N}$. But this fact comes at price: at each next step one has to calculate inter-overflow distribution (or the derivatives of the corresponding LST) which becomes harder and harder as $N$ grows large. We were unable to find the general solution of the system for the $N$-server case but note that for moderate values of $N$ it can be solved using one of the many available numerical software. One can also use the structure of the equations in the system (20)–(26) to elaborate different approximations for the distribution of time between overflows $G(t)$ (see, for example, [30]). Clearly these heavily depend on initial parameters and, for example, exponential approximation with rate $(\lambda + \mu_1 + \mu_2)$ sometimes is rough and sometimes is accurate (if the value of $\lambda$ greatly exceeds the value of $\mu_1 + \mu_2$). Here one must point the fact that matrix-analytic (QBD) approach suits the purpose of obtaining joint stationary distribution in the $N$-server ordered entry queue and by applying it one gets the solution at once (although computational issues arise here also). Without dwelling on the comparison of computational efficiency of the QBD solution for the joint distribution $p_{i,j}$ with the recursive one, proposed in the paper, we note that the latter possesses several nice properties. Firstly it uncovers the interrelationships between the stationary probabilities of the states and secondly it allows one to analytically study some allocation problems previously treated only numerically (or by simulation). For example, in [4] one notices that there is a general rule which says that maximum balance and overall efficiency of the system can best be obtained by allocating extra waiting room to the last server in the ordered entry queueing system. From the expression for $p_{R_1,R_2}$ one can hypothesize that $p_{R_1,R_2+1} < p_{R_1+1,R_2}$, when initial values $\lambda$, $\mu_1$ and $\mu_2$ are held fixed. But the proof is yet to be done.

The proposed algebraic method for recursive computation of the joint distribution $p_{i,j}$ without resorting to matrix approach *apparently fails* for the $N$-server system ($N \geq 3$). Already with the 3-server system we were unable to find the way to compute recursively the whole joint stationary distribution. This results (that the method will fail) might have seem obvious, but our research shows that method fails purely due to the structure of the state-space of the Markov chain. Thus one can hypothesise that there exists a class of multi-dimensional Markov chains which admits recursive computation of the whole joint stationary distribution. The description of this class remains an open issue.

**Acknowledgments.** This work was supported by the Russian Foundation for Basic Research (grants 15-07-03007, 15-07-02354, 15-07-02341).

# References

1. Disney, R.L.: Some multichannel queueing problems with ordered entry. J. Ind. Eng. **14**, 105–108 (1963)
2. Elsayed, E.A., Proctor, C.L.: Ordered entry and random choice conveyors with multiple Poisson Input. Int. J. Prod. Res. **15**, 439–451 (1977)
3. Nawijn, W.M.: Stochastic Conveyor Systems. Ph.D. Dissertation, Technische Hogeschool Twente, Afdeling der Werktuigbouwkunde (1983a)
4. Phillips, D.T., Skeith, R.W.: Ordered entry queueing networks with multiple services and multiple queues. AIEE Trans. **1**, 333–342 (1969)
5. Matsui, M.: Manufacturing and Service Enterprise with Risks: A Stochastic Management. Springer, New York (2008)
6. Gregory, G., Litton, C.D.: A conveyor model with exponential service times. Int. J. Prod. Res. **13**(1), 1–7 (1975)
7. Cooper, R.B.: Queues with ordered servers that work at different rates. Opsearch **13**, 69–78 (1976)
8. Matsui, M., Fukuta, J.: On a multichannel queueing system with ordered entry and heterogeneous servers. AIIE Trans. **9**, 209–214 (1977)
9. Nawijn, W.M.: On a two-server finite queuing system with ordered entry and deterministic arrivals. Eur. J. Oper. Res. **18**, 388–395 (1984)
10. Nawijn, W.M.: A note on many-server queueing systems with ordered entry, with an application to conveyor theory. J. Appl. Prob. **20**, 144–152 (1983)
11. Pourbabai, B., Sonderman, D.: Service utilization factors in queueing loss systems with ordered entry and heterogeneous servers. J. Appl. Prob. **23**, 236–242 (1986)
12. Yao, D.D.: Convexity properties of the overflow in an ordered-entry system with heterogeneous servers. Oper. Res. Lett. **5**, 145–147 (1986)

13. Karlof, J.K., Jenkins, J.: The Behavior of a Multichannel Queueing System under Three Queue Disciplines (2002). http://people.uncw.edu/karlof/publications/jenkins.pdf
14. Jain, M., Shekhar, C.: State-dependent multi-channel queuing system with ordered entry. Int. J. Eng. **15**(3), 275–286 (2002)
15. Elsayed, E.A., Lin, B.W.: Transient behaviour of ordered-entry multichannel queueing systems. Int. J. Prod. Res. **18**, 491–501 (1980)
16. Elsayed, E.A.: Multichannel queueing systems with ordered entry and finite source. Comput. Ops. Res. **10**, 213–222 (1983)
17. Kodama, M., Fukuta, J.: On a mutlichannel queueing system with semi-ordered entry. J. Inf. Optim. Sci. **15**, 65–88 (1994)
18. Isguder, H.O., Celikoglu, C.C.: Minimizing the loss probability in GI/M/3/0 queueing system with ordered entry. Sci. Res. Ess. **7**, 963–968 (2012)
19. Isguder, O., Uzunoglu-Kocer, U.: Analysis of GI/M/n/n queueing system with ordered entry and no waiting line. App. Math. Model. **38**(3), 1024–1032 (2014)
20. Pritsker, A., Alan, B.: Application of multichannel queueing results to the analysis of conveyor systems. J. Ind. Eng. **17**(1), 14–21 (1966)
21. Shanthikumar, J.G., Yao, D.D.: Comparing ordered-entry queues with heterogeneous servers. Queueing Syst. **2**, 235–244 (1987)
22. Zaryadov, I., Meykhanadzhyan, L., Milovanova, T., Razumchik, R.: On the method of calculating the stationary distribution in the finite a two-channel system with ordered input. Syst. Means Inf. **25**(3), 44–59 (2015)
23. Miyazawa, M.: Complementary generating functions for the $M^X/GI/1/k$ and $GI/M^Y/1/k$ queues and their application to the comparison for loss probabilities. J. App. Prob. **27**, 684–692 (1990)
24. Erdelyi, A., Bateman, H.: Higher transcendental functions, vol. II. Robert E. Krieger Publishing Company, Malabar (1985)
25. van Doorn, E.A.: On the overflow process from a finite Markovian queue. Perform. Eval. **4**(4), 233–240 (1984)
26. Medhi, J.: Stochastic Models in Queueing Theory. Academic, Amsterdam (2003)
27. Razumchik, R.V.: Analysis of finite capacity queue with negative customers and bunker for ousted customers using chebyshev and gegenbauer polynomials. Asia Pac. J. Oper. Res. **31**(4) (2014)
28. Avrachenkov, K.E., Vilchevsky, N.O., Shevljakov, G.L.: Priority queueing with finite buffer size and randomized push-out mechanism. In: ACM International Conference on Measurement and Modeling of Computer, San Diego, pp. 324–335 (2003)
29. Ilyashenko, A., Zayats, O., Muliukha, V., Laboshin, L.: Further investigations of the priority queuing system with preemptive priority and randomized push-out mechanism. In: Balandin, S., Andreev, S., Koucheryavy, Y. (eds.) NEW2AN/ruSMART 2014. LNCS, vol. 8638, pp. 433–443. Springer, Heidelberg (2014)
30. Whitt, W.: Approximating a point process by a renewal process: two basic methods. Oper. Res. **30**(1), 125–147 (1982)

# On Polynomial Bounds of Convergence
# for the Availability Factor

Alexander Veretennikov[1,2,3] and Galina Zverkina[4(✉)]

[1] University of Leeds, Leeds, UK
a.veretennikov@leeds.ac.uk
[2] National Research University Higher School of Economics, Moscow, Russia
[3] Institute for Information Transmission Problems, Moscow, Russia
[4] Moscow State University of Railway Engineering, Moscow, Russia
zverkina@gmail.com

**Abstract.** A computable estimate of the readiness coefficient for a standard binary-state system is established in the case where both working and repair time distributions possess heavy tails.

**Keywords:** Readiness coefficient · Restorable system · Heavy tails · Polynomial convergence rate

## 1 Introduction

Let us consider a restorable system, which may be either in the working state during a random time $\xi$ with a distribution function $F_1(s) \stackrel{\text{def}}{=} \mathbf{P}\{\xi \leqslant s\}$, or it may be broken down and being restored by some service during another random time $\eta$ with a distribution function $F_2(s) \stackrel{\text{def}}{=} \mathbf{P}\{\eta \leqslant s\}$. All periods of working and repairing are alternate and independent. The readiness coefficient (or availability factor) $A(t)$ is defined as the probability that at time $t$ the system is in the working (= serviceable) state.

Often in the literature it is accepted that at initial time $t = 0$ the system is serviceable and that it is in the beginning of its working period. We consider a more general case assuming that the activity of the system may have started earlier so that at $t = 0$ the system can be in one of the two states: perfect functionality or complete failure; and further that *before* $t = 0$ the system already spent time $x$ in its current state.

Let us formalize the definition of readiness coefficient (availability factor). We assume that $\xi_i$ are random variables with a common distribution function $F_1(x) = \mathbf{P}\{\xi_i \leqslant x\}$; likewise, $\eta_i$ are random variables with a (another) common distribution function $F_2(x) = \mathbf{P}\{\eta_i \leqslant x\}$; all of them are mutually independent.

Both authors are supported by the RFBR, project No 14-01-00319 A. For the first author the article was prepared within the framework of a subsidy granted to the HSE by the Government of the Russian Federation for the implementation of the Global Competitiveness Program.

© Springer International Publishing Switzerland 2016
V. Vishnevsky and D. Kozyrev (Eds.): DCCN 2015, CCIS 601, pp. 358–369, 2016.
DOI: 10.1007/978-3-319-30843-2_37

If at time $t = 0$ our system is working and its elapsed working time before $t = 0$ equals $x$, then the *residual time* of this working period is a random variable denoted by $\xi^{(x)}$; its distribution function is denoted by

$$F_1^{(x)}(s) \overset{\text{def}}{=\!=} \mathbf{P}\{\xi^{(x)} \leqslant s\} = \mathbf{P}\{\xi \leqslant s + x | \xi > x\} = 1 - \frac{1 - F_1(x + s)}{1 - F_1(x)}.$$

Correspondingly, if at time $t = 0$ the system is under repair and the duration of this repair before $t = 0$ equals $x$, then the residual time of this repair period is a random variable denoted by $\eta^{(x)}$ with a distribution function

$$F_2^{(x)}(s) \overset{\text{def}}{=\!=} \mathbf{P}\{\eta^{(x)} \leqslant s\} = \mathbf{P}\{\eta \leqslant s + x | \eta > x\} = 1 - \frac{1 - F_2(x + s)}{1 - F_2(x)}.$$

In the first case we will use notations

$$t_0 \overset{\text{def}}{=\!=} 0, \quad t_1 \overset{\text{def}}{=\!=} \xi^{(x)} + \eta_1, \quad t_i \overset{\text{def}}{=\!=} \xi^{(x)} + \eta_1 + \sum_{j=2}^{i}(\xi_j + \eta_j),$$

and $t_0' \overset{\text{def}}{=\!=} \xi^{(x)}, \quad t_i' \overset{\text{def}}{=\!=} \xi^{(x)} + \sum_{j=1}^{i-1}(\eta_j + \xi_{j+1}).$

In the second case

$$t_1 \overset{\text{def}}{=\!=} \eta^{(x)}, \quad t_i \overset{\text{def}}{=\!=} \eta^{(x)} + \sum_{j=2}^{i}(\xi_{j-1} + \eta_j),$$

and

$$t_1' \overset{\text{def}}{=\!=} 0, \quad t_1' \overset{\text{def}}{=\!=} \eta^{(x)} + \xi_1, \quad t_i' \overset{\text{def}}{=\!=} \eta^{(x)} + \xi_1 + \sum_{j=2}^{i}(\eta_j + \xi_j).$$

In this notation $A(t) \overset{\text{def}}{=\!=} \mathbf{P}\left\{t \in \bigcup_i [t_i, t_i')\right\}.$

It is well known that if distributions of $\xi + \eta$ are non-arithmetical and $\mathbf{E}\xi + \mathbf{E}\eta < \infty$, then there exists a limiting value

$$\lim_{t \to \infty} A(t) \overset{\text{def}}{=\!=} A = \frac{\mathbf{E}\xi}{\mathbf{E}\xi + \mathbf{E}\eta}.$$

Moreover, if $\mathbf{E}\xi^n + \mathbf{E}\eta^n < \infty$ for some $n > 1$, then

$$\limsup_{t \to \infty} |A(t) - A| t^{n-1} < \infty$$

(see e.g. [1, Theorem 3, Appendix 1] or [2, Theorem 10.7.4]).

In other words, for any $\alpha \in (0, n - 1]$ there exists a constant $C(\alpha)$ such that for all $t > 0$

$$|A(t) - A| \leqslant \frac{C(\alpha)}{(1 + t)^\alpha}.$$

However, the general theory provides neither the value of $C(\alpha)$, nor any bound for it. Any knowledge of the value $C(\alpha)$ or its bound is rather important in applications. Also, in the case where nothing was known earlier about such a constant at all, even rough estimates could be useful. The goal of this paper is to give explicit estimates to this constant.

This paper is an extended version of the conference publication [10]. The Sect. 2 contains assumptions and notations; the Sect. 3 presents the main result; the last Sect. 4 provides the full proof.

## 2   Assumptions and Notations

### 2.1   Assumptions

We suppose that $\quad F_1(x) \;=\; 1 - e^{-\int_0^x \lambda(s)\,dx}$, i.e., almost everywhere $\lambda(s) = \dfrac{F_1'(s)}{1 - F_1(s)}$, and for some $\Lambda > K_1 > 3$, $K_2 > 3$

$$\Lambda \geqslant \lambda(s) \geqslant \frac{K_1}{1+s} \quad \text{when } s > 0, \tag{1}$$

$$F_2(s) \geqslant 1 - \frac{1}{(1+s)^{K_2}} \quad \text{when } s > 0 ; \tag{2}$$

*we do not assume continuity of $F_2(s)$.*

Note that from (1) it follows that $F_1(s) \geqslant 1 - \dfrac{1}{(1+s)^{K_1}}$ and there is

$$f_1'(s) \stackrel{\text{def}}{=\!=} F_1'(s) \in \left( \frac{K_1 e^{-\Lambda s}}{1+s}, \frac{\Lambda}{(1+s)^{K_1}} \right).$$

So, (1) and (2) imply that for all $a \in [1, K_1 - 1]$ and $b \in [1, K_2 - 1]$ we have,

$$\mathbf{E}\,\xi^a \geqslant \int_0^\infty s^a \frac{K_1 e^{-\Lambda s}}{1+s}\,d s \geqslant K_1 \int_0^\infty s^{a-1} e^{-\Lambda s}\,d s = \frac{K_1 \Gamma(a)}{\Lambda^a} \stackrel{\text{def}}{=\!=} \mu_a; \tag{3}$$

$$\mathbf{E}\,\xi^a = a \int_0^\infty s^{a-1}(1 - F_1(s))\,d s \leqslant a \int_0^\infty \frac{s^{a-1}}{(1+s)^{K_1}}\,d s < \frac{a}{K_1 - a} \stackrel{\text{def}}{=\!=} m_1(a), \tag{4}$$

similarly,

$$\mathbf{E}\,\eta^b < \frac{b}{K_2 - b} \stackrel{\text{def}}{=\!=} m_2(b), \tag{5}$$

which suffices for the existence of $A < \infty$.

Notice that $\lambda(s)$ is called *intensity* of failure of the recoverable system, of course, while it is working.

## 2.2    Notations

1. Denote $K \stackrel{\text{def}}{=} \min(K_1, K_2)$.

2. The behaviour of the system under consideration may be presented by the random process

$$X_t = (n_t, x_t) = \begin{cases} (1, t - t_i), & \text{if } t \in [t_i, t'_i); \\ (2, t - t'_i), & \text{if } t \in [t'_i, t_{i+1}); \end{cases} \qquad n(X_t) \stackrel{\text{def}}{=} n_t, \; x(X_t) \stackrel{\text{def}}{=} x_t.$$

The state space of the process $X_t$ is a set $\mathcal{X} \stackrel{\text{def}}{=} \{\{1, 2\} \times \mathbb{R}_+\}$ with a standard $\sigma$-algebra.

Denote $\mathcal{S}_j \stackrel{\text{def}}{=} \{(j, x), x \in \mathbb{R}_+\} \subset \mathcal{X} \quad (j = (1, 2))$.

Let $X_0 = (n_0, x_0)$.

3. Denote (here $j = (1, 2)$):

$$\mathrm{M}_j(k) \stackrel{\text{def}}{=} k \int_0^\infty \frac{s^{k-1}}{(1 + s)^{K_j}} \, ds; \quad \mathrm{M}_j^{(x)}(k) \stackrel{\text{def}}{=} \frac{k}{1 - F_j(x)} \int_0^\infty \frac{s^{k-1}}{(1 + s + x)^{K_j}} \, ds;$$

$$\varkappa(T) \stackrel{\text{def}}{=} \int_0^\infty \frac{K_1 e^{-\Lambda s}}{1 + T + s} \, ds; \quad F_j^{(a)}(s) \stackrel{\text{def}}{=} 1 - \frac{1 - F_j(s + a)}{1 - F_j(a)};$$

$$f_j^{(a)}(s) \stackrel{\text{def}}{=} \left( F_j^{(a)}(s) \right)'; \quad \varphi_{x,y}(s) \stackrel{\text{def}}{=} \min \left( f_1^{(x)}(s), f_1^{(y)}(s) \right);$$

$$\varkappa_{x,y} \stackrel{\text{def}}{=} \int_0^\infty \varphi_{x,y}(s) \, ds; \quad \Phi_{x,y}(s) \stackrel{\text{def}}{=} \int_0^s \varphi_{x,y}(u) \, du;$$

$$\widehat{\Phi}_{x,y}(s) \stackrel{\text{def}}{=} F_1^{(x)}(s) - \Phi_{x,y}(s).$$

4. Let us choose (using (3)–(5))

$$R > \Theta_0 \stackrel{\text{def}}{=} \frac{\mathbf{E}(\xi + \eta)^2}{2(\mathbf{E}\,\xi + \mathbf{E}\,\eta)} \left[ \leq \frac{m_1(2) + 2m_1(1)m_2(1) + m_2(2)}{2\mu_a} \right];$$

let $N$ be such that $e^{-\Lambda R} > (1 + NR)^{-K_1}$, and let

$$q \stackrel{\text{def}}{=} 1 - \left( 1 - \frac{\Theta_0}{R} \right) \left( e^{-\Lambda R} - \frac{1}{(1 + NR)^{K_1}} \right) \varkappa(NR);$$

$$\Psi(\alpha, X_0) \stackrel{\text{def}}{=} \left( \sum_{i=0}^\infty (2i + 4)^\alpha q^i \right) \left( 1 + \mathbf{1}(n_0 = 1) 2^{\alpha - 1} \left( \mathrm{M}_1^{(x_0)}(\alpha) + \mathrm{M}_2(\alpha) \right) \right.$$

$$+ \mathbf{1}(n_0 = 2) \mathrm{M}_2^{(x_0)}(\alpha) + 2^{\alpha - 1} A \left( \frac{\alpha}{(K_1 - \alpha)(K_1 - \alpha - 1)\mathbf{E}\,\xi} + \mathrm{M}_2(\alpha) \right)$$

$$+ \frac{(1 - A)\alpha}{(K_2 - \alpha)(K_2 - \alpha - 1)\mathbf{E}\,\eta} + (i + 1)\mathrm{M}_1(\alpha) + i\,\mathrm{M}_2(\alpha) \Big).$$

## 3   Main Result

**Theorem 1.** *Let $K > 3$ and let the conditions (1), (2) be satisfied. Then for the process described earlier with initial state $X_0 = (n_0, x_0)$, for every $\alpha \in (1, K-1)$ there exists a constant $C(\alpha, X_0) < \Psi(\alpha, X_0)$ such that for all $t \geq 0$ the following inequality is true:*

$$|A(t) - A| \leq \frac{C(\alpha, X_0)}{(1+t)^\alpha}.$$

## 4   Proof

### 4.1   Properties of the Process $X_t$

The process $X_t$ defined in the Subsect. 2.2 (point 2.) is Markov. Moreover, it possesses a strong Markov property. We skip the standard proof of both claims. Note that trajectories of the process $X_t$ are right continuous.

### 4.2   On the Stationary Distribution of $X_t$

In terms of [3,4], the process $X_t$ is a linear-type (piecewise linear) Markov process, and it satisfies the conditions of ergodic theorem from [5, Sect. 2.6] (see also [6, Theorem 2]): there exists a stationary distribution $\mathcal{P}$ on $\mathcal{X}$ such that there is a limit

$$\lim_{t \to \infty} \mathbf{P}\{n_t = j, x_t \leq s\} = \mathcal{P}(\{j\} \times [0, s])$$

for any initial state $X_0$ (again and always in the sequel $j = (1, 2)$);

$$\mathcal{P}(\{j\} \times (s, \infty)) = \frac{\mathbf{1}\{j = 1\} \int_s^\infty (1 - F_1(s))\, ds + \mathbf{1}\{j = 2\} \int_s^\infty (1 - F_2(s))\, ds}{\mathbf{E}\,\xi + \mathbf{E}\,\eta}, \quad (6)$$

and $\mathcal{P}(n_t = 1) = \dfrac{\mathbf{E}\,\xi}{\mathbf{E}\,\xi + \mathbf{E}\,\eta} = A.$

### 4.3   Coupling Method

To prove the Theorem 1 we will use the *coupling method*, which will be now briefly recalled (for details see [7]).

Suppose some strong Markov process $X_t$ weakly converges to its (unique) stationary regime; denote its marginal distribution by $\mathcal{P}$.

Suppose that on some probability space it is possible to construct two *independent* versions $X_t'$ and $X_t''$ of this Markov process – i.e., both with the same generator but possibly with different initial distributions – such that the stopping time

$$\tau(X_0', X_0'') \overset{\text{def}}{=} \inf\{t > 0 : X_t' = X_t''\}$$

has a finite expectation.

If, further, we have an estimate $\mathbf{E}\,\phi(\tau(X_0', X_0'')) \leqslant C(X_0', X_0'')$ where $\phi(s) \uparrow$ and $\phi(s) > 0$ as $s > 0$, then we can use a strong Markov property and *coupling inequality*: for all set $\mathcal{M} \in \mathcal{B}(\mathcal{X})$

$$\left|\mathbf{P}\{X_t' \in \mathcal{M}\} - \mathbf{P}\{X_t'' \in \mathcal{M}\}\right| \leqslant \mathbf{P}\{t \leqslant \tau(X_0', X_0'')\} = \mathbf{P}\{\phi(t) \leqslant \phi(\tau(X_0', X_0''))\}.$$

Hence, due to Markov's inequality,

$$\left|\mathbf{P}\{X_t' \in \mathcal{M}\} - \mathbf{P}\{X_t'' \in \mathcal{M}\}\right| \leqslant \frac{\mathbf{E}\,\phi(\tau(X_0', X_0''))}{\phi(t)} \leqslant \frac{C(X_0', X_0'')}{\phi(t)}. \qquad (7)$$

Once the inequality (7) is established for the pair of processes, we may conclude that for the stationary process $\widetilde{X}_t$ with the initial distribution $\mathcal{P}$ and for the process $X_t$ starting from an arbitrary initial state $X_0$ we get,

$$\left|\mathbf{P}\{X_t \in \mathcal{M}\} - \mathbf{P}\{\widetilde{X}_t \in \mathcal{M}\}\right| = |\mathbf{P}\{X_t \in \mathcal{M}\} - \mathcal{P}(\mathcal{M})\}|$$

$$\leqslant \frac{\displaystyle\int_{\mathcal{X}} C(X_0, Y)\mathcal{P}(\mathrm{d}Y)}{\phi(t)} = \frac{\widetilde{C}(X_0)}{\phi(t)}. \qquad (8)$$

Note that since the right hand side here does not depend on $\mathcal{M} \subseteq \mathcal{X}$, this inequality, of course, provides an estimate in total variation, that is,

$$\sup_{\mathcal{M} \in \mathcal{X}} \left|\mathbf{P}\{X_t \in \mathcal{M}\} - \mathcal{P}(\mathcal{M})\}\right| \leqslant \frac{\widetilde{C}(X_0)}{\phi(t)}.$$

Also, if $\mathcal{M} = \{n(X_t) = 1\}$, then $\mathbf{P}\{X_t \in \mathcal{M}\} = A(t)$. Hence, in particular, the inequality (8) implies that

$$\left|A(t) - A\right| \leqslant \frac{\widetilde{C}(X_0)}{\phi(t)}.$$

Now, the goal is to give an estimate of $\widetilde{C}(X_0)$ for the function $\phi(t) = (1+t)^\alpha$.

## 4.4  Coupling, Continued

We will be using a procedure first suggested in [8]. On some probability space we construct a "paired" Markov process $Z_t = (Z_t', Z_t'')$ in the state space $\mathcal{X} \times \mathcal{X}$ so that the marginal distributions of the processes $Z_t'$ and $Z_t''$ coincide with the distributions of the processes $X_t'$ and $X_t''$, respectively:

$$(Z_t', t \geqslant 0) \overset{\mathcal{D}}{=} (X_t', t \geqslant 0) \quad \text{and} \quad (Z_t'', t \geqslant 0) \overset{\mathcal{D}}{=} (X_t'', t \geqslant 0); \qquad (9)$$

$Z_0' = X_0'$ and $Z_0'' = X_0''$.

In addition, we suppose, that if at some moment $\bar{\tau}$ the random variable $Z'_t$ coincides with $Z''_t$, i.e. $Z'_{\bar{\tau}} = Z''_{\bar{\tau}}$, then for all $t \geqslant \bar{\tau}$, $Z'_t = Z''_t$. This pair $(Z', Z'')$ is called coupling. Of course, in general, the processes $Z'_t$ and $Z''_t$ will be dependent.

Assuming that the process $Z_t = (Z'_t, Z''_t)$ is already constructed, let us denote

$$\bar{\tau}(X'_0, X''_0) \left[= \bar{\tau}(Z'_0, Z''_0)\right] \stackrel{\text{def}}{=} \inf\{t > 0 : Z'_t = Z''_t\}.$$

The coupling is called *successful* if $\mathbf{P}\left\{\bar{\tau}(X'_0, X''_0) < \infty\right\} = 1$. Our coupling constructed below will be successful.

Then, we can use the coupling inequality (7) for the processes $Z'_t$ and $Z'_t$:

$$\left|\mathbf{P}\{Z'_t \in \mathcal{M}\} - \mathbf{P}\{Z''_t \in \mathcal{M}\}\right| \leqslant \mathbf{P}\{t \leqslant \bar{\tau}(X'_0, X''_0)\}.$$

Due to (9) the same inequality holds true for $X'_t$ and $X''_t$.

## 4.5   Construction of the Process $Z_t$

For any distributional function $F(s)$ denote $F^{-1}(s) \stackrel{\text{def}}{=} \inf\{u : F(u) = s\}$; it is well known that on the probability space $(\Omega^{\mathcal{L}}, \mathcal{F}^{\mathcal{L}}, \mathbf{P}^{\mathcal{L}}) \stackrel{\text{def}}{=} ([0,1], \mathcal{B}([0,1]), \mathcal{L})$ (where $\mathcal{L}$ is a Lebesgue measure) the random variable $\xi \stackrel{\text{def}}{=} F^{-1}(\mathfrak{U}_{\Omega^{\mathcal{L}}})$ has a distribution function $F(s)$ if $\mathfrak{U}_{\Omega^{\mathcal{L}}}$ is a uniformly distributed random variable on the space $\Omega^{\mathcal{L}}$.

We will construct the process $Z_t$ on the probability space

$$(\Omega, \mathcal{F}, \mathbf{P}) \stackrel{\text{def}}{=} \prod_{i=0}^{\infty} \left((\Omega^{\mathcal{L}}_{i,1}, \mathcal{F}^{\mathcal{L}}_{i,1}, \mathbf{P}^{\mathcal{L}}_{i,1}) \times (\Omega^{\mathcal{L}}_{i,2}, \mathcal{F}^{\mathcal{L}}_{i,2}, \mathbf{P}^{\mathcal{L}}_{i,2})\right),$$

where the probability spaces $\Omega^{\mathcal{L}}_{i,j}$ are the copies of the described above space $\Omega^{\mathcal{L}}$. The construction of $Z_t$ is based on a sequence of stopping times $t_k$, at which

$$\mathbf{1}\left\{n(Z'_{t-0}) \neq n(Z'_{t+0})\right\} + \mathbf{1}\left\{n(Z''_{t-0}) \neq n(Z''_{t+0})\right\} > 0,$$

i.e., of (random) times $t_k$ where one of the processes $Z'_t$ and $Z''_t$ – or both of them – changes its first component.

Let $t_0 = 0$ and denote

$$m'_t \stackrel{\text{def}}{=} n(Z'_t), \quad m''_t \stackrel{\text{def}}{=} n(Z''_t), \quad z'_t \stackrel{\text{def}}{=} x(Z'_t), \quad z''_t \stackrel{\text{def}}{=} x(Z''_t).$$

The sequence $(t_k)$ will be built by induction. Assume that $t_k$ is already determined for some $k$ and consider three cases.

**Case 1.** Suppose that $Z'_{t_k} \neq Z''_{t_k}$ and $m'_{t_k} + m''_{t_k} > 2$ (that is, at least one of the processes is in the set $\mathcal{S}_2$). Then on the probability space $\left(\Omega^{\mathcal{L}}_{k,1} \times \Omega^{\mathcal{L}}_{k,2}\right)$ we take an independent random variables $\theta'_k \stackrel{\text{def}}{=} \left(F^{(z'_{t_k})}_{m'_{t_k}}\right)^{-1}\left(\mathfrak{U}_{\Omega^{\mathcal{L}}_{k,1}}\right)$   and

$\theta_k'' \overset{\text{def}}{=} \left( F_{m_{t_k}''}^{(z_{t_k}'')} \right)^{-1} \left( \mathrm{u}_{\Omega_{k,2}^{\mathcal{L}}} \right)$, they have a distribution functions $F_{m_{t_k}'}^{(z_{t_k}')}(s)$ and $F_{m_{t_k}''}^{(z_{t_k}'')}(s)$ respectively: they are residual times of stay of the processes $Z_t'$ and $Z_t''$ in the sets $\mathcal{S}_{m_{t_k}'}$ and $\mathcal{S}_{m_{t_k}''}$ correspondingly.

Denote $\theta_k \overset{\text{def}}{=} \min(\theta_k', \theta_k'')$ and $t_{k+1} \overset{\text{def}}{=} t_k + \theta_k$. For $t \in [t_k, t_{k+1})$ define,

$$Z_t' \overset{\text{def}}{=} (m_{t_k}', z_{t_k}' + t - t_k); \qquad Z_t'' \overset{\text{def}}{=} (m_{t_k}'', z_{t_k}'' + t - t_k);$$

$$Z_{t_{k+1}}' \overset{\text{def}}{=} 1\{\theta_k' = \theta_k\} \left( m_{t_k}' - (-1)^{m_{t_k}'}, 0 \right) + 1\{\theta_k' \neq \theta_k\} \left( m_{t_k}', z_{t_k}' + t_{k+1} - t_k \right);$$

$$Z_{t_{k+1}}'' \overset{\text{def}}{=} 1\{\theta_k'' = \theta_k\} \left( m_{t_k}'' - (-1)^{m_{t_k}''}, 0 \right) + 1\{\theta_k'' \neq \theta_k\} \left( m_{t_k}'', z_{t_k}'' + t_{k+1} - t_k \right).$$
$$(10)$$

**Case 2.** Suppose now that $Z_{t_k}' \neq Z_{t_k}''$ and $m_{t_k}' = m_{t_k}'' = 1$. In this case, using the idea of the "Lemma about three random variables" (see [9]) we construct on *one* space $\Omega_{k,1}^{\mathcal{L}}$ the pair of *dependent* random variables $(\theta_k', \theta_k'')$ such that:

$$\mathbf{P}\{\theta_k' \leqslant s\} = F_1^{(z_{t_k}')}(s); \quad \mathbf{P}\{\theta_k'' \leqslant s\} = F_1^{(z_{t_k}'')}(s);$$

$$\mathbf{P}\{\theta_k' = \theta_k''\} = \int_0^\infty \min\left( f_1^{(z_{t_k}')}(s), f_1^{(z_{t_k}'')}(s)' \right) ds$$

$$= \varkappa_{z_{t_k}', z_{t_k}''} \geqslant \int_0^\infty \frac{K_1 e^{-\Lambda s}}{1 + s + \max\left( z_{t_k}', z_{t_k}'' \right)} ds = \varkappa\left( \max\left( z_{t_k}', z_{t_k}'' \right) \right).$$
$$(11)$$

Note that clearly $\varkappa(T) \downarrow 0$ if $T \uparrow +\infty$.
The construction of $(\theta_k', \theta_k'')$ is as follows. Let

$$\Xi_{x,y}(s) \overset{\text{def}}{=} \begin{cases} \Phi_{x,y}^{-1}(s) & s \in [0, \varkappa_{x,y}); \\ \widehat{\Phi}_{x,y}^{-1}(s - \varkappa_{x,y}) & s \in [\varkappa_{x,y}, 1); \end{cases}$$

$$\theta_k' \overset{\text{def}}{=} \Xi_{z_{\vartheta_k}', z_{\vartheta_k}''} \left( \mathrm{u}_{\Omega_{k,1}^{\mathcal{L}}} \right), \qquad \theta_k' \overset{\text{def}}{=} \Xi_{z_{\vartheta_k}'', z_{\vartheta_k}'} \left( \mathrm{u}_{\Omega_{k,1}^{\mathcal{L}}} \right).$$

It is easy to see that in this case the formulas (11) for $\theta_k'$ and $\theta_k''$ are true.

Next, we again denote $t_{k+1} \overset{\text{def}}{=} t_k + \min(\theta_k', \theta_k'')$ and apply the same construction given in the formulae (10). This definition and (11) imply the inequality

$$\mathbf{P}\left\{ Z_{t_{k+1}}' = Z_{t_{k+1}}' \right\} \geqslant \varkappa\left( \max\left( z_{t_k}', z_{t_k}'' \right) \right).$$

**Case 3.** Now, suppose $Z'_{t_k} = Z''_{t_k} = (m_{t_k}, z_{t_k})$. In this case we construct random variables $\theta'_k = \theta''_k \stackrel{\text{def}}{=} \left(F^{z_{t_k}}_{m_{t_k}}\right)^{-1}(\mathfrak{U}_{\Omega^{\mathcal{L}}_{k,1}})$ (i.e., they are identical) with distribution function $F^{(z_{t_k})}_{m_{t_k}}(s)$ on the space $\Omega^{\mathcal{L}}_{k,1}$, and $t_{k+1} \stackrel{\text{def}}{=} t_k + \theta'_k$. Here for $t \in [t_k, t_{k+1})$,

$$Z'_t = Z''_t = \left(m'_{t_k}, z'_{t_k} + t - t_k\right); \quad Z'_{t_{k+1}} = Z''_{t_{k+1}} = \left(m'_{t_k} - (-1)^{m'_{t_k}}, 0\right).$$

*This construction gives us the desired pair $Z_t = (Z'_t, Z''_t)$, which satisfies (9) and which is suitable for the successful coupling procedure.*

Indeed, each of the processes $Z'_t$ and $Z''_t$ is an alternating, wherein periods when these processes are in the sets $\mathcal{S}_1$ or $\mathcal{S}_2$ have the distribution functions $F_1(s)$ and $F_2(s)$, respectively; the first period of their stay in the set $\mathcal{S}_{n'_0}$ or $\mathcal{S}_{n''_0}$ has a distribution function $F^{(x'_0)}_{n'_0}(s)$ and $F^{(x''_0)}_{n''_0}(s)$ – these properties are guaranteed by the construction of processes. Moreover, for each of the processes $Z'_t$ and $Z''_t$ considered separately periods of its stay in the sets $\mathcal{S}_1$ and $\mathcal{S}_2$ are mutually independent.

## 4.6   Using Coupling Method

Let us fix two initial values $X'_0 \equiv Z'_0 \neq Z''_0 \equiv X''_0$. In this step of the proof we will show the coupling inequality for the process $Z = (Z', Z'')$; hence, the same inequality will be established for the couple $(X', X'')$.

**First Hits to the Set. $\mathcal{S}_1$.** For $t > 0$ denote,

$$\tau'(t) \stackrel{\text{def}}{=} \inf\left\{s > t : Z'_t = (1,0)\right\}, \quad \tau''(t) \stackrel{\text{def}}{=} \inf\left\{s > t : Z''_t = (1,0)\right\}.$$

These are the moments of the beginning of regeneration periods for the processes $Z'$ and $Z''$ after the nonrandom $t$.

Denote also

$$\tau(Z'_0, Z''_0) \stackrel{\text{def}}{=} \max\left(\tau'(0), \tau''(0)\right).$$

At $\tau'(0)$ the regeneration period of the process $Z'_t$ begins.

Its length equals $\theta \stackrel{\mathcal{D}}{=} \xi + \eta$ where $\xi$ and $\eta$ were introduced in the Sect. 1. After that, the behaviour of $Z'$ does not depend on the initial state $Z'_0$ (given $\tau'(0)$). The same can be said about the process $Z''$.

Let $t > \tau(Z'_0, Z''_0)$. Then, there was at least one beginning of the regeneration period of each of the processes $Z'$ and $Z''$ before $t$.

Denote $\vartheta'(t) \stackrel{\text{def}}{=} (\tau'(t) - t)$ – the residual time of the last regeneration period of $Z'_t$, which started before time $t$. From the corollary of W. Smith's Key Renewal Theorem (cf. [6, Lemma 1]), the following inequality holds true: if $t > \tau(Z'_0, Z''_0)$, then

$$\mathbf{E}\left(\vartheta'(t)\Big|t > \tau\big(Z_0', Z_0''\big)\right) \leqslant \frac{\mathbf{E}\,\theta^2}{2\mathbf{E}\,\theta} = \frac{\mathbf{E}\,(\xi + \eta)^2}{2(\mathbf{E}\,\xi + \mathbf{E}\,\eta)}\,[=\Theta_0]. \tag{12}$$

The same statement applies to the process $Z_t''$.

Note that $\tau\big(Z_0', Z_0''\big) \leqslant \tau'(0) + \tau''(0)$, and, by virtue of Jensen's inequality, for all $\alpha \in (1, K-1)$

$$\mathbf{E}\left(\tau\big(Z_0', Z_0''\big)\right)^{\alpha} \leqslant 2^{\alpha-1}\left(\mathbf{E}\left(\tau'(0)\right)^{\alpha} + \mathbf{E}\left(\tau''(0)\right)^{\alpha}\right).$$

**Coupling After Common Hit to the Set $S_1$.** Without loss of generality we can assume that $\tau\big(Z_0', Z_0''\big) = \tau''(0)$.

Let $\tau_1 \overset{\text{def}}{=} \tau''(0)$, $\tau_{k+1} \overset{\text{def}}{=} \min\{\tau''(t), t > \tau_{k+1}\}$; $\{\tau_k\}$ is a sequence of beginnings of regeneration periods of $Z''$.

And let $\widetilde{\tau}_k \overset{\text{def}}{=} \inf\{t > \tau_k : Z_t'' = (2,0)\}$ – time of the (first) jump of $Z_t''$ to the set $S_2$ after time $\tau_k$.

Denote the event $\mathcal{E}_k \overset{\text{def}}{=} \{\vartheta'(\tau_k) < R \,\&\, (\widetilde{\tau}_k - \tau_k) \in (R, NR)\}$, i.e. at time $\widehat{\tau}_k \overset{\text{def}}{=} \tau_k + \vartheta'(\tau_k)$ $Z_{\widehat{\tau}_k}' = (1,0)$, $Z_{\widehat{\tau}_k}'' = (1,z)$, and $z < R$.

Using (12) and condition (1) by Markov inequality we can estimate $\mathbf{P}\{\mathcal{E}_k\}$ :

$$\mathbf{P}\{\mathcal{E}_k\} \geqslant \left(1 - \frac{\Theta_0}{R}\right)\left(e^{-\Lambda R} - \frac{1}{(1+NR)^{K_1}}\right) \overset{\text{def}}{=} \pi(R, N).$$

Now, using (11), we have: $\mathbf{P}\left\{Z_{\tau_{k+1}}' = Z_{\tau_{k+1}}''\right\} \geqslant \pi(R, N)\varkappa(RN) \overset{\text{def}}{=} p.$

## 4.7  Completion of the Proof

The number of regeneration periods of $Z_t''$ before the processes $Z_t'$ and $Z_t''$ meet each other according to the scheme from the step 4.6. (that is, any meeting outside this scheme is ignored) is a random variable $\nu$ dominated by another one with a geometric distribution with parameter $p$ ($\nu$ itself has a more complicated distribution). Denote $q \overset{\text{def}}{=} 1-p$, and $\varsigma\left(Z_0', Z_0''\right) \overset{\text{def}}{=} \inf\{t > 0 : Z_t' = Z_t''\}$. Obviously, $\varsigma\left(Z_0', Z_0''\right) \leqslant \tau_\nu$.

Since we know the distribution of $\tau = \tau\left(Z_0', Z_0''\right)$ and $\theta \overset{D}{=} \xi + \eta$, we can obtain an estimation of $\mathbf{E}\left(1 + \varsigma\left(Z_0', Z_0''\right)\right)^{\alpha}$ for all $\alpha \in (1, K-1)$ : by Jensen's inequality we get,

$$\mathbf{E}\left(1 + \varsigma\left(Z_0', Z_0''\right)\right)^{\alpha}$$

$$\leqslant \mathbf{E}\left(1 + \tau\left(Z_0', Z_0''\right) + \xi + \sum_{i=1}^{\infty}\left(\mathbf{P}\{\nu = i\}\sum_{k=1}^{i}(\xi_k + \eta_k)\right)\right)^{\alpha}$$

$$\leqslant \sum_{i=1}^{\infty} q^{i-1}\mathbf{E}\left(1 + \tau'(0) + \tau''(0) + \xi + \sum_{k=1}^{i}(\xi_k + \eta_k)\right)^{\alpha}$$

$$\leqslant \sum_{i=1}^{\infty} q^{i-1}(2i+4)^{\alpha-1}\left(1 + \mathbf{E}\big(\tau'(0)\big)^{\alpha} + \mathbf{E}\big(\tau''(0)\big)^{\alpha} + (i+1)\mathbf{E}\,\xi^{\alpha} + i\,\mathbf{E}\,\eta^{\alpha}\right)$$

$$\leqslant \sum_{i=1}^{\infty} q^{i-1}(2i+4)^{\alpha-1}\left(1 + \mathbf{1}\,(n_0'=1)\,2^{\alpha-1}\left(\mathrm{M}_1^{(x_0')}(\alpha) + \mathrm{M}_2(\alpha)\right)\right.$$

$$+ \mathbf{1}\,(n_0'=2)\,\mathrm{M}_2^{(x_0')}(\alpha) + \mathbf{1}\,(n_0''=1)\,2^{\alpha-1}\left(\mathrm{M}_1^{(x_0'')}(\alpha) + \mathrm{M}_2(\alpha)\right)$$

$$\left. + \mathbf{1}\,(n_0''=2)\,\mathrm{M}_2^{(x_0'')}(\alpha) + (i+1)\mathrm{M}_1(\alpha) + i\mathrm{M}_2(\alpha)\right)$$

$$= C\,(\alpha,Z_0',Z_0'') = C\,(\alpha,X_0',X_0'')\,.$$

Now, considering (8), it is necessary to estimate the integral $\int_{\mathcal{X}} C(\alpha,X_0',X_0'')\mathcal{P}\,(\mathrm{d}\,X_0'')$.

First, we make estimations of the integrals:

$$\int_{\mathcal{X}} (\mathbf{1}\,(n_0''=1))\,\mathcal{P}\,(\mathrm{d}\,X_0'') = \mathcal{P}(\{1\}\times(0,+\infty)) = A; \tag{13}$$

$$\int_{\mathcal{X}} (\mathbf{1}\,(n_0''=2))\,\mathcal{P}\,(\mathrm{d}\,X_0'') = \mathcal{P}(\{1\}\times(0,+\infty)) = 1-A; \tag{14}$$

$$\int_{\mathcal{X}} \left(\mathbf{1}\,(n_0''=1)\,\mathrm{M}_1^{(x_0'')}(\alpha)\right)\mathcal{P}\,(\mathrm{d}\,X_0'')$$

$$= \int_{\mathcal{X}} \left(\frac{\mathbf{1}\,(n_0''=1)\,\alpha}{1-F_1(x_0'')}\int_0^{\infty}\frac{s^{\alpha-1}}{(1+s+x_0'')^{K_1}}\mathrm{d}\,s\right)\times\mathrm{d}\left(1-\frac{\int_{x_0''}^{\infty}(1-F_1(s))\,\mathrm{d}\,s}{\mathbf{E}\,\xi+\mathbf{E}\,\eta}\right)$$

$$< \int_0^{\infty}\left(\frac{\alpha}{1-F_1(x_0'')}\int_0^{\infty}\frac{(1+s+x_0)^{\alpha-1}}{(1+s+x_0'')^{K_1}}\,\mathrm{d}\,s\right)\frac{1-F_1(x_0'')}{\mathbf{E}\,\xi+\mathbf{E}\,\eta}\,\mathrm{d}\,x_0''$$

$$\leqslant \int_0^{\infty}\frac{\alpha(1+x_0'')^{\alpha-K_1}}{(K_1-\alpha)(\mathbf{E}\,\xi+\mathbf{E}\,\eta)}\,\mathrm{d}\,x_0'' = \frac{\alpha}{(K_1-\alpha)(K_1-\alpha-1)(\mathbf{E}\,\xi+\mathbf{E}\,\eta)}$$

$$= \frac{\alpha A}{(K_1-\alpha)(K_1-\alpha-1)\mathbf{E}\,\xi}\left[<\frac{\alpha A}{(K_1-\alpha-1)\mathbf{E}\,\xi}\right]; \tag{15}$$

analogously $\int_{\mathcal{X}}\left(\mathbf{1}\,(n_0''=2)\,\mathrm{M}_1^{(x_0'')}(\alpha)\right)\mathcal{P}\,(\mathrm{d}\,X_0'')$

$$< \frac{\alpha(1-A)}{(K_2-\alpha)(K_2-\alpha-1)\mathbf{E}\,\eta}\left[<\frac{\alpha(1-A)}{(K_2-\alpha-1)\mathbf{E}\,\eta}\right]. \tag{16}$$

Considering, that for all constant $\Upsilon$ and set $\mathcal{M} \in \mathcal{B}(\mathcal{X})$ we have $\int_{\mathcal{M}} \Upsilon \mathcal{P}(\mathrm{d}X_0'') = \Upsilon \mathcal{P}(\mathcal{M})$, and collecting formulas (13)–(16), we can estimate the integral $\int_{\mathcal{X}} C(\alpha, X_0', X_0'') \mathcal{P}(\mathrm{d}X_0'')$, and this estimation give as an inequality

$$\int_{\mathcal{X}} C(\alpha, X_0', X_0'') \mathcal{P}(\mathrm{d}X_0'') \leqslant \Psi(\alpha, X_0'), \qquad (17)$$

this completes the proof of the theorem.

**Remark.** The estimate (17) is better than the similar estimate in [10], but it could be improved; moreover, a more careful choice of parameters $R$ and $N$ may provide some enhancement of this bound.

**Acknowledgments.** The authors are grateful to L. G. Afanasieva and V. V. Kozlov for very useful consultations.

# References

1. Borovkov, A.A.: Stochastic Processes in Queueing Theory. Springer-Verlag, New York (1976)
2. Thorisson, H.: Coupling, Stationarity, and Regeneration. Springer, New York (2000)
3. Gnedenko, B.V., Kovalenko, I.N.: Introduction to Queueing Theory. Birkhauser, Boston (1989)
4. Kalashnikov, V.V.: Some properties of piecewise linear Markov processes. Teor. Veroyatnost. i Primenen. **20**(3), 571–583 (1975)
5. Klimov, G.P.: Probability Theory and Mathematical Statistics. Mir Publishers, Moscow (1986)
6. Zverkina, G.A.: On some limit theorems following from Smith's Theorem (2015). arxiv:1509.06178
7. Lindvall, T.: Lectures on the Coupling Method. Wiley, New York (1992)
8. Griffeath, D.: A maximal coupling for Markov chains. Zeitschrift für Wahrscheinlichkeitstheorie und Verwandte Gebiete **31**(2), 95–106 (1975)
9. Veretennikov, A.: Coupling method for Markov chains under integral Doeblin type condition. Theory Stoch. Process. **8**(24(3–4)), 383–390 (2002)
10. Veretennikov, A., Zverkina, G.: On polynomial convergence rate of the availability factor to its stationary value. In: Proceedings of the Eighteenth International Scientific Conference on Distributed Computer and Communication Networks: Control Computation, Communications (DCCN-2015), pp. 168–175. ICS RAS, Moscow (2015)

# Feedback in Infinite-Server Queuing Systems

Svetlana Moiseeva$^{(\boxtimes)}$ and Lyubov Zadiranova

Tomsk State University, Tomsk, Russia
{smoiseeva,zhidkovala}@mail.ru

**Abstract.** Queuing systems with infinite servers $M|M|\infty$, $MMPP|M|\infty$, $GI|M|\infty$ with feedback are considered. Asymptotic characteristic functions of the flow of repeating requests in the systems are obtained by using method of asymptotic analysis under a condition of increasing service time. Results of numeric experiments are presented.

**Keywords:** Queuing system with infinite servers · Asymptotic analysis · Feedback

## 1 Introduction

Due to development of the telecommunication and information-computing systems, assessment of transmission capacity and of service quality is a very important design stage. Queuing systems (QS) with feedback are adequate mathematic models of many real situations in which part of processed requests needs to return to systems for repeating service.

In queuing theory there are two types of the models with feedback: models with delayed feedback (with an orbit) [1,2]; models with instantaneous feedback [3,4].

Article by D'Avignon G.R., Disney R.L [5] has covered queuing system that has one server, two independent Poisson input flows which give two types of requests, and an instantaneous feedback. Each type of requests has its own distribution for service time. Whether there is feedback (allowing for a repeating service) or there isn't decided depending on request type. This work has covered conditions of existence of stationary mode and conclusion about joint and marginal distributions for the length of a queue.

Articles by I.S. Zaryadov [6,7] have covered the research of single-channel QS with a recurrent input flow of requests, the service time for each call distributed by an exponential law, unlimited capacity storage system and the renewal mechanism with repeating service.Request that is in the server will either leave system with probability of $p$, or stay with probability of 1-$p$, while discarding all other requests are rejected.

Real technical systems have limited number of servers, but in the case when probability of request loss could be disregarded such systems could also be approximated by infinite-server QS [8].

© Springer International Publishing Switzerland 2016
V. Vishnevsky and D. Kozyrev (Eds.): DCCN 2015, CCIS 601, pp. 370–377, 2016.
DOI: 10.1007/978-3-319-30843-2_38

QS with infinite number of servers, stationary Poisson arrival process and repeating requests were considered in articles [9].

It's worth noting that using of Poisson arrival process follows a greater inaccuracy when calculating characteristics of service quality [10,11], thus it is necessary to complicate models by using non-Poisson arrival process.

This paper has covered models of queuing with infinite number of servers, Poisson process ($M$), Markov-modulated Poisson process ($MMPP$) and a renewal process ($GI$) as arrival process. A flow of requests which are returned to the system for repeating service is an object of study.

## 2    Problem Statement

Let us consider with the queuing systems with infinite number of servers, Poisson process ($M$), Markov-modulated Poisson process ($MMPP$) and a renewal process ($GI$) as models of arrivals.

Service time of a request is a random variable and is distributed by an exponential law with parameter $\mu$. The incoming call occupies any of free server, and after finishing the service leaves the system with probability 1-$r$, or stays for additional service with it probability $r$.

The object of the study is a number of requests that have come into reviewed systems for additional service during time interval $t$ (the flow of repeating requests).

## 3    The Flow of Repeating Requests in $M|M|\infty$ System with Feedback

Let us start with the queuing system with infinite number of servers, and Poisson arrival process with an intensity $\lambda$.

Let us denote the following: $i(t)$ is a number of occupied servers at the instant $t$, $n(t)$ is a number of requests that have come into system for an additional service during time interval $t$. Two-dimensional flow $\{i(t), n(t)\}$ is Markovian.

For its probability distribution $P(i, n, t) = P\{i(t) = i, n(t) = n\}$ we write down Kolmogorov differential equation system in the form

$$\frac{\partial P(i, n, t)}{\partial t} = -\lambda P(i, n, t) - i\mu P(i, n, t) + \lambda P(i - 1, n, t)$$
$$+ i\mu r P(i, n - 1, t) + \mu(1 - r)(1 + i)P(i + 1, n, t), \qquad (1)$$

where i=0,1,..., n=0,1,....

Let us introduce characteristic functions in the form of

$$H(u, w, t) = \sum_{i=0}^{\infty} \sum_{n=0}^{\infty} e^{jui} e^{jwn} P(i, n, t),$$

then we get the following differential equation from (1)

$$\frac{\partial H(u,w,t)}{\partial t} = \lambda(e^{ju}-1)H(u,w,t)+j\mu(1-re^{jw}-(1-r)e^{-ju})\frac{\partial H(u,w,t)}{\partial u}. \quad (2)$$

The resulting equation lets us determine main statistically distributed characteristics of the considered system, as well as for the flow of repeating requests into the system. Let us research the flow of repeating requests into the system during time interval $t$, by using method of asymptotic analysis under condition of increasing service time [12]. To do this, let us denote the following

$$\mu = \epsilon, u = \epsilon y, H(u,w,t) = F(y,w,t,\epsilon).$$

We can rewrite (2) in the following form

$$\frac{\partial F(y,w,t,\epsilon)}{\partial t} = j(1-re^{jw}-(1-r)e^{-j\epsilon y})\frac{\partial F(y,w,t,\epsilon)}{\partial y}+\lambda(e^{j\epsilon y}-1)F(y,w,t,\epsilon). \quad (3)$$

**Theorem 1.** *Limit value, where $\epsilon \to 0$, of the function $F(y,w,t)$ of solution $F(y,w,t,\epsilon)$ of the Eq. (3) has the following form*

$$F(y,w,t) = exp\left\{\frac{\lambda rt(e^{jw}-1)}{1-r}+\frac{j\lambda y}{1-r}\right\}. \quad (4)$$

*Proof.* After executing the transmission to the limit in Eq. (3) we get the first-order partial differential equation

$$\frac{\partial F(y,w,t,\epsilon)}{\partial t} = jr(1-e^{jw})\frac{\partial F(y,w,t,\epsilon)}{\partial y}. \quad (5)$$

General solution of that equation is

$$F(y,w,t) = \varphi\left(t+\frac{jy}{r(e^{jw}-1)}\right),$$

where $\varphi(y)$ is some function the form of which is determinable by using initial condition.

Let us consider function $F(y,w,t)$ at time zero, obviously, this function doesnt depend on number of repeating requests $w$, so the initial condition will have the form

$$F(y,w,0) = \Phi(y), \quad (6)$$

where $\Phi(y)$ is an asymptotic approximation of characteristic function of distribution for a number of occupied servers in the system under condition of increasing service time of requests. Its easy to derive that the function has the following form

$$\Phi(y) = exp\left\{\frac{j\lambda y}{1-r}\right\}.$$

Thus, solution of the Eq. (3) has the following form

$$F(y,w,t) = exp\left\{\frac{\lambda rt(e^{jw}-1)}{1-r}+\frac{j\lambda y}{1-r}\right\}.$$

which is the same as the Eq. (4). Theorem is proved.

By assuming the number of occupied devices at (4) as $y = 0$, we have an asymptotic approximation of characteristic function of the number of repeating requests that have come into system during time interval $t$ under a condition of increasing service time

$$h(w, t) = exp\left\{\frac{\lambda rt(e^{jw} - 1)}{1 - r}\right\}.$$

## 4    The Flow of Repeating Requests in $MMPP|M|\infty$ System with Feedback

Let us consider the queuing system with infinite number of servers, and Markov-modulated Poisson arrival process ($MMPP$), underlying by Markov chain with a finite number of states $k(t) = 1, 2, \ldots, K$, given matrix of infinitesimal characteristics $\mathbf{Q}$, $i, j = 1, 2, \ldots, K$, and a matrix of conditional densities $\Lambda$. Let us denote $i(t)$ is a number of occupied devices at the instant $t$, $n(t)$ is a number of repeating requests that have come during the time interval $t$, $k(t)$ is a state of the Markov chain. Three-dimensional process $\{k(t), i(t), n(t)\}$ is Markovian. For probability distribution $P(k, i, n, t) = P\{k(t) = k, i(t) = i, n(t) = n\}$ we can write down Kolmogorov differential equation system for partial characteristic function in form of differential matrix equation

$$\frac{\partial \mathbf{H}(u, w, t)}{\partial t} + j\mu(re^{jw} - 1 + (1-r)e^{-ju})\frac{\partial \mathbf{H}(u, w, t)}{\partial u} = \mathbf{H}(u, w, t)[(e^{ju} - 1)\Lambda + \mathbf{Q}], \quad (7)$$

where

$$\mathbf{H}(u, w, t) = [H(1, u, w, t), H(2, u, w, t), \ldots, H(K, u, w, t)],$$

$$\Lambda = \begin{pmatrix} \lambda_1 & 0 & \cdots & 0 \\ 0 & \lambda_2 & \cdots & 0 \\ \vdots & \vdots & \ddots & \vdots \\ 0 & 0 & \cdots & \lambda_K \end{pmatrix}, \mathbf{Q} = \begin{pmatrix} q_{11} & q_{12} & \cdots & q_{1K} \\ q_{21} & q_{22} & \cdots & q_{2K} \\ \vdots & \vdots & \ddots & \vdots \\ q_{K1} & q_{K2} & \cdots & q_{KK} \end{pmatrix}.$$

Let us find the asymptotic characteristic function of the number of repeating requests in $MMPP|M|\infty$ system during time interval $t$ under condition of increasing time. We will denote

$$\mu = \epsilon, u = \epsilon y, \mathbf{H}(u, w, t) = \mathbf{F}(y, w, t, \epsilon),$$

Let us rewrite the Eq. (7) considering all introduced designations

$$\frac{\partial \mathbf{F}(y, w, t, \epsilon)}{\partial t} + j(re^{jw} - 1 + (1-r)e^{-jy\epsilon})\frac{\partial \mathbf{F}(y, w, t, \epsilon)}{\partial y} = \mathbf{F}(y, w, t, \epsilon)[(e^{jy\epsilon} - 1)\Lambda + \mathbf{Q}]. \quad (8)$$

Let us formulate the theorem, proof of which could be carried out the same as the one of the Theorem 1.

**Theorem 2.** *Sum of the components of limit value where $\epsilon \to 0$ of the vector-function $\boldsymbol{F}(y, w, t)$ of the solution $\boldsymbol{F}(y, w, t, \epsilon)$ of the Eq. (8) has the following form*

$$\boldsymbol{F}(y, w, t)\boldsymbol{E} = exp\left\{\frac{\lambda rt(e^{jw} - 1)}{1 - r} + \frac{j\lambda y}{1 - r}\right\}, \tag{9}$$

*where the value of $\lambda$ is determined by expression*

$$\lambda = \boldsymbol{R}\boldsymbol{\Lambda}\boldsymbol{E},$$

*vector $\boldsymbol{R}$ is a stationary probability distribution of states of the Markov chain, which is determined by the linear equation system* $\begin{cases} \boldsymbol{R}\boldsymbol{Q} = 0, \\ \boldsymbol{R}\boldsymbol{E} = 1. \end{cases}$

By assuming that $y = 0$ at (9), we obtain the asymptotic approximation of characteristic function of the number of repeating requests that have come into the system during time interval $t$ under a condition of increasing service time:

$$h(w, t) = exp\left\{\frac{\lambda rt(e^{jw} - 1)}{1 - r}\right\}.$$

## 5    The Flow of Repeating Requests in $GI|M|\infty$ System with Feedback

Now, let us start with the queuing system with infinite number of servers and reneval arrival process with the distribution function $A(x)$ for the lengths of intervals between arrivals.

Let us denote the following: $i(t)$ is a number of occupied servers in the system at the moment of time $t$, $n(t)$ is a number of requests that have come into system for repeating service during time interval $t$. Because of the resulting random process $\{i(t), n(t)\}$ being non-Markovian, we have to Markovize it by introducing additional variable $z(t)$ which is equal to the length of interval between the instant $t$ and the epoch the next. Then the three-dimensional process $\{z(t), i(t), n(t)\}$ will be a Markov process.

Let us denote the probability distribution for values of the process $P(z, i, n, t) = P\{z(t) < z, i(t) = i, n(t) = n\}$. Then we can write the Kolmogorov differential equation system for partial characteristic function

$$\frac{\partial H(z, u, w, t)}{\partial t} = \frac{\partial H(z, u, w, t)}{\partial z} - j\mu(re^{jw} - 1 + (1 - r)e^{-ju})\frac{\partial H(z, u, w, t)}{\partial u}$$
$$+ (e^{ju}A(z) - 1)\frac{\partial H(0, u, w, t)}{\partial z}. \tag{10}$$

Let us define characteristics of the flow of repeating requests in the system by means of method of asymptotic analysis. We consider the $GI|M|\infty$ system with feedback under a condition of increasing service time. Denote the following:

$$\mu = \epsilon, u = \epsilon y, H(z, u, w, t) = F(z, y, w, t, \epsilon),$$

then we can rewrite (10) considering all introduced designations

$$\frac{\partial F(z, u, w, t, \epsilon)}{\partial t} = \frac{\partial F(z, u, w, t, \epsilon)}{\partial z} - j(re^{jw} - 1 + (1 - r)e^{-jy\epsilon})\frac{\partial F(z, u, w, t, \epsilon)}{\partial y}$$

$$+ (e^{jy\epsilon}A(z) - 1)\frac{\partial F(0, u, w, t, \epsilon)}{\partial z}. \tag{11}$$

Let us formulate the theorem, proof of which could be carried out the same as the one of the Theorem 1.

**Theorem 3.** *The limit value where of the $\epsilon \to 0$, of the function $F(y, w, t)$ of the solution $F(y, w, t, \epsilon)$ of the Eq. (11) has the following form*

$$F(y, w, t) = exp\left\{\frac{\lambda rt(e^{jw} - 1)}{1 - r} + \frac{j\lambda y}{1 - r}\right\}, \tag{12}$$

*where*

$$\lambda = R'(0),$$

$R(z)$ *is a stationary probability distribution of values of the random process $z(t)$.*

By assuming that $y = 0$ at (12), we have the asymptotic approximation of the characteristic function of the number of repeating requests that have come into system during time interval $t$ under condition of increasing service time

$$h(w, t) = exp\left\{\frac{\lambda rt(e^{jw} - 1)}{1 - r}\right\}.$$

## 6    Numerical Results

In order to evaluate the assessment of error of the probability distribution for the reviewed process we have to compare the results of simulation with the results obtained by using method of asymptotic analysis.

The domain of applicability of the results of asymptotic analysis will be defined by means of Kolmogorov distance

$$\Delta = \max_i |\widehat{F(i)} - F(i)|.$$

Where $\widehat{F(i)}$ – empirical probability distribution function obtained by means of imitational modeling, $F(i)$ is a Poisson probability distribution function with the parameter $\lambda rt/(1 - r)$.

**Example 1.** Consider the system with $MMPP$ arrival process with the following parameters:

$$\Lambda = \begin{pmatrix} 1 & 0 & 0 \\ 0 & 5 & 0 \\ 0 & 0 & 10 \end{pmatrix}, Q = \begin{pmatrix} -0.5 & 0.3 & 0.2 \\ 0.2 & -0.6 & 0.4 \\ 0.5 & 0.3 & -0.8 \end{pmatrix}.$$

Let probability of the return of a request back into the system for repeating service is $r = 0.1$. Let us notice that for these parameters the rate of arrivals is $\lambda = 4,73$, and the parameter of Poisson distribution will be equal to $\lambda r/(1-r) = 0.525$.

By increasing the value of an average service time $1/\mu$ we can define when the Kolmogorov distance will become acceptable.

Table 1 has covered values of Kolmogorov distances for the system at various values of a service parameter $\mu$.

**Table 1.** Error of approximation of the flow of repeating requests into the system by the Poisson process at various values of $\mu$

| $\mu$ | 1 | 0.5 | 0.1 | 0.01 |
|---|---|---|---|---|
| $\Delta$ | 0.081 | 0.038 | 0.008 | 0.002 |

**Example 2.** Let us consider $GI|M|\infty$ QS with renewal arrival process. Lengths of intervals between the epochs of arrivals have Gamma-distribution with a parameter of form $\alpha = 0,5$ and a parameter of scale $\beta = 2,5$, probability of return of a request into the system $r = 0,1$. Then the rate of the arrival process is $\lambda = 2$, and a parameter of the Poisson distribution is equal to $\lambda r/(1 - r) = 0.22$.

Values of Kolmogorov distances for the system at various values of a service parameter $\mu$ are given in a Table 2.

**Table 2.** Error of approximation of the flow of repeating requests into the system by the Poisson flow at various values of $\mu$

| $\mu$ | 1 | 0.5 | 0.1 | 0.01 |
|---|---|---|---|---|
| $\Delta$ | 0.0540 | 0.0252 | 0.0090 | 0.0082 |

The analysis demonstrates that probability distribution for a number of requests in the flow of repeating requests in the system with increasing service time is coming close to Poisson distribution. Even at $\mu = 0.5$ the Kolmogorov distances is under 0.03 which is an acceptable error.

The same results were obtained for other values of parameters of arrival process.

## 7    Conclusion

Thus, this work has covered mathematical models of $M|M|\infty$, $MMPP|M|\infty$ and $GI|M|\infty$ systems with feedback. Using the method of asymptotic analysis under condition of increasing service time, we obtain asymptotic approximation of characteristic functions of the flow of repeating requests for each system. Its proven that the flow of repeating requests could be approximated by the Poisson process with a parameter $\lambda rt/(1-r)$, where $\lambda$ is a rate of the arrival process, $r$ is a probability of repeating service. The numerical analysis of the obtained results is presented; the domain of applicability of approximation is defined.

**Acknowledgments.** This work is performed under the state order No. 1.511.2014/K of the Ministry of Education and Science of the Russian Federation.

# References

1. Pekoz, E.A., Joglekar, N.: Poisson traffic flow in a general feedback. J. Appl. Probab. **39**(3), 630–636 (2002)
2. Foley, R.D., Disney, R.L.: Queues with delayed feedback. Adv. Appl. Probab. **15**(1), 162–182 (1983)
3. Neuts, F.: Matrix-Geometric Solutions in Stochastic Models: An Algorithmic Approach. John Hopkins University Press, Baltimore (1981)
4. Melikov, A.Z., Ponomarenko, L.A., Kuliyeva, K.H.N.: Calculation of the characteristics of multichannel queuing system with pure losses and feedback. J. Autom. Inf. Sci. **47**(6), 19–29 (2015)
5. D'Avignon, G.R., Disney, R.L.: Queues with instantaneous feedback. Manag. Sci. **24**(2), 168–180 (1977)
6. Zaryadov, I.S., Scherbanskaya, A.A.: Time characteristics of queuing system with renovation and reservice. Physics **2**, 61–65 (2014). (in Russian), Bulletin of PFUR. Serires Mathematics. Information Sciences
7. Zaryadov, I.S., Korolkov, A.V., Milovanov, T.A., Sherbanskaya, A.A.: Mathematical model of calculating and analysis of characteristics of systems with generalized and repeating service. T-Comm Telecommun. Transp. **6**(8), 16–20 (2014). (in Russian)
8. Moiseeva, S.P., Zakhorol'naya, I.A.: Mathematical model of parallel retrial queueing of multiple requests. Optoelectronics Instrum. Data Process. **47**(6), 567–572 (2011)
9. Morozova, A.S., Moiseeva, S.P., Nazarov, A.A.: Research of QS with repeating requests and uunlimited number of servers by means of a method of limit decomposing. Comput. Technol. **13**(5), 88–92 (2005). (in Russian)
10. Leland, W.E., Willinger, W., Taqqu, M.S., Wilson, D.V.: On the self-similar nature of ethernet traffic. ACM SIGCOMM Comput. Commun. Rev. **47**(6), 202–213 (1995)
11. Klemm, A., Lindemann, C., Lohmann, M.: Modeling IP traffic using the batch Markovian arrival process (extendend version). Perform. Eval. **54**, 149–173 (2003)
12. Nazarov, A.A., Moiseeva, S.P.: Methods of Asymptotic Analysis in Queuing Theory. NTL, Tomsk (2006). (In Russian)

# Author Index

Printed in the United States
by Baker & Taylor Publisher Services